Symmetry in Engineering Sciences

Symmetry in Engineering Sciences

Special Issue Editors

Raúl Baños Navarro
Francisco G. Montoya

MDPI • Basel • Beijing • Wuhan • Barcelona • Belgrade

MDPI

Special Issue Editors

Raúl Baños Navarro   Francisco G. Montoya
University of Almería   University of Almería
Spain   Spain

Editorial Office
MDPI
St. Alban-Anlage 66
4052 Basel, Switzerland

This is a reprint of articles from the Special Issue published online in the open access journal *Symmetry* (ISSN 2073-8994) from 2018 to 2019 (available at: https://www.mdpi.com/journal/symmetry/special_issues/Symmetry_Engineering_Sciences)

For citation purposes, cite each article independently as indicated on the article page online and as indicated below:

LastName, A.A.; LastName, B.B.; LastName, C.C. Article Title. *Journal Name* **Year**, *Article Number, Page Range.*

ISBN 978-3-03921-874-5 (Pbk)
ISBN 978-3-03921-875-2 (PDF)

Contents

About the Special Issue Editors

Raul Baños Navarro (Ph.D.) is Associate Professor at the Department of Engineering, University of Almeria (Spain). He received his first Bachelor's degree in Computer Science at the University of Almeria and his second Bachelor's degree in Economics by the National University of Distance Education (UNED). He completed his PhD dissertation on computational methods applied to optimization of energy distribution in power networks and water distribution networks. His research activity includes computational optimization, power systems, renewable energy systems, and energy economics. The research is being carried out at Napier University (Edinburgh, UK) and at the Universidade do Algarve (Portugal). As a result of his research, he has published more than 150 papers in peer-reviewed journals, books, and conference proceedings.

Francisco G. Montoya (Ph.D.) is Professor at the Engineering Department and the Electrical Engineering Section in the University of Almeria (Spain), received his MS from the University of Malaga and his PhD from the University of Granada (Spain). He has published over 70 papers in JCR journals and is the author or coauthor of books published by MDPI, RA-MA, and others. His main interests are power quality, smart metering, smart grids and evolutionary optimization applied to power systems, and renewable energy. Recently, he has become passionately interested in geometric algebra as applied to power theory.

Preface to "Symmetry in Engineering Sciences"

Symmetry is a frequent issue widely studied in different research fields, but with particular implications unique to each of them. For example, in mathematics, symmetry is often considered a type of invariance because the objects are invariant under a set of transformations. However, other particular meanings and implications are considered in different fields, such as physics, chemistry, biology, etc.

Complex systems with symmetry also arise in a wide range of engineering disciplines. This is the case of mechanical engineering, where symmetric and synchronized systems are considered to analyze stability criteria for rotating structures, vibration and noise, fault diagnosis, etc. The study of symmetrical and asymmetrical faults is also a critical issue in the study of power systems. Data speed or quantity are the same in both directions as averaged over time in some telecommunications systems. Civil engineering often considers that the strength of the objects depends on the symmetry. Symmetric network structures and symmetric algorithms are often studied in computer science, and many other examples can be found in these and other engineering fields.

Due to the high complexity of engineering applications, inherent symmetry is not easily recognizable so that although in certain cases, certain symmetry properties can be detected, these may be partial while others may not be perceived. Furthermore, some systems have imperfect symmetry characteristics that can be measured in terms of similarity, while nonsymmetry is a measure of difference. Therefore, there are many open research areas in engineering which need further work to determine symmetrical and asymmetrical properties. Therefore, this book includes recent theoretical or practical advances of symmetry in multidisciplinary engineering applications so that readers can familiarize themselves with the new problems and methods explained directly by experts in the field.

Raúl Baños Navarro, Francisco G. Montoya
Special Issue Editors

symmetry

MDPI

Editorial

Symmetry in Engineering Sciences

Francisco G. Montoya *, Raúl Baños, Alfredo Alcayde and Francisco Manzano-Agugliaro

Department of Engineering, University of Almeria, ceiA3, 04120 Almeria, Spain; rbanos@ual.es (R.B.); aalcayde@ual.es (A.A.); fmanzano@ual.es (F.M.-A.)
* Correspondence: pagilm@ual.es; Tel.: +34-950-015791; Fax: +34-950-015491

Received: 13 June 2019; Accepted: 13 June 2019; Published: 15 June 2019

Abstract: The symmetry concept is mainly used in two senses. The first from the aesthetic point of view of proportionality or harmony, since human beings seek symmetry in nature. Or the second, from an engineering point of view to attend to geometric regularities or to explain a repetition process or pattern in a given phenomenon. This special issue dedicated to geometry in engineering deals with this last concept, which aims to collect both the aspects of geometric solutions in engineering, which may even have a certain aesthetic character, and the aspect of the use of patterns that explain observed phenomena.

Keywords: asymmetry; synchronization; topology; electrical circuits; electronic devices; mechanical structures; robots; graphic modelling; complex networks; optimization; computing applications

1. Introduction

Symmetry is a frequent pattern widely studied in different research fields. In particular, complex systems with symmetry arise in engineering science (e.g., in mechanical engineering, symmetric and synchronized systems are often used to satisfy stability criteria for rotating structures; in electrical engineering, the study of symmetrical and asymmetrical faults in power systems is a critical issue; in telecommunications engineering, many systems are symmetrical since data speed or quantity is the same in both directions; in civil engineering, the strength of the objects depend on the symmetry; in computer engineering, symmetric network structures and symmetric algorithms are often studied, etc.).

This Special Issue invites researchers to submit original research papers and review articles related to any engineering discipline where theoretical or practical issues of symmetry are considered. The topics of interest include, but are not limited to:

- Symmetry in electrical and electronic engineering
- Symmetry in mechanical engineering
- Symmetry in automation and robotic engineering
- Symmetry in computer engineering
- Symmetry in telecommunications engineering
- Symmetry in civil engineering (transportation, hydraulics, etc.)
- Symmetry in chemical engineering
- Symmetry and topology of complex networks in engineering
- Symmetry and optimization in engineering applications

2. Statistics of the Special Issue

The statistics of the call for papers for this special issue related to published or rejected items was: Total submissions (19), Published (12; 73%), and Rejected (7; 27%).

The authors' geographical distribution by country for published papers is shown in Table 1, where it is possible to observe 45 authors from five different countries. Note that it is usual for an article to be signed by more than one author and for authors to collaborate with others of different affiliations.

Table 1. Geographic distribution by the country of author.

Country	Number of Authors
China	31
Spain	8
Pakistan	3
Czech Republic	2
Korea	1
Total	45

3. Authors of this Special Issue

The authors of this special issue and their main affiliations are summarized in Table 2, where there are four authors on average per manuscript.

Table 2. Affiliations and bibliometric indicators for the authors.

Author	Main Affiliation	Reference
Cristina Velilla	Universidad Politécnica de Madrid	[1]
Alfredo Alcayde	University of Almeria	[1]
Carlos San-Antonio-Gómez	Universidad Politécnica de Madrid	[1]
Francisco G. Montoya	University of Almeria	[1]
Ignacio Zavala	Universidad Politécnica de Madrid	[1]
Francisco Manzano-Agugliaro	University of Almeria	[1]
José Ignacio Rojas-Sola	University of Jaen	[2]
Eduardo De la Morena-De la Fuente	University of Jaen	[2]
Yu Zhang	South China University of Technology	[3]
Yuanpeng Zhu	South China University of Technology	[3]
Xuqiao Li	South China University of Technology	[3]
Xiaole Wang	South China University of Technology	[3]
Xutong Guo	South China University of Technology	[3]
Nasar Iqbal	University of Engineering and Technology	[4]
Sadiq Ali	University of Engineering and Technology	[4]
Imran Khan	University of Engineering and Technology	[4]
Byung Moo Lee	Sejong University	[4]
Ling Wang	Henan Agricultural University	[5]
Dongfang Zhou	National Digital Switching System Engineering and Technology R&D Center (NDSC)	[5]
Hui Tian	Henan Agricultural University	[5]
Hao Zhang	Henan Agricultural University	[5]
Wei Zhang	Henan Agricultural University	[5]
Yanrong Wang	Beihang University	[6]
Hang Ye	Beihang University	[6]
Xianghua Jiang	Beihang University	[6]
Aimei Tian	Beihang University	[6]
Daniel Chalupa	Brno University of Technology	[7]
Jan Mikulka	Brno University of Technology	[7]
Ke Ruan	Xi'an University of Architecture and Technology	[8]
Qi Zhang	Xi'an University of Architecture and Technology	[8]
Han-ye Zhang	Jiujiang University	[9]
Wei-ming Lin	Jiujiang University	[9]
Ai-xia Chen	Jiujiang University	[9]
Siqi Liu	Beijing Jiaotong University	[10]
Boliang Lin	Beijing Jiaotong University	[10]
Jianping Wu	Beijing Jiaotong University	[10]

Table 2. *Cont.*

Author	Main Affiliation	Reference
Yinan Zhao	Beijing Jiaotong University	[10]
Jianjie Zheng	Dalian Jiaotong University	[11]
Yu Yuan	Dalian Jiaotong University	[11]
Li Zou	Dalian Jiaotong University	[11]
Wu Deng	Dalian Jiaotong University	[11]
Chen Guo	Dalian Jiaotong University	[11]
Huimin Zhao	Dalian Jiaotong University	[11]
Zihan Qu	Beijing Jiaotong University	[12]
Shiwei He	Beijing Jiaotong University	[12]

4. Brief Overview of the Contributions to This Special Issue

The analysis of the topics (Table 3) identifies or summarizes the research undertaken. This section classifies the manuscripts according to the topics proposed in the special issue. It was observed that there are four topics that have dominated the others: Symmetry in electrical and electronic engineering; Symmetry in mechanical engineering; Symmetry in computer engineering; and Symmetry in civil engineering (transportation).

Table 3. Topic analysis.

Topic	Number of Manuscripts
Symmetry in electrical and electronic engineering	2
Symmetry in mechanical engineering	2
Symmetry in computer engineering	2
Symmetry in civil engineering (transportation, hydraulics, etc.)	2
Symmetry in automation and robotic engineering	1
Symmetry in telecommunications engineering	1
Symmetry and topology of complex networks in engineering	1
Symmetry and optimization in engineering applications	1
Total	12

Author Contributions: All authors contributed equally to this work.

Conflicts of Interest: The authors declare no conflict of interest.

References

1. Velilla, C.; Alcayde, A.; San-Antonio-Gómez, C.; Montoya, F.G.; Zavala, I.; Manzano-Agugliaro, F. Rampant Arch and Its Optimum Geometrical Generation. *Symmetry* **2019**, *11*, 627. [CrossRef]
2. Rojas-Sola, J.I.; la Morena-De la Fuente, D. The Hay Inclined Plane in Coalbrookdale (Shropshire, England): Geometric Modeling and Virtual Reconstruction. *Symmetry* **2019**, *11*, 589. [CrossRef]
3. Zhang, Y.; Zhu, Y.; Li, X.; Wang, X.; Guo, X. Anomaly Detection Based on Mining Six Local Data Features and BP Neural Network. *Symmetry* **2019**, *11*, 571. [CrossRef]
4. Iqbal, N.; Ali, S.; Khan, I.; Lee, B.M. Adaptive Edge Preserving Weighted Mean Filter for Removing Random-Valued Impulse Noise. *Symmetry* **2019**, *11*, 395. [CrossRef]
5. Wang, L.; Zhou, D.; Tian, H.; Zhang, H.; Zhang, W. Parametric Fault Diagnosis of Analog Circuits Based on a Semi-Supervised Algorithm. *Symmetry* **2019**, *11*, 228. [CrossRef]
6. Wang, Y.; Ye, H.; Jiang, X.; Tian, A. A Prediction Method for the Damping Effect of Ring Dampers Applied to Thin-Walled Gears Based on Energy Method. *Symmetry* **2018**, *10*, 677. [CrossRef]
7. Chalupa, D.; Mikulka, J. A Novel Tool for Supervised Segmentation Using 3D Slicer. *Symmetry* **2018**, *10*, 627. [CrossRef]
8. Ruan, K.; Zhang, Q. Accessibility Evaluation of High Order Urban Hospitals for the Elderly: A Case Study of First-Level Hospitals in Xi'an, China. *Symmetry* **2018**, *10*, 489. [CrossRef]

9. Zhang, H.Y.; Lin, W.M.; Chen, A.X. Path Planning for the Mobile Robot: A Review. *Symmetry* **2018**, *10*, 450. [CrossRef]
10. Liu, S.; Lin, B.; Wu, J.; Zhao, Y. Modeling the Service Network Design Problem in Railway Express Shipment Delivery. *Symmetry* **2018**, *10*, 391. [CrossRef]
11. Zheng, J.; Yuan, Y.; Zou, L.; Deng, W.; Guo, C.; Zhao, H. Study on a Novel Fault Diagnosis Method Based on VMD and BLM. *Symmetry* **2019**, *11*, 747. [CrossRef]
12. Qu, Z.; He, S. A Time-Space Network Model Based on a Train Diagram for Predicting and Controlling the Traffic Congestion in a Station Caused by an Emergency. *Symmetry* **2019**, *11*, 780. [CrossRef]

symmetry

MDPI

Article

Feature Selection with Conditional Mutual Information Considering Feature Interaction

Jun Liang [1,2,*], Liang Hou [2], Zhenhua Luan [1,2] and Weiping Huang [2]

[1]   State Key Lab of Nuclear Power Safety Monitoring Technology and Equipment, Shenzhen 518124, China
[2]   State Key Lab of Industrial Control Technology, College of Control Science and Engineering,
     Zhejiang University, Hangzhou 310027, China
*   Correspondence: jliang@zju.edu.cn

Received: 30 May 2019; Accepted: 25 June 2019; Published: 2 July 2019

Abstract: Feature interaction is a newly proposed feature relevance relationship, but the unintentional removal of interactive features can result in poor classification performance for this relationship. However, traditional feature selection algorithms mainly focus on detecting relevant and redundant features while interactive features are usually ignored. To deal with this problem, feature relevance, feature redundancy and feature interaction are redefined based on information theory. Then a new feature selection algorithm named CMIFSI (Conditional Mutual Information based Feature Selection considering Interaction) is proposed in this paper, which makes use of conditional mutual information to estimate feature redundancy and interaction, respectively. To verify the effectiveness of our algorithm, empirical experiments are conducted to compare it with other several representative feature selection algorithms. The results on both synthetic and benchmark datasets indicate that our algorithm achieves better results than other methods in most cases. Further, it highlights the necessity of dealing with feature interaction.

Keywords: feature selection; conditional mutual information; feature interaction; classification; computer engineering

1. Introduction

In an era of growing data complexity and volume, high dimensional data brings a huge challenge for data processing, as it increases the computational complexity in computer engineering. Feature selection is a widely used technique to address this issue. Theoretically, the more features are used, the more information is provided, however this is not always true in practical experience. Excessive features not only bring high computation complexity, but also cause the learning algorithm to over-fit the training data. Since feature selection could provide many advantages, such as avoiding over-fitting, resisting noise, reducing computation complexity and increasing predictive accuracy, it has attracted increasing interest in the field of machine learning and a large amount of feature selection algorithms have been proposed during recent years.

Feature selection could be broadly categorized into three types, i.e., wrapper, filter, and embedded methods according to whether the selection algorithm is independent of the specified learning algorithm [1]. Wrapper methods use a predetermined classifier to evaluate the candidate feature subset. Therefore, they usually achieve a higher predictive accuracy than other methods, like some heuristic algorithms that excessively depend on hyper-parameters, with a heavy computational burden and a high risk of being overly specific to the classifier. One of the typical wrapper methods is shown in reference [2]. For the embedded methods, feature selection is integrated into the training process for a given learning algorithm. They are less computationally expensive, but need strict model structure assumptions. In contrast, filter methods are independent of learning algorithms because they involve defining a heuristic evaluation criterion to provide a proxy measure of the classification accuracy. Compared with wrapper and embedded methods, due to the computational efficiency and generalization ability, filter methods are gaining more interest and many contributions have been made in feature selection since 2008 [3]. Filter methods could be further divided according to different kinds of evaluation criterions, such as distance, information, dependency and consistency [4]. Among these evaluation criterions, the information metric has gained more attention and is more comprehensively studied because of its ability to quantify the nonlinear relevance among features and classes.

Traditional feature selection algorithms mainly focus on the removing of irrelevant and redundant features. Irrelevant features provide no useful information and redundant features provide overlapped information about the selected features. However, feature interaction is usually ignored. Feature interaction was first proposed by Jakulin, et al. [5] and some recent research has pointed out its effect on classification. Interactive features could provide more information when combined together than the sum of information provided individually. Unintentional removal of interactive features would result in poor classification performance. An extreme example of feature interaction is the XOR problem. Suppose we defined label C based on two features f1, f2, C= f1⊕f2, then each feature is independent of the label C and provides no information about the class individually. However, these two features completely determine the class together.

Wrapper methods could deal with feature interaction implicitly to some extent. However, the heavy computational burden makes wrapper methods intractable for large scale classification tasks. Some newly proposed filter methods have considered feature interaction [6–8]. However, it's still a challenge for most filter methods to handle interaction and more work is needed on an explicit treatment of this issue. These challenges include sensitivity to data noise and data transformation [9].

Many feature selection algorithms have been proposed and widely used. Genetic Algorithm (GA) is a heuristic algorithm with global optimization. However, "pre-mature" outcomes can occur with expected hyper-parameters. The Symmetric Uncertainty (SU) algorithm assumes that the evaluated feature is independent of other features and reflects only the single feature and category. The Relief algorithm takes samples randomly while the number of samples greatly affects the results. Correlation-based feature selection (CFS) is a filter method that selects features by measuring the correlation between features and categories and the redundancy between different features, but its result may not be the global optimum. The Minimum-Redundancy Maximum-Relevance (MRMR) method searches for the most closely related features with objective category, or a subset of features that are least redundant. It can meticulously characterize feature correlation and redundancy weights. Conditional Mutual Information Maximization (CMIM) uses conditional mutual information to measure distance, which makes a tradeoff between the predictive power of the candidate feature and its independence from previously selected features. However, it may be difficult to calculate the multidimensional probability density in high dimensional space. Those methods have achieved good performance in some cases. However, they ignored the significance of feature interaction. Feature interaction is very significant and can be used in many fields like object detection and recognition, and neurocomputing and so on. Reference [10] integrated feature interaction into their proposed linear regression model to capture the nonlinear property of data. Reference [11] proposed a method to remove relevant features by considering the feature interaction and reducing the weakly relevant features.

In this paper, a new feature selection method based on conditional mutual information (named CMIFSI) is proposed. Firstly, some basic information-theoretic concepts and related work are reviewed, then a new information metric is proposed to evaluate the redundancy and interaction of candidate features. With the aid of this metric, CMIFSI could restrain the redundant features and redress interactive ones in the feature ranking process. To verify its performance, CMIFSI is compared with several of the state-of-the-art feature selection methods mentioned above.

2. Basic Information-Theoretic Concepts

In this section, we give a brief introduction to information-theoretic concepts, followed by a summary of applications used for feature selection. Information theory was initially developed by

Shannon to deal with communication problems, and entropy is the key measure. Because of its capability to quantify the uncertainty of random variables and the amount of information shared by different random variables, information theory has also been widely applied to feature selection [12].

Let X be a random variable with m discrete values and $p(x_i)$ represents the probability of x_i, x_i is the i-th value of X, then its uncertainty measured by entropy $H(X)$ is defined as

$$H(X) = -\sum_{i=1}^{m} p(x_i) \log p(x_i) \tag{1}$$

It's worth noting that entropy doesn't depend on actual values but just the probability distribution of discrete values. Then the joint entropy $H(X, Y)$ of X and Y, a random variable with n discrete values is defined as

$$H(X, Y) = -\sum_{i=1}^{m}\sum_{j=1}^{n} p(x_i, y_j) \log p(x_i, y_j) \tag{2}$$

When $p(x_i, y_j)$ is the joint distribution probabilities of x_i and y_i, and variable Y is known, y_i is the j-th of Y, then the reserved uncertainty of X is measured by conditional entropy $H(X|Y)$ which is defined as

$$H(X|Y) = -\sum_{i=1}^{m}\sum_{j=1}^{n} p(x_i, y_j) \log p(x_i|y_j) \tag{3}$$

where $p(x_i|y_j)$ is the posterior probabilities of X given Y. And it could be proven that

$$H(X|Y) = H(X, Y) - H(Y) \tag{4}$$

To quantify the information shared by two random variables X and Y, a new concept termed as mutual information (MI) is defined as

$$I(X; Y) = \sum_{i=1}^{m}\sum_{j=1}^{n} p(x_i, y_j) \log \frac{p(x_i|y_j)}{p(x_i)} \tag{5}$$

MI could quantify the relevance between variables, whether liner or nonlinear, and plays a key role in feature selection based on information metric.

Additionally, the MI and the entropy could be related by the following formula

$$I(X; Y) = H(X) - H(X|Y) \tag{6}$$

In addition, conditional mutual information (CMI) of X and Y when given a new random variable Z is defined as

$$I(X; Y|Z) = H(X|Z) - H(X|Y, Z) \tag{7}$$

CMI represents the quantity of information shared by X and Y when Z is known. It implies Y brings information about X which is not already contained in Z.

3. Related Work

Evaluation criterion is the key role in filter methods, which is intended to measure how potentially useful a feature or feature subset should be when used in a classifier. The general evaluation criterion of feature selection based on information metric could be represented as

$$J(f) = I(C; f) - g(C, S, f) \tag{8}$$

where f is a candidate feature, S is the selected feature subset, C is the class vector that evaluates the candidate feature f and $g(C, S, f)$ is a deviated function which is used to penalize or compensate the first part, i.e., $I(C; f)$. Different feature selection methods were proposed by designing modified evaluation criterions according to Equation (8).

A simple method termed as Mutual Information Maximization (MIM) is proposed in [13], which simplifies Equation (8) by removing the deviated function

$$J(f) = I(C; f) \tag{9}$$

Since mutual information tends to favor features with more discrete values, a normalized mutual information criterion named symmetrical uncertainty (SU) [14] is then introduced into the feature selection.

$$J(f) = \frac{2I(C; f)}{H(C) + H(f)} \tag{10}$$

where $H(C)$ and $H(f)$ is defined as Equation (1), $I(C; f)$ is defined as Equation (6). This criterion compensates mutual information's bias towards features with more discrete values and restricts its value to the range of [0,1].

In general, it is widely accepted that an optimal feature set should not only be relevant with the class individually, but also consider feature redundancy. Therefore, other modified criterions have been proposed to pursue the "relevancy-redundancy" goal.

Battiti [15] proposed the Mutual Information Feature Selection (MIFS) criterion:

$$J(f) = I(C; f) - \beta \sum_{f_i \in S} I(f; f_i) \tag{11}$$

This criterion uses mutual information to identify the relevant features, and a penalty to ensure low redundancy within selected features. β is a configurable parameter to determine the trade-off between relevance and redundancy. However, β is set experimentally, which results in unstable performance.

A Minimum-Redundancy Maximum-Relevance (MRMR) criterion was proposed by Peng et al. [16].

$$J(f) = I(C; f) - \frac{1}{|S|} \sum_{f_i \in S} I(f; f_i) \tag{12}$$

where $|S|$ is the number of features in selected feature subset S In this criterion, the deviated function $g(C, S, f) = \frac{1}{|S|} \sum_{f_i \in S} I(f; f_i)$ acts as a penalty to feature redundancy.

Another similar criterion is called Joint Mutual Information (JMI) [17].

$$J(f) = \sum_{f_i \in S} I(f, f_i; C) \tag{13}$$

This criterion could be re-written in the form of Equation (8) by using some relatively simple manipulations.

$$J(f) = I(C; f) - \frac{1}{|S|} \sum_{f_i \in S} [I(f; f_i) - I(f; f_i | C)] \tag{14}$$

In this criterion, $I(f; f_i) - I(f; f_i | C)$ represents the amount of information about C shared by f and f_i. Therefore, the second part of this criterion is another modified deviated function to penalize feature redundancy.

Fleuret [18] proposed the Conditional Mutual Information Maximization (CMIM) criterion

$$J(f) = \min_{f_i \in S} [I(C; f | f_i)] \tag{15}$$

This criterion could also be re-written in the form of Equation (8)

$$J(f) = I(C; f) - \min_{f_i \in S}[I(C; f) - I(C; f|f_i)] \tag{16}$$

Actually, the initial form of this criterion is $J(f) = I(C; f|S)$, since $I(C; f|S)$ is difficult to calculate, it should be approximated by some simplified form. When only taking feature redundancy into consideration, the following inequality is established

$$I(C; f|S) \leq \mathop{I}_{S_i \in S}(C; f|S_i) \leq \mathop{I}_{f_i \in S}(C; f|f_i) \tag{17}$$

Therefore, we could estimate $I(C; f|S)$ by using the minimum value, i.e.,

$$I(C; f|S) \approx \min_{f_i \in S}[I(C; f|f_i)] \tag{18}$$

Many other criterions based on information metric have also been proposed, such as FCBF [19], AMIFS [20], CMIFS [21]. Reviewing these criterions, it is easy to find that almost all of these information based criterions focus on selecting relevant features and penalizing feature redundancy by a deviated function, while feature interaction is ignored. As stated above, feature interaction does exist and unintentional ignoring of this feature interaction may result in poor classification performance. Therefore an appropriate deviated function in Equation (8) should not only penalize feature redundancy but also compensate for feature interaction. After taking feature interaction into account, many of the presented criterions would be ill-considered or even improper. Taking CMIM as an example, the inequality (17) would be not tenable once feature interaction is considered, then the final criterion $\min[I(C; f|f_i)]$ would be improper as well. However, little work has been conducted to deal with feature interaction using the information metric.

4. Some Definitions about Feature Relationships

In this section, we first present some classic definitions of feature relevance and redundancy, then provide our formal definitions of feature irrelevance, redundancy and interaction based on information theory.

John et al. [22] classifies features into three disjoint categories, namely, strong relevance, weak relevance and irrelevant features. Then Yu and Liu [18] proposed the definition of redundancy base on the concept of Markov blanket.

Let F be a full set of features, f_i a feature and $S_i = F - \{f_i\}$, C the class vector. These definitions are as follows.

Definition 1. *(Strong relevance) A feature fi is strong relevant if and only if*

$$P(C|F) \neq P(C|S_i) \tag{19}$$

Definition 2. *(Weak relevance) A feature fi is weak relevant if and only if*

$$P(C|F) = P(C|S_i)$$
$$\exists S_i' \subset S_i, \text{ such that } P(C|f_i, S_i') \neq P(C|S_i') \tag{20}$$

Corollary 1. *(Irrelevance) A feature f_i is irrelevant if and only if*

$$\forall S_i' \subset S_i, \ P(C|f_i, S_i') = P(C|S_i') \tag{21}$$

Definition 3. *(Markov blanket) Given a feature f_i, let $M_i \subset F(f_i \notin M_i)$, M_i is said to be a Markov blanket for f_i if and only if*

$$P(F - M_i - \{f_i\}, C | f_i, M_i) = P(F - M_i - \{f_i\}, C | M_i) \tag{22}$$

Definition 4. *(Redundancy) A feature f_i is redundant if and only if it's weakly relevant and has a Markov blanket M_i within F.*

It's important to note that strong relevance, weak relevance and irrelevance are paratactic, while redundancy is a part of weak relevance. The objective of feature selection is to select an optimal or relatively suboptimal feature subset which contains the strongly relevant and weakly relevant but non-redundant features. However, the above definitions for relevance and redundancy rely on a whole probability distribution, which is intractable to guide feature selection. Moreover, feature interaction has not been defined specifically either.

In the following sections, we will provide some definitions for relevance, redundancy, irrelevance and interaction based on information theory. It's worth noting that the following definitions are based on a candidate feature and a selected feature subset, which is different from the above definitions but could be directly used to guide feature selection especially those using greedy search strategy.

Let S be the subset of features which has been selected, f_i is a candidate feature, C is the class vector.

Definition 5. *(Irrelevance) Feature f_i is irrelevant if and only if*

$$I(f_i; C) = I(f_i; C | S) = 0 \tag{23}$$

According to definition 5, an irrelevant feature f_i couldn't provide any information about C.

Definition 6. *(Strong redundancy) Feature f_i is strongly redundant with S if and only if*

$$I(f_i; C) > I(f_i; C | S) = 0 \tag{24}$$

Where $I(f_i; C) > 0$ suggests that f_i could provide some information about C individually, while $I(f_i; C | S) = 0$ indicates that f_i provides no more information when S is given. Therefore, the information provided by f_i has already been contained in the selected feature subset S, thus f_i should not be selected.

Definition 7. *(Weak redundancy) Feature f_i is weakly redundant with S if and only if*

$$I(f_i; C) > I(f_i; C | S) > 0 \tag{25}$$

According to definition 7, $I(f_i; C) > I(f_i; C | S)$ means the information about C provided by f_i would decrease when S is given, i.e., f_i and S share some information about C. But f_i is still useful and may be selected since it could provide more information even S is known according to $I(f_i; C | S) > 0$.

Definition 8. *(Independent relevance) Feature f_i is independently relevant with S if and only if*

$$I(f_i; C) = I(f_i; C | S) > 0 \tag{26}$$

According to definition 8, the information provided by an independently relevant feature has totally not been contained in S, therefore an independently relevant feature should be selected.

Definition 9. *(Interaction) Feature f_i is interactive with S if and only if*

$$I(f_i; C | S) > I(f_i; C) \geq 0 \tag{27}$$

According to definition 9, $I(f_i; C | S) > I(f_i; C)$ suggests that the information provided by f_i would increase when S is given. Thus feature f_i and feature subset S have a synergy.

In the process of feature selection, when a feature subset is selected, a candidate feature belonging to weak redundancy, independent relevance and interaction should be selected, while irrelevant or strong redundant features would be eliminated.

5. A New Feature Smethod Considering Feature Interaction

The main goal of feature selection is to select a small number of features that can carry as much information as possible. When using information metric, the objective function is defined as $\max I(S;C)$, where S is the selected feature subset. Based on this objective function, the evaluation criterion for a candidate feature fi in a greedy search strategy could be represented as [23]

$$J(f_i) = I(C; f_i|S) \tag{28}$$

Re-write this criterion in the form of Equation (8)

$$J_{CMI}(f_i) = I(C; f_i) + [I(f_i; C|S) - I(f_i; C)] \tag{29}$$

To directly calculate JCMI(fi), we need to compute the complex joint probability, which would be computationally intractable. To address this issue, we would like to evaluate $JCMI(f_i)$ by using some approximation technique without the involvement of complex joint probability.

The second part in JCMI (f_i) (termed as $DF = I(f_i; C|S) - I(f_i; C)$) acts as a deviated function, to penalized or compensate the first part. Therefore, a proper approximation of DF should consider both redundancy and interaction. One feasible method is to consider redundancy and interaction severally.

For a candidate feature f_i, the selected S could be divided into three kinds of subsets, which are redundant, independently relevant and interactive with fi respectively, denoted as S_{redu}, S_{inde} and S_{inte}.

The redundancy between S and f_i could be represented by the subset with the highest redundancy degree, denoted as

$$S_{redu} = \arg\min_{S_{redu} \subset S} [I(f_i; C|S_{redu}) - I(f_i; C)] \tag{30}$$

Similarly, the interaction between S and f_i could be represented by the subset with the highest interaction degree, denoted as

$$S_{inte} = \arg\max_{S_{inte} \subset S} [I(f_i; C|S_{inte}) - I(f_i; C)] \tag{31}$$

In addition, according to Definition 8, $I(f_i; C) = I(f_i; C|S_{inde})$, so independently relevant subsets S_{inde} doesn't influence the selection of candidate feature f_i and could be ignored. Therefore, the deviated function DF could be replaced by

$$DF = [I(f_i; C|S_{redu}) - I(f_i; C)] + [I(f_i; C|S_{inte}) - I(f_i; C)] \tag{32}$$

For features in S_{redu}, more features would intensify redundancy, thus

$$I(f_i; C|S_{redu}) \le \underset{f_j \in S_{redu}}{I} (f_i; C|f_j) \tag{33}$$

We estimate $I(f_i; C|S_{redu})$ by their minimum value, i.e., $I(f_i; C|S_{redu}) \approx \min_{f_j \in S_{redu}} I (f_i; C|f_j)$.

Similarly, for features in S_{inte}, more features would intensify interaction

$$I(f_i; C|S_{inte}) \ge \underset{f_j \in S_{inte}}{I} (f_i; C|f_j) \tag{34}$$

We estimate $I(f_i; C|S_{inte})$ by their maximum value, i.e., $I(f_i; C|S_{inte}) \approx \max_{f_j \in S_{inte}} I (f_i; C|f_j)$.

Therefore, DF could be approximated by

$$DF \approx [\min_{\substack{f_j \in S_{redu}}} I\,(f_i; C|f_j) - I(f_i; C)] + [\max_{\substack{f_j \in S_{inte}}} I\,(f_i; C|f_j) - I(f_i; C)] \tag{35}$$

Since the subsets S_{redu}, S_{inte} are all implicit, we should generalize the above DF into the whole subset S. It's easy to prove that $\min_{\substack{f_j \in S_{redu}}} I\,(f_i; C|f_j) = \min_{\substack{f_j \in S}} I\,(f_i; C|f_j)$. Therefore, the first part of DF is denoted as $\min[[\min_{\substack{f_j \in S}} I\,(f_i; C|f_j) - I(f_i; C)], 0]$, comparing with zero in case of $S_{redu} = \varnothing$. Similarly, the second part of DF is denoted as $\max[[\max_{\substack{f_j \in S}} I\,(f_i; C|f_j) - I(f_i; C)], 0]$.

Finally, the deviated function DF is represented as

$$DF = \min[[\min_{\substack{f_j \in S}} I\,(f_i; C|f_j) - I(f_i; C)], 0] + \max[[\max_{\substack{f_j \in S}} I\,(f_i; C|f_j) - I(f_i; C)], 0] \tag{36}$$

And the evaluation criterion of a candidate f_i is defined as

$$
\begin{aligned}
J_{CMI}(f_i) &= I(C; f_i) + DF \\
&= I(C; f_i) + \min[[\min_{\substack{f_j \in S}} I\,(f_i; C|f_j) - I(f_i; C)], 0] + \max[[\max_{\substack{f_j \in S}} I\,(f_i; C|f_j) - I(f_i; C)], 0],
\end{aligned} \tag{37}
$$

It's important to note that this new evaluation criterion makes use of a similar idea with CMIM, except taking feature interaction into consideration. And it would degenerate to the CMIM criterion once interaction is ignored.

With this newly defined evaluation criterion, a new feature selection algorithm termed as CMIFSI is proposed in this paper, the details of the Algorithm 1 are as follows

---

Algorithm 1 CMIFSI algorithm

Input: A training dataset D with a full feature set $F = \{f_1, f_2, \ldots f_n\}$ and class vector C
　　A predefined threshold K
Output: The selected feature sequence
1. Initialize parameters: the selected feature subset $S = \varnothing$, $k = 0$, deviated function $DF(f_i, S) = 0$ for all candidate features;
2. for $i = 1$ to n do
3. 　Calculate $I(f_i; C)$
4. end
5. While $k < K$ do
6. 　For each candidate feature $f_i \in F$ do
7. 　　Update $DF(f_i, S)$ according to Equation (36)
8. 　　Calculate the evaluation value
　　$J_{CMI}(f_i) = I(f_i; C) + DF(f_i, S)$
9. 　End
10. 　Select the feature f_j with the largest $J_{CMI}(f_i)$
　　$S = S \cup f_j$
　　$F = F - f_j$
11. 　$k = k + 1$
12. end

As shown in Algorithm 1, a sequential forward search strategy was adopted in this algorithm and the procedure was terminated by a predefined threshold K. The low K value can achieve low computation complexity but may lose many effective features that are useful, while a high K value can achieve better classification accuracy but also entails high computation complexity. Actually, when the threshold is exceeded, the classification accuracy doesn't increase much but the computation complexity still increases. As a result, an appropriate K value is needed. The features are ranked in descending order according to the new evaluation criterion.

Now we analyze the time complexity of our algorithm. Suppose the total number of candidate feature is n, and the predefined threshold is K. When k features have been selected, the complexity of updating the deviated function is O(n-k) for each while loop. Therefore, the complexity for K selected features is n + (n − 1) + (n − 2)+ ... + (n − K + 1) = nK − 1/2(K2 − K), namely its time complexity is O(nK). In the worst case, when all candidate features are selected, i.e., K = n, the time complexity is O(n2).

6. Experiments and Results

In this section, we empirically evaluate the effectiveness of the proposed algorithm by comparing it with some other representative feature selection algorithms using both synthetic and benchmark datasets. The experiment setup is described in Section 6.1, while the results and discussion on synthetic and benchmark datasets are shown in Sections 6.2 and 6.3, respectively.

6.1. Experiment Setup

To evaluate the effectiveness of a feature selection algorithm, a simple and direct criterion is the similarity degree between the selected subset and the optimal subset, but it can only be measured using synthetic data whose optimal subset is known beforehand. For real-world data like some benchmark datasets, such prior knowledge in unavailable and we usually use the predictive accuracy on selected subset of features as an indirect measure.

Six representative feature selection algorithms (GA, SU, Relief [24], CFS [25], MRMR, CMIM) were selected to compare with CMIFSI using both synthetic and benchmark datasets. All of these methods could effectively identify irrelevant features and some of them (e.g. CFS, MRMR, CMIM) could detect redundant features as well.

In the experiment on benchmark datasets, two different learning algorithms (C4.5 and SVM) were used to evaluate the predictive accuracy on the selected feature subsets. This meant we could verify whether the performance of our new algorithm would be limited to the specified learning algorithm.

The experiment is mainly conducted in the WEKA (WEKA is a software and can be available at http://www.cs.waikato.ac.nz/~{}ml) environment with the default settings. Some of these feature selection algorithms and learning algorithms (like SU, Relief, CFS, C4.5) could be found in the WEKA environment, other algorithms were implemented in MATLAB. To achieve impartial results, five-fold validation and ten-fold cross validation were adopted for each step in selecting features and verifying the classification capability, respectively.

6.2. Experiment on Synthetic Datasets

6.2.1. Synthetic Datasets

In this section, five synthetic datasets were employed to evaluate the effectiveness of different feature selection algorithms. These synthetic datasets were generated by the data generation RDG1 in a WEKA environment. The description of these five datasets is as follows:

Data1

There are 100 instances and 10 Boolean features denoted as $a_0, a_1, a_2 \dots , a_9$ with 2 classes. Six of these ten features are irrelevant and the target concept was defined as

$$c_1 = \bar{a}_5 \vee (\bar{a}_1 \wedge \bar{a}_6).$$

Data2

There are 100 instances and 10 Boolean features denoted as $a_0, a_1, a_2 \ldots, a_9$ with 2 classes. Five of the ten features are irrelevant and the target concept was defined as

$$c_1 = (a_0 \wedge a_1 \wedge \bar{a}_5) \vee (a_8 \wedge a_0 \wedge \bar{a}_6) \vee \bar{a}_0.$$

Data3

There are 100 instances and 10 Boolean features denoted as $a_0, a_1, a_2 \ldots, a_9$ with 2 classes. Six of the ten features are irrelevant and the target concept was defined as

$$c_1 = \bar{a}_5 \vee (\bar{a}_1 \wedge \bar{a}_6 \wedge \bar{a}_8).$$

Data4

There are 200 instances and 15 Boolean features denoted as $a_0, a_1, a_2 \ldots, a_{14}$ with 3 classes. Eleven of the fifteen features are irrelevant and the target concept was defined as

$$c_0 = \bar{a}_2 \vee (a_2 \wedge a_{12} \wedge \bar{a}_8)$$
$$c_1 = a_8 \vee (a_2 \wedge \bar{a}_{12} \wedge \bar{a}_8)$$
$$c_2 = therest$$

Data5

There are 100 instances and 10 Boolean features denoted as $a_0, a_1, a_2 \ldots, a_9$ with 2 classes. Six of the ten features are irrelevant and the target concept was defined as

$$c_1 = (\bar{a}_1 \wedge \bar{a}_5) \oplus (\bar{a}_0 \wedge \bar{a}_6).$$

Features absent in the definitions of the target concepts are redundant or irrelevant, and the relevant and interactive features in each synthetic dataset are shown in Table 1.

Table 1. Relevant and interactive features of the four synthetic datasets.

Dataset	Relevant Features	Interactive Features
Data1	a_1, a_5, a_6	(a_1, a_6)
Data2	a_0, a_1, a_5, a_6, a_8	$(a_0, a_1, a_5), (a_0, a_6, a_8)$
Data3	a_1, a_5, a_6, a_8	(a_1, a_6, a_8)
Data4	a_2, a_8, a_{12}, a_{13}	(a_2, a_8, a_{12})
Data5	a_0, a_1, a_5, a_6	$(a_1, a_5), (a_0, a_6)$

6.2.2. Results on Synthetic Datasets

Feature selection methods could be divided into two types: subset selection and feature ranking [1]. Subset selection preserves relevant features and removes as much irrelevant and redundant features as possible, while feature ranking ranks features in a descending order according to specific evaluation criterions and the number of selected features is predefined. In this experiment, GA, SU, Relief, MRMR, CMIM, CMIFSI belonged to feature ranking and CFS was a kind of subset selection method. The feature selection/ranking results are shown in Table 2 with no threshold predefined. The bold values in entries represent features belong to the optimal subset and the notation" *" denotes correct selected/ranked features.

It can be seen that when comparing with other feature selection algorithms, CMIFSI achieves the best performance. For data1, data2 and data3, CMIFSI ranks the optimal subset in the top, which means the feature ranking result is more accurate than the other results. For data4 and data5, all feature selection algorithms failed to obtain the correct results, but CMIFSI still performed better than other algorithms since its feature ranking sequence is more similar to the optimal subset. This is mainly because feature interaction really exists in these datasets (as shown in Table 1), but all algorithms except for CMIFSI just focus on detecting irrelevant and redundant features while feature interaction is ignored, which results in poor performance. Among the other feature selection algorithms, CMIM and Relief performed with approximately suboptimal results on all the four synthetic datasets, and other methods failed to obtain satisfying results.

Table 2. Feature selection results on synthetic datasets.

Algorithm	Data1	Data2
GA	$a_1,a_2,a_5,a_0,a_6,a_9,a_4,a_7,a_3,a_8$	$a_8,a_0,a_6,a_5,a_7,a_3,a_2,a_1,a_9,a_4$
SU	$a_1,a_5,a_2,a_0,a_7,a_9,a_8,a_4,a_3,a_6$	$a_8,a_0,a_6,a_7,a_1,a_5,a_4,a_3,a_2,a_9$
Relief	$a_1,a_5,a_6,a_2,a_7,a_7,a_4,a_9,a_0,a_3^*$	$a_8,a_0,a_6,a_1,a_5,a_7,a_2,a_3,a_4,a_9^*$
CFS	a_1,a_2,a_5	a_0,a_6,a_7,a_8
MRMR	$a_1,a_0,a_7,a_8,a_6,a_5,a_2,a_9,a_4,a_3$	$a_8,a_6,a_5,a_3,a_2,a_0,a_7,a_1,a_9,a_4$
CMIM	$a_1,a_6,a_5,a_2,a_7,a_9,a_0,a_8,a_3,a_4^*$	$a_8,a_0,a_6,a_5,a_7,a_1,a_4,a_3,a_9,a_2$
CMIFSI	$a_1,a_6,a_5,a_4,a_2,a_9,a_7,a_8,a_3,a_0^*$	$a_8,a_0,a_1,a_6,a_5,a_7,a_4,a_2,a_9,a_3^*$
Optimal subset	a_1,a_5,a_6	a_0,a_1,a_5,a_6,a_8

Algorithm	Data3	Data4
GA	$a_5,a_6,a_9,a_1,a_3,a_2,a_0,a_7,a_8,a_4$	$a_2,a_{13},a_{11},a_8,a_1,a_{14},a_5,a_{12},a_0,a_6,a_{10},a_7,a_3,a_4,a_9$
SU	$a_5,a_1,a_6,a_7,a_0,a_3,a_4,a_9,a_8,a_2$	$a_2,a_8,a_{13},a_{10},a_6,a_{11},a_0,a_1,a_{14},a_4,a_7,a_5,a_3,a_9,a_{12}$
Relief	$a_5,a_1,a_6,a_2,a_0,a_9,a_4,a_8,a_7,a_3$	$a_2,a_8,a_{13},a_{11},a_{14},a_{10},a_7,a_4,a_{12},a_1,a_6,a_3,a_5,a_0,a_9$
CFS	a_0,a_1,a_5,a_7	$a_0,a_2,a_8,a_{10},a_{11},a_{13},a_{14}$
MRMR	$a_5,a_6,a_3,a_2,a_9,a_1,a_0,a_4,a_7,a_8$	$a_2,a_{13},a_{11},a_1,a_8,a_7,a_5,a_{12},a_{10},a_6,a_0,a_{14},a_4,a_3,a_9$
CMIM	$a_5,a_6,a_1,a_0,a_7,a_2,a_4,a_8,a_9,a_3$	$a_2,a_8,a_{13},a_4,a_{10},a_1,a_0,a_{14},a_{11},a_6,a_{12},a_3,a_5,a_7,a_9$
CMIFSI	$a_5,a_6,a_1,a_8,a_2,a_4,a_0,a_9,a_7,a_3^*$	$a_2,a_8,a_{13},a_4,a_{12},a_{11},a_5,a_1,a_{14},a_{10},a_3,a_0,a_6,a_9,a_7$
Optimal subset	a_1,a_5,a_6,a_8	a_2,a_8,a_{12},a_{13}

Algorithm	Data5
GA	$a_1,a_6,a_9,a_5,a_3,a_2,a_8,a_4,a_0,a_7$
SU	$a_1,a_5,a_6,a_2,a_9,a_0,a_7,a_3,a_4,a_8$
Relief	$a_5,a_1,a_6,a_2,a_9,a_4,a_8,a_0,a_7,a_3$
CFS	a_6,a_2,a_1,a_5
MRMR	$a_1,a_6,a_2,a_9,a_5,a_3,a_0,a_8,a_7,a_4$
CMIM	$a_5,a_6,a_1,a_9,a_0,a_2,a_4,a_7,a_8,a_3$
CMIFSI	$a_5,a_6,a_1,a_2,a_0,a_9,a_4,a_8,a_7,a_3$
Optimal subset	a_0,a_1,a_5,a_6

The above results and discussion demonstrate the necessity of taking feature interaction into consideration in feature selection. For datasets that involved interactive features, most of the traditional feature selection algorithms would fail to achieve an optimal result. Take data4 for example, a_{12} is a relevant feature in the optimal subset which is interactive with a_2, a_8, traditional algorithms remove it or rank it at the back of the feature sequence mainly because its correlation with the class is low individually and its interaction with other features is ignored. However, during the selection process of CMIFSI, once a_2 and a_8 are selected, the evaluation criterion value of a_{12} would be compensated because of its interaction with a_2, a_8, which would increase its probability of being selected. Therefore, the result of CMIFSI is found to be superior to others.

6.3. Experiment on Benchmark Datasets

6.3.1. Benchmark Datasets

Ten datasets from the UCI Machine Learning Repository [26] are adopted in our simulation experiments. These datasets contain various numbers of features, instances, and classes, as shown in Table 3. At the same time, the distribution of each class in terms of number of instances is shown as Figure 1.

Data preprocess was applied before feature selection. Missing values were replaced by the most frequently used values and means for nominal and numeric features, respectively. For algorithms based on information metric, the MDL discretization method was applied to transform the numerical features into discrete ones. Algorithms conducted in WEKA are set with default parameters. For SVM, the grid searching method was adopted to obtain its relatively optimal parameters. In addition, we selected the top K features that produce the highest accuracy and limited their maximum to 20 (used in GA, SU, Relief MRMR, CMIM and CMIFSI), since the objective of feature selection is to reduce the original feature dimension.

Table 3. Summary of UCI benchmark datasets.

No.	Datasets	Features	Instances	Classes
1	Wine	13	178	3
2	Kr-vs-kp	36	3196	2
3	SPECTF-heart	44	267	2
4	Zoo	16	101	7
5	Credit Approval	15	690	2
6	Optical Recognition of Handwritten Digits	64	1797	10
7	Contraceptive Method Choice	9	1473	3
8	Congressional Voting Records	16	435	2
9	Waveform	21	5000	3
10	Waveform+noise	40	5000	3

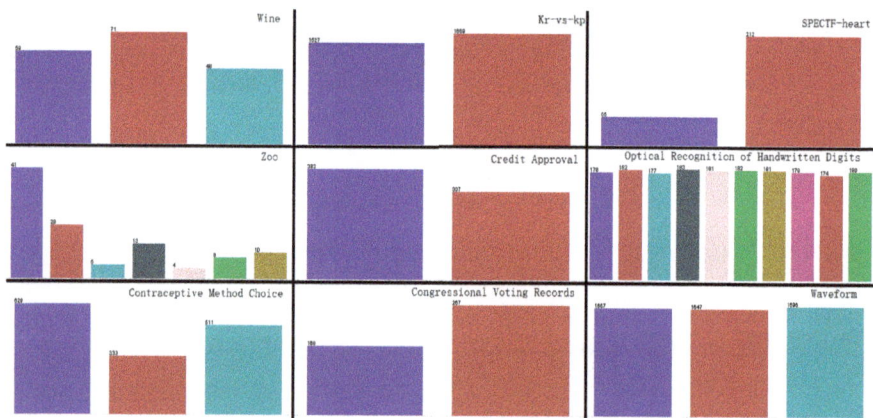

Figure 1. Distribution of each class in terms of number of instances.

6.3.2. Results on Benchmark Datasets

Tables 4 and 5 record the number of features selected by different feature selection algorithms using C4.5 and SVM, respectively. It is shown that all these feature selection algorithms achieve reduction of dimensionality by selecting only a portion of the original features. Furthermore, CMIFSI tends to obtain smaller feature subsets than those of other feature selection algorithms. AVE is the average of the same selected features of the same dataset. From the table, we can see that in most cases CMIFSI outperform the other algorithms.

Tables 6 and 7 show the 10-fold cross-validation accuracies of C4.5 and SVM respectively, where "Unselected" depicts the accuracies on datasets with original features. The bold values indicate the highest accuracies among these six feature selection algorithms using the same classifier. Notation"*" denotes the highest accuracies in each dataset corresponding to a specific classifier, including "Unselected". The last row "W/T/L" in each table summarizes the wins/ties/losses in accuracy over all datasets by comparing various feature sets with those selected by CMIFSI. At the same time, we computed the confidence intervals of all our results to evaluate the algorithms. Here, we used the bootstrapping method with 1000 as the sampling number and a 95% confidence level. The results are shown in Tables 8 and 9.

Table 4. Number of features selected by different feature selection algorithms using C4.5.

No.	C4.5							
	Total	GA	SU	Relief	CFS	MRMR	CMIM	CMIFSI	AVE.
1	13	6	3	5	11	3	3	3	4.86
2	36	15	13	15	7	14	19	19	14.57
3	44	3	2	1	21	1	2	3	4.71
4	16	10	14	9	9	10	4	4	8.57
5	15	8	6	9	6	4	3	7	6.14
6	64	20	15	20	38	17	20	20	21.43
7	9	8	7	7	8	8	7	4	7.0
8	16	12	13	9	5	12	10	6	9.57
9	21	17	17	16	16	17	16	9	15.43
10	40	16	13	11	15	16	11	13	13.57

Table 5. Number of features selected by different feature selection algorithms using SVM.

No.	SVM							
	Total	GA	SU	Relief	CFS	MRMR	CMIM	CMIFSI	AVE.
1	13	2	5	8	11	9	6	8	7.0
2	36	13	13	20	7	12	15	15	13.57
3	44	15	18	5	21	14	8	19	14.29
4	16	9	9	8	9	7	9	5	8.0
5	15	6	5	1	6	6	4	3	4.43
6	64	20	19	13	38	17	11	11	18.43
7	9	8	5	5	8	9	3	7	6.43
8	16	6	2	1	5	3	5	4	3.71
9	21	19	19	18	16	18	20	19	18.43
10	40	20	20	19	15	18	17	16	18.29

Table 6. Accuracy of selected features using C4.5.

No.	C4.5							
	Unselected	GA	SU	Relief	CFS	MRMR	CMIM	CMIFSI
1	93.80	92.67	96.07	94.38	93.80	**97.19***	**97.19***	**97.19***
2	99.31*	95.48	96.62	97.70	94.09	96.65	96.53	**97.74**
3	74.90	79.40	79.78	79.40	77.90	79.40	79.78	**80.15***
4	92.08	**95.05***	**95.05***	**95.05***	93.07	**95.05***	94.06	94.06
5	84.93	86.24	86.09	86.81	86.81	**86.96***	85.65	85.94
6	87.42	86.16	**87.48***	86.94	87.42	87.20	86.92	87.26
7	53.22	55.53	55.53	55.53	54.58	45.96	55.53	**56.21***
8	96.32	95.86	96.32	96.32	94.94	**96.78***	96.32	96.32
9	75.94	76.82	77.04	76.92	76.76	77.04	77.16	**77.22***
10	75.08	77.40	77.82	**77.92***	77.30	76.58	77.82	77.82
Ave.	83.30	84.06	84.78	84.71	83.67	83.88	84.70	**84.99**
W/T/L	1/1/8	2/0/8	3/2/5	3/2/5	2/0/8	3/1/6	0/4/6	

Table 7. Accuracy on selected features using SVM.

No.	SVM							
	Unselected	GA	SU	Relief	CFS	MRMR	CMIM	CMIFSI
1	94.44	97.85	97.22	98.89	95.56	98.89	98.33	99.40*
2	94.91	95.03	95.68	97.80*	94.87	96.21	95.03	96.94
3	79.78	82.57	83.70	83.33	81.48	82.22	82.59	84.81*
4	93.07	97.27	94.54	96.36	88.18	97.27	98.18*	97.27
5	85.51	87.79	88.10	87.82	88.50	87.39	87.10	88.55*
6	95.11*	88.28	90.78	88.39	88.28	89.44	88.00	88.00
7	51.28	53.85	54.32	53.85	54.79	50.81	54.46	55.34*
8	96.09	94.31	95.90	95.45	93.86	96.59	96.59	97.27*
9	84.08	85.59	86.22	85.70	85.20	86.12	86.56*	86.24
10	86.02	86.12	86.18	86.12	85.50	86.42*	85.82	86.18
Ave.	86.03	86.87	87.26	87.37	85.62	87.14	87.27	88.00
W/T/L	1/0/9	1/1/8	1/1/8	1/0/9	1/0/9	2/1/7	2/2/6	

Table 8. Confidence intervals on selected features using C4.5.

No.	C4.5							
	Unselected	GA	SU	Relief	CFS	MRMR	CMIM	CMIFSI
1	[93.41,94.35]	[92.04,92.96]	[95.73,96.70]	[93.50,94.46]	[93.20,94.16]	[96.77,97.70]	[95.53,97.70]	[96.84,97.86]
2	[98.62,99.58]	[94.92,95.81]	[95.95,97.00]	[97.55,98.50]	[93.49,94.36]	[95.59,96.62]	[95.81,96.76]	[97.68,98.66]
3	[74.45,75.44]	[78.76,79.71]	[79.84,80.90]	[79.10,80.08]	[77.59,78.66]	[78.78,79.80]	[77.96,79.80]	[79.68,80.65]
4	[91.92,92.86]	[94.48,95.46]	[94.55,95.55]	[94.68,95.59]	[92.53,93.52]	[94.19,95.23]	[93.58,94.61]	[93.56,94.51]
5	[84.03,85.05]	[85.80,86.73]	[85,58,86.40]	[86.50,87.55]	[86.04,87.17]	[86.84,87.72]	[85.13,86.10]	[85.14,86.00]
6	[86.87,87.73]	[85.57,86.60]	[87.20,88.09]	[86.40,87.49]	[87.05,87.92]	[86.71,87.74]	[86.53,87.44]	[86.36,87.40]
7	[52.47,53.93]	[54.29,55.75]	[55.06,56.35]	[55.13,56.60]	[53.82,55.66]	[45.45,46.71]	[53.85,55.38]	[55.34,56.33]
8	[96.32,96.74]	[95.56,96.37]	[95.81,96.72]	[95.56,96.38]	[94.08,94.87]	[96.16,96.95]	[96.06,96.87]	[96.16,97.05]
9	[75.19,76.16]	[76.63,77.52]	[76.67,77.56]	[76.58,77.51]	[76.33,77.30]	[76.51,77.47]	[76.45,77.31]	[77.09,78.09]
10	[74.58,75.58]	[77.01,77.97]	[76.92,77.70]	[77.86,78.77]	[76.97,77.86]	[75.96,76.91]	[77.10,78.06]	[77.27,78.15]

Table 9. Confidence intervals on selected features using SVM.

No.	SVM							
	Unselected	GA	SU	Relief	CFS	MRMR	CMIM	CMIFSI
1	[93.95,94.87]	[97.71,98.52]	[96.80,97.64]	[98.59,99.35]	[95.14,95.95]	[98.48,99.27]	[97.61,98.36]	[98.47,99.52]
2	[94.40,95.27]	[94.81,95.60]	[95.14,95.92]	[97.50,98.32]	[94.22,94.96]	[95.35,96.23]	[94.46,95.31]	[96.83,97.60]
3	[78.68,79.49]	[82.39,83.31]	[83.05,83.84]	[82.94,83.73]	[80.81,81.55]	[81.98,82.88]	[82.50,83.39]	[84.32,85.13]
4	[92.38,93.33]	[96.71,97.42]	[94.31,95.20]	[95.56,96.35]	[87.56,88.47]	[96.77,97.61]	[96.78,98.55]	[96.88,97.77]
5	[85.26,86.08]	[87.52,88.27]	[87,58,88.46]	[87.01,87.85]	[88.22,89.14]	[86.77,87.52]	[86.49,87.38]	[88.56,89.40]
6	[94.68,95.60]	[87.69,88.44]	[90.39,91.22]	[87.87,88.77]	[87.88,88.65]	[89.07,89.98]	[87.66,88.46]	[87.20,88.03]
7	[51.24,52.02]	[53.41,54.22]	[53.74,54.59]	[53.63,54.46]	[54.34,55.17]	[50.53,51.38]	[54.02,54.79]	[55.18,55.98]
8	[95.82,96.60]	[93.64,94.47]	[95.34,96.21]	[95.30,96.18]	[93.60,94.38]	[96.13,96.97]	[96.37,97.23]	[96.53,97.28]
9	[83.79,84.59]	[85.13,86.00]	[85.43,86.40]	[85.06,85.80]	[84.89,85.80]	[85.77,86.63]	[86.18,86.98]	[86.10,86.90]
10	[85.69,86.58]	[86.33,87.13]	[86.68,87.60]	[87.45,88.21]	[84.86,85.69]	[86.94,87.77]	[86.80,87.62]	[87.71,88.56]

The results in Tables 6 and 7 show that CMIFSI tends to outperform other feature selection algorithms on these ten benchmark datasets, regardless of whether C4.5 or SVM is used. The proposed method gets the highest classification accuracy on five datasets of ten. The average classification accuracies of CMIFSI are higher than others in both tables. From the view of "W/T/L", CMIFSI is also relatively superior to other selectors. More specifically, CMIFSI obtains 5 maximal classification accuracies (denoted by bold value) over ten datasets whether use C4.5 or SVM.

CMIFSI and CMIM are similar to some extent, since they apply the same method to evaluate feature redundancy, except that the feature interaction is ignored by CMIM. Actually, CMIFSI could be regarded as a modification to the original CMIM. Therefore, it's worth comparing these two algorithms to verify whether the modification to CMIM is worthwhile. The results in Tables 6 and 7 show that CMIFSI outperforms CMIM in almost all of the datasets. In contrast, MRMR, which uses a different way to evaluate feature redundancy, achieves higher classification accuracies than CMIFSI in some cases, even though its average accuracy is lower than CMIM and CMIFSI in both tables.

Apart from comparison among feature selectors, it's interesting to find that for some cases such as the 6th dataset using C4.5 and SVM, the accuracies based on feature selection tend to be lower than the ones based on the original features. This doesn't mean that feature selection deteriorated the classification performance. Actually it mainly resulted from a limitation in this experiment. Since we restricted the maximal number of features selected to 20, when the optimal feature subset consisted of more than 20 features, feature selector resulted in poor performance.

To further evaluate the effectiveness of our new algorithm, another experiment was conducted to compare different feature ranking algorithms (GA, SU, Relief, MRMR, CMIM and CMIFSI) by adding features for learning one by one in the order that the features are ranked. This experiment was applied to two datasets with more than 20 features (i.e., kr-vs-kp and waveform). These two datasets with different number of selected features were tested on both C4.5 and SVM, and the average classification accuracies of these features are shown in Figure 2.

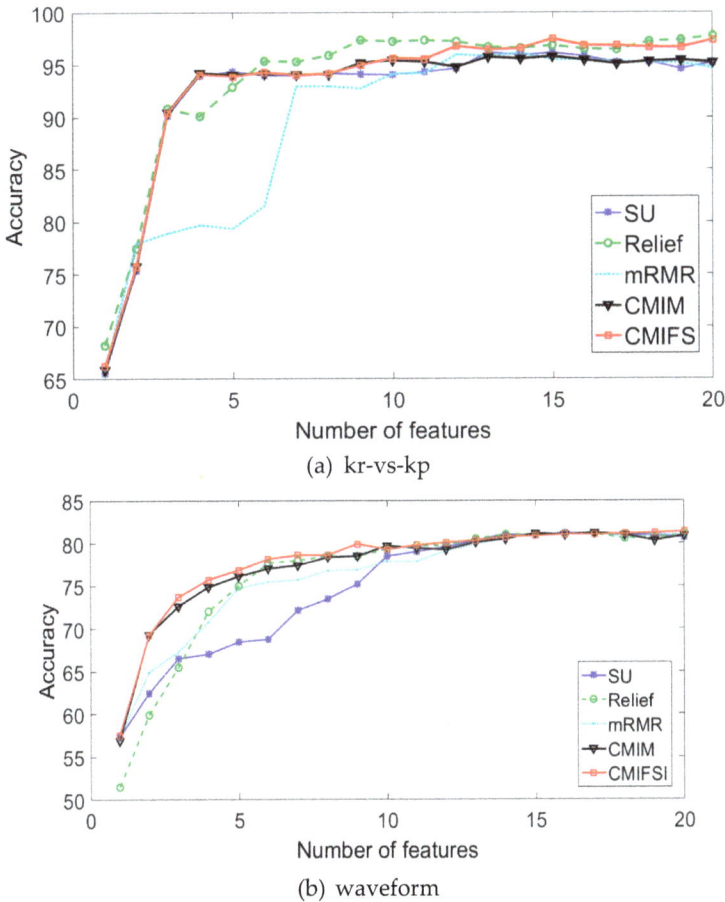

(a) kr-vs-kp

(b) waveform

Figure 2. Average accuracy with different number of selected features using different selector.

The result in Figure 2a shows that Relief seems to outperform other selectors in the majority of cases for dataset "kr-vs-kp". The result may be confusing since Relief does not consider feature redundancy and interaction. A similar result was obtained by reference [27] on this dataset and this may be due to the properties inherent to this dataset. For other selectors which are all based on information metric, CMIFSI achieves the best performance, which is comparable to Relief to some extent. And the superiority of CMIFSI increases as more features are added for learning. For dataset "waveform" (shown in Figure 2b), CMIFSI outperforms other selectors in most cases. For instance, all plots of CMIFSI are higher than others when feature number is less than ten especially in the number of 9. As more features are added, all selectors tend to perform comparably. It is worth noting that with the increase of the number of features, it more sources may needed for calculation as part of the process of prediction.

7. Discussion and Conclusions

In this paper, several feature selection algorithms based on information metric were reviewed in the framework of a general evaluation criterion. Then feature interaction was introduced and some new definitions were proposed to better analyze feature relationships. The state-of-the art methods like CMIM based on conditional mutual information are not rigorous enough to select features, because they don't consider the relationship between features beyond relevance and redundancy. To address the drawback that most of these traditional feature selection methods ignore feature interaction, a new algorithm CMIFSI based on conditional mutual information was proposed to take interaction into consideration. The main idea of CMIFSI is to penalize feature redundancy and compensate feature interaction in the evaluation criterion. Experiments were conducted to compare CMIFSI with other 6 up-to-data feature selection algorithms on both synthetic and benchmark datasets, and the results showed that CMIFSI works well and outperforms other algorithms in most cases. Many features are related to each other, so if we only select features by considering relevant or redundant features while ignoring feature interactions, some feature interactions may cause bad performance on the results. Therefore, the methods that consider feature interaction perform better on some datasets than the methods that do not. Based on the mutual information methods, we proposed exploiting feature interaction to capture more information, making the classifier more efficient for prediction.

Further work is still needed to improve the performance stability of this new algorithm. Furthermore, while CMIFSI adopts an approximation method to estimate feature redundancy and interaction, other new methods are called for to better handle feature interaction. It is still a challenging task to deal with feature interaction.

Author Contributions: Methodology, L.H.; software, Z.L.; writing—original draft preparation, W.H.; writing—review and editing, L.H.; supervision, conceptualization, formal analysis, and editing, J.L.

Funding: This research was funded by the National Natural Science Foundation of China (U1664264,U1509203).

Conflicts of Interest: The authors declare no conflict of interest.

References

1. Guyon, I.; Elisseeff, A. An Introduction to Variable and Feature Selection. *J. Mach. Learning Res.* **2002**, *3*, 1157–1182.
2. Maldonado, S.; Weber, R. A wrapper method for feature selection using support vector machines. *Inf. Sci.* **2009**, *179*, 2208–2217. [CrossRef]
3. Bolón-Canedo, V.; Sánchez-Maroño, N.; Alonso-Betanzos, A. Recent advances and emerging challenges of feature selection in the context of big data. *Knowledge-Based Syst.* **2015**, *86*, 33–45. [CrossRef]
4. Dash, M.; Liu, H. Feature Selection for Classification. *Intel. Data Anal.* **1997**, *1*, 131–156. [CrossRef]
5. Jakulin, A.; Bratko, I. Testing the Significance of Attribute Interactions. In Proceedings of the twenty-first international conference on Machine learning, Banff, AB, Canada, 4–8 July 2004.
6. Zhao, Z.; Liu, H. Searching for interacting features in subset selection. *Intel. Data Anal.* **2009**, *13*, 207–228. [CrossRef]

7. Zeng, Z.; Zhang, H.; Zhang, R.; Yin, C. A novel feature selection method considering feature interaction. *Pattern Recogn.* **2015**, *48*, 2656–2666. [CrossRef]
8. Tao, H.; Hou, C.; Nie, F.; Jiao, Y.; Yi, D. Effective Discriminative Feature Selection with Nontrivial Solution. *IEEE Trans. Neural Netw. Learning Syst.* **2016**, *27*, 3013–3017. [CrossRef] [PubMed]
9. Murthy, C.A.; Chanda, B. Generation of Compound Features based on feature Interaction for Classification. *Expert Syst. Appl.* **2018**, *108*, 61–73.
10. Chang, X.; Ma, Z.; Lin, M.; Yang, Y.; Hauptmann, A.G. Feature Interaction Augmented Sparse Learning for Fast Kinect Motion Detection. *IEEE Trans. Image Process.* **2017**, *26*, 3911–3920. [CrossRef] [PubMed]
11. Yin, Y.; Zhao, Y.; Zhang, B.; Li, C.; Guo, S. Enhancing ELM by Markov Boundary Based Feature Selection. *Neurocomputing* **2017**, *261*, 57–69. [CrossRef]
12. Brown, G.; Pocock, A.; Zhao, M.J.; Luján, M. Conditional Likelihood Maximisation: A Unifying Framework for Information Theoretic Feature Selection. *J. Mach. Learning Res.* **2012**, *13*, 27–66.
13. Lewis, D.D. Feature Selection and Feature Extraction for Text Categorization. In Proceedings of the workshop on Speech and Natural Language, New York, NY, USA, 23–26 February 1992; pp. 212–217.
14. Press, W. Numerical Recipes in FORTRAN: The Art of Scientific Computing. Available online: https://doi.org/10.5860/choice.30-5638a (accessed on 29 May 2019).
15. Battiti, R. Using mutual information for selecting features in supervised neural net learning. *IEEE Trans. Neural Netw.* **1994**, *5*, 537–550. [CrossRef] [PubMed]
16. Peng, H.; Long, F.; Ding, C. Feature selection based on mutual information: criteria of max-dependency, max-relevance, and min-redundancy. *IEEE Trans. Pattern Anal. Mach. Intel.* **2005**, *27*, 1226–1238. [CrossRef] [PubMed]
17. Yang, H.H.; Moody, J. Data visualization and feature selection: New algorithms for non-gaussian data. *Advances in Neural Information Processing Systems.* 2000, pp. 687–693. Available online: https://www.researchgate.net/publication/2460722 (accessed on 29 May 2019).
18. Fleuret, F. Fast Binary Feature Selection with Conditional Mutual Information. *J. Mach. Learn. Res.* **2004**, *5*, 1531–1555.
19. Yu, L.; Liu, H. Efficient Feature Selection via Analysis of Relevance and Redundancy. *J. Mach. Learn. Res.* **2004**, *5*, 1205–1224.
20. Tesmer, M.; Estevez, P.A. AMIFS: Adaptive feature selection by using mutual information. In Proceedings of the 2004 IEEE International Joint Conference on Neural Networks, Budapest, Hungary, 25–29 July 2004; pp. 303–308.
21. Cheng, G.; Qin, Z.; Feng, C.; Wang, Y.; Li, F. Conditional Mutual Information-Based Feature Selection Analyzing for Synergy and Redundancy. *Etri J.* **2011**, *33*, 210–218. [CrossRef]
22. John, G.H.; Kohavi, R.; Pfleger, K. Irrelevant Features and the Subset Selection Problem. In Proceedings of the Machine Learning Proceedings, New Brunswick, NJ, USA, 10–13 July 1994; pp. 121–129.
23. Wang, G.; Lochovsky, F.H. Feature selection with conditional mutual information maximin in text categorization. In Proceedings of the 13th ACM international conference on Information and knowledge management, Washington, DC, USA, 8–13 November 2004; pp. 342–349.
24. Kira, K.; Rendell, L.A. A Practical Approach to Feature Selection. In Proceedings of the Ninth International Workshop on Machine Learning, Aberdeen, Scotland, 1–3 July 1992; pp. 249–256.
25. Hall, M.A. Correlation-based Feature Selection for Discrete and Numeric Class Machine Learning. In Proceedings of the Seventeenth International Conference on Machine Learning, Stanford, CA, USA, 29 June–2 July 2000; pp. 359–366.
26. UCI Repository of Machine Learning Databases. Available online: http://www.ics.uci.edu/~{}mlearn/MLRepository.html (accessed on 29 May 2019).
27. Liu, H.; Sun, J.; Liu, L.; Zhang, H. Feature selection with dynamic mutual information. *Pattern Recognit.* **2009**, *42*, 1330–1339. [CrossRef]

symmetry

MDPI

Article

A Time-Space Network Model Based on a Train Diagram for Predicting and Controlling the Traffic Congestion in a Station Caused by an Emergency

Zihan Qu [1] and Shiwei He [2,*]

[1] Key Laboratory of Transport Industry of Big Data Application Technologies for Comprehensive Transport, Ministry of Transport, Beijing Jiaotong University, Beijing 100044, China; 17120866@bjtu.edu.cn
[2] School of Transportation, Beijing Jiaotong University, Beijing 100044, China
* Correspondence: shwhe@bjtu.edu.cn

Received: 23 May 2019; Accepted: 10 June 2019; Published: 12 June 2019

Abstract: Timely predicting and controlling the traffic congestion in a station caused by an emergency is an important task in railway emergency management. However, traffic forecasting in an emergency is subject to a dynamic service network, with uncertainty surrounding elements such as the capacity of the transport network, schedules, and plans. Accurate traffic forecasting is difficult. This paper proposes a practical time-space network model based on a train diagram for predicting and controlling the traffic congestion in a station caused by an emergency. Based on the train diagram, we constructed a symmetric time-space network for the first time by considering the transition of the railcar state. On this basis, an improved A* algorithm based on the railcar flow route was proposed to generate feasible path sets and a dynamic railcar flow distribution model was built to simulate the railcar flow distribution process in an emergency. In our numerical studies, these output results of our proposed model can be used to control traffic congestion.

Keywords: railway transportation; time-space network; A* algorithm; traffic congestion; traffic forecasting; traffic control; railcar flow distribution

1. Introduction

1.1. The Description of the Subject of Research and the Motives to Take

When the station in a transportation network stops operation due to natural disasters and other emergencies, trains cannot pass through the station. Thus, trains arriving or passing through the station have to be rerouted. As there are capacity limitation stations in the transportation network, after a period of time, the railcar flow in some of the other stations will become overstocked, thus forming congestion. The congestion will greatly reduce the network's transport capacity. Failure to timely predict the congestion will result in propagation of congestion and even the paralysis of the transportation network.

Effectively predicting and controlling the traffic congestion in a station caused by an emergency must have the following characteristics:

- Strong timeliness. Because of the transmissibility of congestion, it is of great significance to predict congestion in time for maintaining and improving the capacity of the transport network in an emergency.
- Time-space. The final output of the model should be which station will become congested first and when the station begins to become congested. These output results will become the basis of the decision of the railway transportation managers.

- Strongly controllable. Rail transportation is a strongly controlled mode of transport, which means that the turnover process of railcar flow and the distribution of railcar flow in the transport network are subject to schedules and plans.
- Multiple highly dynamic constraints. Congestion prediction and control in an emergency is a complicated system involving many components. First, the capacity in the network will be reduced in an emergency. Second, the state of railcar flow will change dynamically with the operation process and the state of some railcar flows changes more than twice within the decision-making period horizon. Third, the completion rate of the transportation plan and scheduled train departure will be affected by the emergency.

Congestion prediction and control problem is a new research field with strong exploration, which is not only the premise of rational utilization and enhancement of transport capacity, but also the premise of adjustment of rail transit. However, large-scale railcar flow, the complex railcar flow turnover process, and dynamic constraints (schedule constraints, loading plan constraints, emptying plan constraints, capacity constraints) make congestion prediction and control in an emergency very difficult.

To summarize, reasonable prediction and control of traffic congestion in a station caused by an emergency has already become a theoretically and practically urgent problem.

1.2. Literature Review

Recent research efforts have shown the ability to model congestion prediction and control of road transportation [1–5] and the metro [6–10]. Sun et al. [11] explored the hazardous materials route problem in the road-rail multimodal transportation network with a hub-and-spoke structure. Jiang et al. [12] proposed an emergency material vehicle dispatching and routing model which has multi-objective and multi-dynamic-constraint. Lu et al. [13] presented an approach to measure and analyze the vulnerability of urban rail transit network based on accessibility, which explicitly considers the characteristics of rail transit passenger flow, travel cost changes and alternative modes of transportation. Deng et al. [14] designed for passenger bus bridge circuit in the case of urban rail traffic interruption. The results of these models have provided a basis for emergency traffic control planning and evacuation order in time for transportation disruptions caused by equipment failures, accidents, peak traffic, and natural disasters.

The core problem of prediction and control of traffic congestion in a station caused by an emergency is the railcar management problem (this problem mainly concerns the distribution of railcar flow in the transportation network). White et al. [15] generated the network from the time-space graph and solved the transshipment problem by using the improved out-of-kilter algorithm. Static formulations solvable as transportation problems were proposed by Allman et al. [16]. Spieckermann et al. [17] formulated the problem of empty railcar assignment as a scheduling problem with machine to represent railcar and job to represent railcar process request. Holmberg et al. [18] proposed a multi-commodity network flow model for operational distribution of empty railcars. Narisetty et al. [19] proposed an optimization model to implement the union Pacific railway (UP) to allocate empty freight railcars based on demand. Gorman et al. [20] discussed the practice of optimizing the allocation of empty freight railcars for shippers by BNSF and CSX, two major U.S. freight companies. Heydari et al. [21] proposed a new solution to the problem of empty railcars distribution, which not only meets the needs of customers, but also takes into account the practical technical and business needs. Jiang et al. [22] proposed an integrated scheduling coordination model, which aimed to maximize the production connection time (PCT) and production delivery time (PDT). Sun et al. [23] explored the impact of demand uncertainty on the problem of the run-level cargo path of the multi-modal transport network based on rail transit based on time arrangement and road traffic based on time flexibility. Aiming at the design problem of railway express freight service network, Lin et al. [24] proposed a bi-level programming model, and transformed the lower-level model into a linear model by using linearization technology, which can be solved directly by standard optimization solver. Lawley et al. [25] proposed a time-space network flow model for scheduling regular bulk rail traffic from suppliers to customers. Sayarshad et al. [26]

proposed a solution to optimize the fleet size and the configuration of freight railcars under uncertain demand conditions. However, the traditional railcar management model such as the empty railcar distribution model does not consider the transition of the railcar state (empty railcar to loaded railcar and loaded railcar to empty railcar) [27]. The comparisons between a few published studies on the railcar management model and this work are presented in Table 1. The innovation points of this paper are as follows:

- Based on the train diagram, we constructed a time-space network for the first time by considering the transition of the railcar state.
- The problem is split into two sub-problems: feasible path set generation and railcar flow distribution. An improved A* algorithm based on railcar flow route was used to generate a feasible path set of any pair of transportation demands. A dynamic railcar flow distribution model in an emergency which can be solved by mathematical programming software was established to simulate the process of railcar flow operation in emergency, and the railcar flow distribution in the future period was estimated based on the distribution results of railcar flow in the time-space network, so as to predict where there will be traffic congestion and when the traffic congestion will begin in the station.

Table 1. Comparisons between a few published studies and this work.

Studies	Holmberg et al. (1998)	Narisetty et al. (2008)	Heydari et al. (2017)	This Model
Combines railcar routing and distribution?	Yes	No	No	Yes
Formulation	Capacitated network design	Transportation problem	Path-based capacitated network	Multi-commodity flow problem
Considers transition of railcar state?	No	No	No	Yes
Capacitated?	Yes	No	Yes	Yes
Solution method	Tabu Search	LP	LP	Heuristic decomposition

The paper is organized as follows: Section 1 introduces the subject of research and the motives to take, and provides a brief survey of the literature devoted to the congestion prediction and control problem. In Section 2, we describe the congestion prediction and control problem in detail. In Section 3, an improved A* algorithm based on railcar flow route and a dynamic railcar flow distribution model in emergency are proposed. Sections 4 and 5 present a numerical example and conclusion respectively.

2. Problem Description

In this section, first, we describe the process resulting in traffic congestion in a station due to the emergency to illustrate the need for timely predicting and controlling the traffic congestion. Then, we analyze the turnover process of railcar and transition of the railcar state in detail. Finally, we abstract the railcar management problem into a multi-commodity flow problem.

2.1. The Process Resulting in Traffic Congestion due to the Emergency

We describe the process resulting in traffic congestion due to the emergency in Figure 1. There are five normal stations and two railcar flow types (a railcar flow type refers to a class of railcar flows with the same origin station, destination station, and flow route) in period 1. In period 2, station B is interrupted due to an emergency, so the railcar flows with type 1 have to adjust the transport path. When a large number of railcar flows of type 1 and type 2 converge at station C simultaneously, it is likely that the number of real-time railcars in station C exceeds the capacity of the station, thus forming traffic congestion in period 3. For both types of railcar flows, the two paths to be traveled are blocked, which will result in the accumulation of railcars at station B and station A in period 4. Stations A and B become congested stations when the number of accumulated railcars exceeds their capacity. The propagation of congestion will greatly decrease the station capacity and transportation efficiency, therefore, it is of great significance to predict and control the traffic congestion in time.

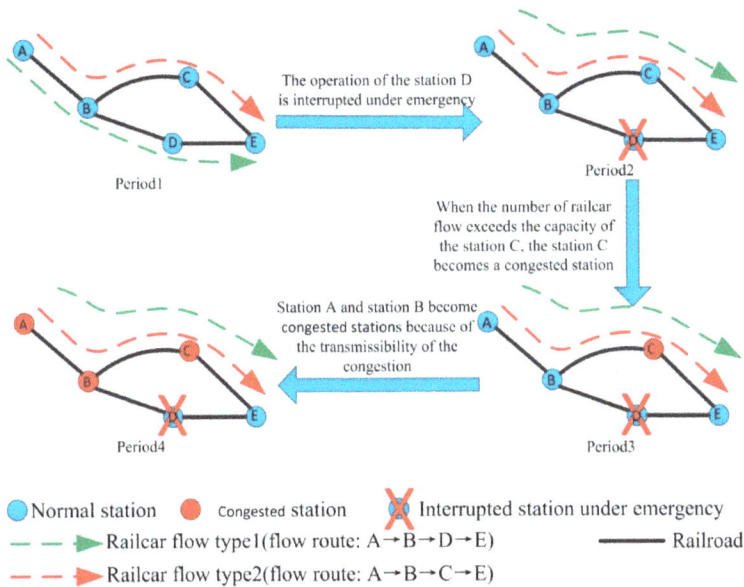

Figure 1. Schematic diagram of the process resulting in traffic congestion due to the emergency.

2.2. The Turnover Process of Railcar and Transition of Railcar State

The railcar flow (a railcar flow refers to a collection of railcars with the same railcar flow type and generated in the same period) starts from the origin station and ends at the destination station through various transfer stations. As shown in Figure 2, each railcar flow is assigned into different trains along the way and each railcar flow can go through 11 activities, including train arrival, train break-up, loaded railcar delivery, empty railcar delivery, railcar loading, railcar unloading, loaded railcar pickup, empty railcar pickup, railcar assembly, train formation, and train departure. The implications of these 11 activities are as follows:

- Train arrival. The activity begins when the inbound train carrying railcars arrives at the station's receiving yard and ends when railcars detached from the inbound train are pushed over the top of the station's hump.
- Train break-up. The activity begins when the railcars starts to slide off the hump and ends when all railcars detached from the inbound train roll down to the appropriate track in the station's switchyard.
- Loaded railcar delivery. The activity begins when all loaded railcars detached from the inbound train roll down to the station's switchyard and ends when the loaded railcars are pulled to the freight yard by the local locomotive.
- Empty railcar delivery. The activity begins when all empty railcars detached from the inbound train roll down to the station's switchyard and ends when the empty railcars are pulled to the freight yard by the local locomotive.

- Railcar loading. The activity begins when all empty railcars detached from the inbound train start to load in the freight yard and ends when the empty railcars have finished loading the goods.
- Railcar unloading. The activity begins when all loaded railcars detached from the inbound train start to unload in the freight yard and ends when the loaded railcars have finished unloading the goods.
- Loaded railcar pickup. The activity begins when the loaded railcars detached from the inbound train start to hang to the local locomotive and ends when all the loaded railcars are pulled back to the station's switchyard from the freight yard by the local locomotive.
- Empty railcar pickup. The activity begins when the empty railcars detached from the inbound train start to hang to the local locomotive and ends when all the empty railcars are pulled back to the station's switchyard from the freight yard by the local locomotive.
- Railcar assembly. The activity begins when all railcars detached from the inbound train are pulled back to the station's switchyard from the freight yard and ends when the railcars are able to form an outbound train.
- Train formation. The activity begins when the railcars detached from many inbound trains are able to form an outbound train and ends when the formation of an outbound train is completed.
- Train departure. The activity begins when the formation of an outbound train is completed and ends when the outbound train carrying railcars starts from the departure yard.

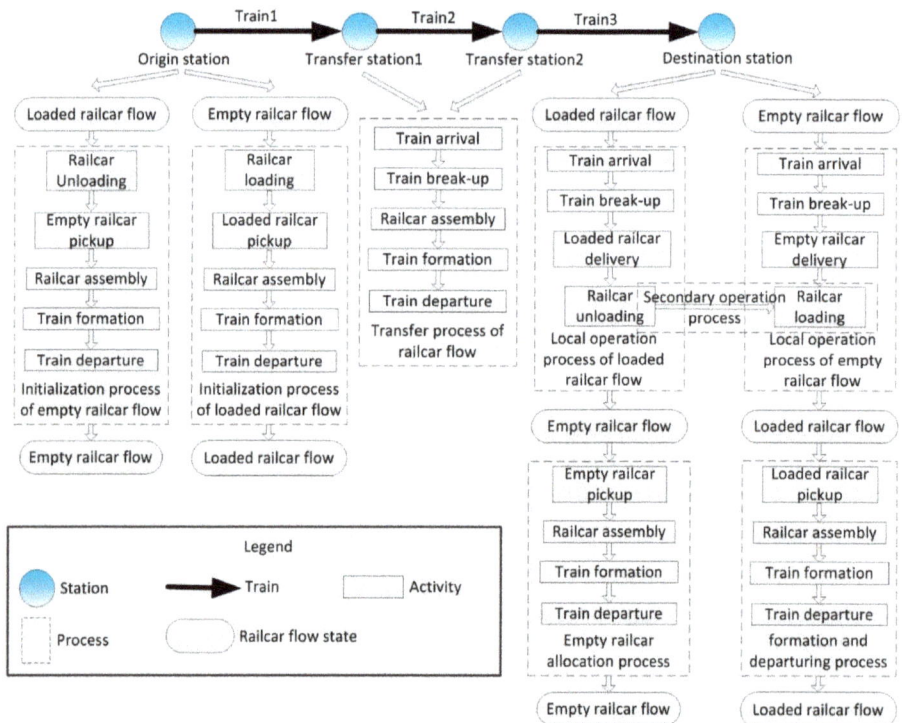

Figure 2. Schematic diagram of the turnover process of railcar and transition of railcar state.

The operation process of a railcar flow is different in its origin station, transfer station, and destination station. The state of each railcar flow changes at the departure and destination stations, but does not change at the transfer stations. The processes of a railcar flow in different stations are as follows:

- Origin station. The operation processes of a railcar flow mainly include the initialization process of loaded railcar flow and the initialization process of empty railcar flow. The initialization process of empty railcar flow consists of railcar unloading, empty railcar pickup, railcar assembly, train formation, and train departure. The process realizes the transition of loaded railcar flow into empty railcar flow. The initialization process of loaded railcar flow consists of railcar loading, loaded railcar pickup, railcar assembly, train formation, and train departure. The process realizes the transition of loaded railcar flow into empty railcar flow.

- Transfer station. The main operation process is the transfer process of railcar flow. The transfer process of railcar flow consists of train arrival, train break-up, railcar assembly, train formation, and train departure.

- Destination station. The operation processes of a railcar flow mainly include the local operation process of loaded railcar flow, the secondary operation process, the local operation process of empty railcar flow, the empty railcar allocation process, and the formation and departure process. The local operation process of loaded railcar flow consists of train arrival, train break-up, loaded railcar delivery, and railcar unloading. The secondary operation process is the process of reloading for the loaded railcar after unloading. The local operation process of empty railcar flow consists of train arrival, train break-up, empty railcar delivery, and railcar loading. The empty railcar allocation process consists of empty railcar pickup, railcar assembly, train formation, and train departure. The formation and departure process consists of loaded railcar pickup, railcar assembly, train formation, and train departure. The local operation process of loaded railcar flow realizes the transition of loaded railcar flow into empty railcar flow. The secondary operation process and the local operation process of empty railcar flow realize the transition of loaded railcar flow into empty railcar flow.

2.3. Problem Abstraction

Kwon et al. [28] defined the railcars dispatching process as the blocks and sequences of railcars s that are scheduled to continue moving, as well as the stations where the railcar will be taken and started during its travel. In addition, they formulated the railcar management problem as a multi-commodity flow problem on a transport network.

We abstract the transportation process of railcar flow in the railroad network into the service network with time attributes. The loaded railcar flow has a fixed origin station and destination station in the transportation process, and cannot be separated [29,30]. Therefore, each loaded railcar flow can be regarded as a commodity in the service network, so the loaded railcar flow management problem can be regarded as a multi-commodity flow problem [31]. The empty railcar flow does not have a fixed origin station and destination station in the transportation process. However, the empty railcar flow comes from the loaded railcar flow after unloading and is consumed by converting to the loaded railcar flow after loading, so the empty railcar flow can also be regarded as a commodity in the service network. In this paper, we abstracted the railcar management problem into a multi-commodity flow problem.

3. Methods

The flowchart of predicting traffic congestion in a station in an emergency is shown in Figure 3. We disassemble the railcar management problem into two sub-problems. One of the sub-problems is the construction of a time-space network based on the turnover process of railcar and transition of railcar state, the other sub-problem is the construction of a dynamic railcar flow distribution model in an emergency. Based on the solution results of the model, we can calculate the real-time number of railcars at each station in each period. Finally, the congested station is determined by comparing the real-time number of railcars at the station and the storage capacity of the station.

We define each OD as a set of railcars flow which has the same origin station and destination station.

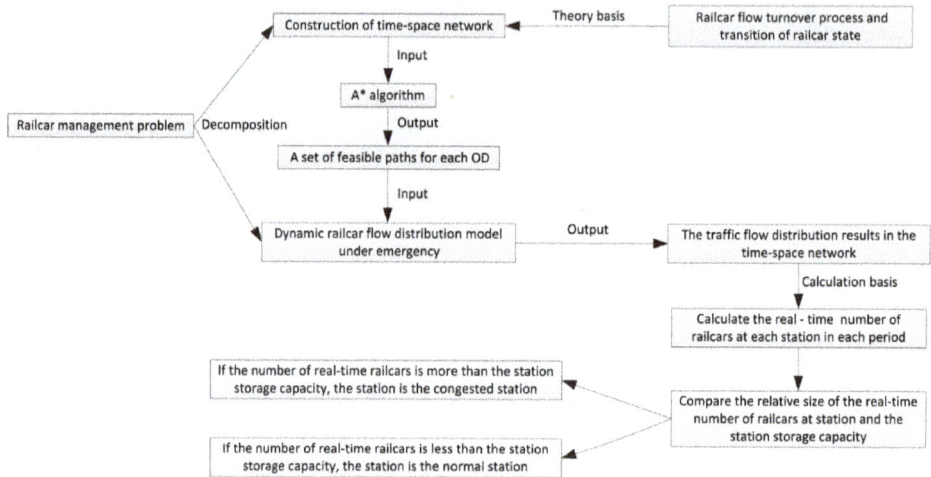

Figure 3. Flowchart of predicting traffic congestion in a station in an emergency.

3.1. Construction of time-space Network Based on the Train Diagram

In order to simplify the scale of the problem, a railway bureau is taken as the research object, and the marshaling station, district station, large freight station, and boundary station of the railway bureau are taken as the main fulcrum stations, while the railcar flow of the intermediate station is merged in accordance with the section center rule [32]. Set the fulcrum station set in the railway bureau as S, and define the time-space network as $G = (N, A)$, where N denotes the set of nodes and A denotes the set of arcs in the network. The network takes 18:00 of the current day to 18:00 of the next day as a decision cycle. The train diagram is the main frame of constructing the time-space network.

Network G takes the Arrival Node (AN) and Departure Node (DN) of as the main time-space nodes. The AN represents the arrival moment of a train and the DN represents the departure moment of a train. The Loaded Railcar State End Node (LRSEN) and the Empty Railcar State End Node (ERSEN) represent the mutual transformation moment of loaded railcar flow and empty railcar flow. Secondary Operation Completion Node (SOCN) represents the completion moment of secondary operation of loaded railcar flow. Each station sets two Initial Nodes (IN), which are divided into loaded railcar layer and empty railcar layer, to represent the start moment of the railcar flow in the decision cycle. Each station sets two End Nodes (EN), which are also divided into loaded railcar layer and empty railcar layer, to represent the end moment of the railcar flow in the decision cycle. The type indexes of time-space nodes are shown in Table 2.

Table 2. Type indexes of time-space nodes.

Time-Space Node Type	Type Index
AN	1
DN	2
LRSEN	3
ERSEN	4
SOCN	5
IN	6
EN	7

Network G takes the train as the Train Arc (TA), and adds the corresponding Local Operation Arc (LOA) and Secondary Operation Arc (SOA) based on the technical operation time of each fulcrum station, and realizes the connection of each arc through the Formation and Departure Arc (FADA),

Empty Railcar Allocation Arc (ERAA), Transfer Operation Arc (TOA), and Stay Arc (SA). The Initial Arc (IA) represents the initialization process of the railcar flow. The End Arc (EA) and the Detention Arc (DA) represent the process of railcar flow detention to the end of the decision cycle. At the same time, hierarchy attributes were added to distinguish the Loaded Railcar Layer (LRL), Empty Railcar Layer (ERL), and Public Layer (PL). At present, the research on dynamic railcar flow operation lacks the consideration of the secondary operation of the loaded railcar flow, and this paper represents the secondary operation process of the loaded railcar flow through the connection between LRSEN and SOCN. The network topology relationship between time-space nodes and time-space arcs is shown in Figure 4. The type indexes of time-space arcs are shown in Table 3 and the hierarchy indexes of the hierarchy set are shown in Table 4.

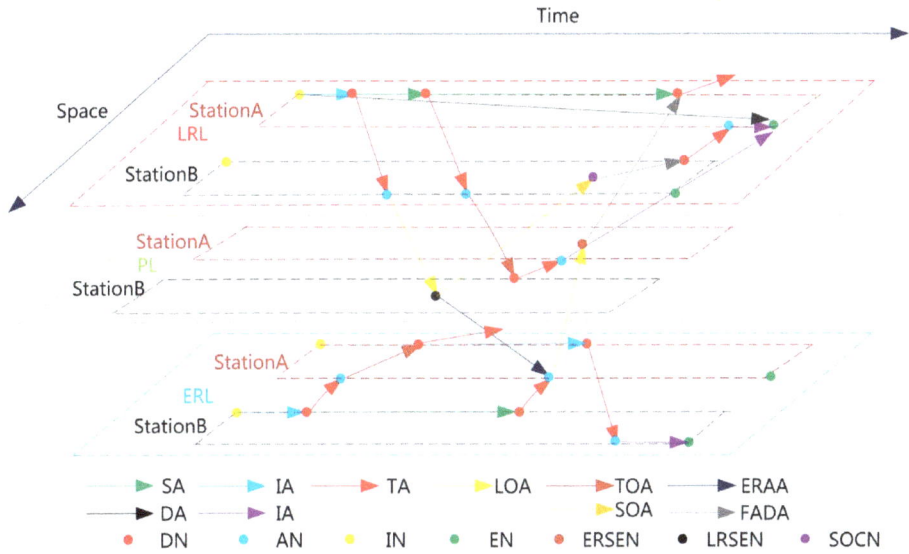

Figure 4. Schematic diagram of time-space network.

Table 3. Type indexes of time-space arcs.

time-space Arc Type	Type Index
TA	1
LOA	2
SOA	3
FADA	4
ERAA	5
TOA	6
SA	7
IA	8
EA	9
DA	10

Table 4. Hierarchy indexes of hierarchy.

Hierarchy	Hierarchy Index
LRL	1
ERL	2
PL	3

Because loaded railcar flow has the characteristics of a tree path, this paper describes the transport process of loaded railcar flow by constructing a time-space path which is a set of time-space arcs that satisfy the constraint of time continuity. The time-space path of loaded railcar flow is divided into four categories: Initial Path (IP), Secondary Operation Path (SOP), Empty to Loaded Path (ETLP), and Detention Path (DP). The IP represents the transport process of the initial loaded railcar flow. The path starts from the IN and ends at the LRSEN. The path takes the LOA as the final arc. The SOP represents the transport process of secondary operation loaded railcar flow. The path starts from the LRSEN and ends at the ETN. The path takes the SOA as the first arc and cannot contain the LOA. ETLP represents the transport process of self-loading railcar flow. The path starts from the ERSEN and ends at the EN. The DP representing the loaded railcar flow remains at the station throughout the decision cycle. The path starts from the IN and ends at the EN. The type indexes of time-space paths are shown in Table 5.

Table 5. Type indexes of time-space paths.

time-space Path Type	Type Index
IP	1
SOP	2
ETLP	3
DP	4

3.2. Feasible Path Set Generation Algorithm Based on Improved A* Algorithm

According to the requirements of a train marshaling plan, each OD (a set of railcars flow which has the same origin station and destination station) has a fixed railcar flow connection mode, thus each OD has a specific route, that is, the railcar flow route. In this paper, the feasible path has a broader meaning in the time-space network. For each OD, a set of time-space arc sets that satisfies the time continuity constraint and the constraint of the railcar flow route station sequence is a feasible time-space path.

A* algorithm is an efficient algorithm for solving K-shortest path [33–35]. We improve the algorithm based on the railcar flow route. Set OD index as f, the flowchart of improved A* algorithm is shown in Figure 5 and the algorithm for generating the feasible path set is as follows:

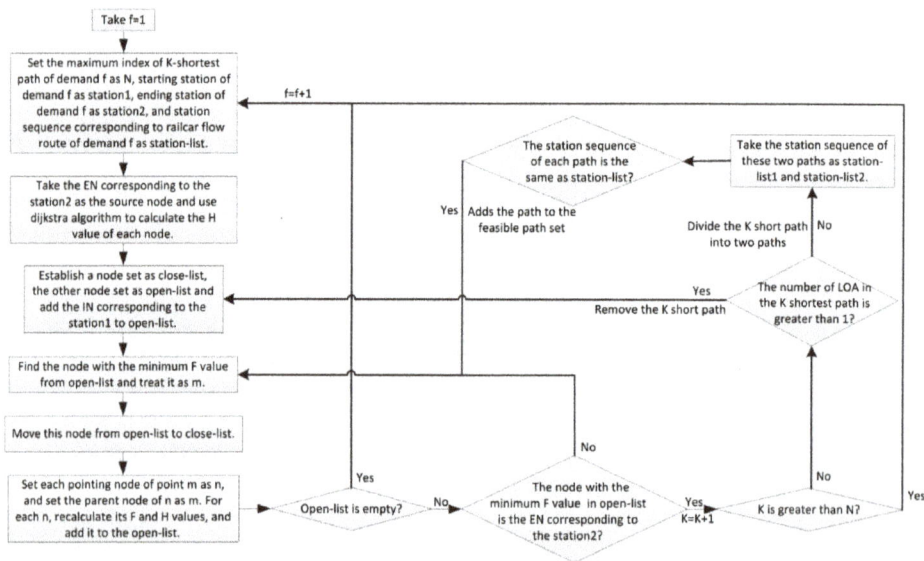

Figure 5. Flowchart of improved A* algorithm.

Step 1: Take f = 1, and turn to Step 2;

Step 2: Set the maximum index of K-shortest path of demand f to N, and initial K = 0, set the starting station of demand f as station1, and the final station as station2. Traverse the railcar flow route set, find the railcar flow route of the starting station as station1 and ending station as station2, and record the station sequence corresponding to the railcar flow route as station-list, and turn to Step 3;

Step 3: Taking the EN corresponding to station2 as the source node, the Dijkstra algorithm is used to find the shortest distance between the source node and other nodes in the time-space network, which is taken as the H value of each node, and turn to Step 4;

Step 4: Establish a node set as close-list which has been estimated and a node set as open-list which will be estimated. Add the IN corresponding to the station1 to the open-list, and turn to Step 5;

Step 5: Traverse the nodes in the open-list to find the node with the lowest F value and treat it as m to be processed, and turn to Step 6;

Step 6: Move this node from the open-list to the close-list, and turn to Step 7;

Step 7: For each pointing node of m as n, set its parent node as m, and recalculate its G and F values, and add them to the open-list, and turn to Step 8;

Step 8: If the open-list is empty, the target K short path is not found, and the search fails, then f = f + 1, and turn to Step 2. If the open-list is not empty, traverse the open-list, if the node with the lowest F value is the EN corresponding to station2, the target K shortest path is found, then K = K + 1, and turn to Step 9; otherwise turn to Step 5;

Step 9: First, judge whether K is greater than N, if K is greater than N, f = f + 1, then turn to Step 2. Otherwise, judge the number of LOA in the K shortest path. If the number is greater than 1, remove the K short path and turn to Step 4; if the number is equal to 1, divide the K short path into two paths with the end node of the LOA as the cutting node, and turn to Step 10;

Step 10: Take the station sequence of these two paths as station-list1 and station-list2. Determine whether the station sequence of each path is the same as the station-list, and if so, add the path to the feasible path set, and turn to Step 5;

Step 11: When all OD demands are traversed to generate a feasible path set, the algorithm is finished.

3.3. The Dynamic Railcar Flow Distribution Model in an Emergency

The key to predict the traffic congestion is to grasp the real-time number of railcars at a station. The train runs according to the train diagram, and the time-space network constructed based on the train diagram conforms to the situation of railway field operation. In this paper, based on the time-space network, the distribution results of railcar flow on the time-space network are obtained by establishing the dynamic railcar flow distribution model in an emergency. According to the arc flow, the real-time number of railcars at each station can be obtained. By comparing the real-time number of railcars at each station with the storage capacity of the station, it can be concluded whether the station belongs to the congestion stage, so as to be able realize the prediction of traffic congestion in the station.

3.3.1. Set Parameter and Variable Definitions

The general notations used in the formulation are listed in Table 6.

Table 6. Notations used in the formulation.

Sets	Descriptions
H	Set of hierarchy index;
N	Set of time-space node;
B_N	Set of type indexes of time-space nodes;
A	Set of time-space arc;
A_{ns}	Set of time-space arcs starting from time-space node $n, n \in N$;
A_{ne}	Set of time-space arcs ending in time-space node $n, n \in N$;

Table 6. *Cont.*

B_A	Set of type indexes of time-space arcs;
P	Set of time-space path;
P_{ns}	Set of time-space paths starting from time-space node $n, n \in N$;
P_{ne}	Set of time-space paths ending in time-space node $n, n \in N$;
B_P	Set of type indexes of time-space paths;
F_1	Set of initial loaded railcar flow;
P_{f1}	Set of feasible paths of initial loaded railcar flow $f_1, f_1 \in F_1$;
F_2	Set of initial empty railcar flow;
A_{f2}	Set of feasible arcs of initial empty railcar flow $f_2, f_2 \in F_2$;
L	Set of loading plan;
P_l	Set of time-space path in line with loading plan $l, l \in L$;
E	Set of emptying plan set;
A_e	Set of time-space arc in line with emptying plan $e, e \in E$;
A_T	Set of train arc affected in an emergency;
G	Set of period;
S	Set of station;
A_g^s	Set of time-space arcs used to calculate the number of railcars at station s in period g

Parameters	Descriptions
b_n	Type index of time-space node n, $b_n \in B_N$, $n \in N$;
h_n	Hierarchy index of time-space node n, $h \in H$, $n \in N$;
b_a	Type index of time-space arc a, $b_a \in B_A$, $a \in A$;
h_a	Hierarchy index of time-space arc a, $h \in H$, $a \in A$;
t_a	Time of time-space arc a, $a \in A$;
c_a	Capacity of time-space arc a, $a \in A$;
b_p	Type index of time-space path p, $b_p \in B_P$, $p \in P$;
h_p	Hierarchy index of time-space path p, $h \in H$, $p \in P$;
t_p	Time of time-space path p, $p \in P$;
A_p	time-space arc set of time-space path p, $p \in P$;
q_{f1}	Number of initial loaded railcar flow $f_1, f_1 \in F_1$;
q_{f2}	Number of initial empty railcar flow $f_2, f_2 \in F_2$;
q_l	Planned loading number for one day of loading plan $l, l \in L$;
r_l	Completed when emergency occurs, $l \in L$;
q_e	Planned emptying number for one day of emptying plan $e, e \in E$;
r_e	Completed when emergency occurs, $e \in E$;
m	Maximum number of marshaling railcars in a train;
st_a	Start moment in minutes of time-space arc a, $a \in A$;
et_a	End moment in minutes of time-space arc a, $a \in A$;
et_g	End moment in minutes of period $g, g \in G$;
s_n	Station of time-space node n, $s_n \in S$, $n \in N$;
n_{as}	Start node of time-space arc a, $a \in A$;
n_{ae}	End node of time-space arc a, $a \in A$;

Decision variables	Descriptions
x_a	Empty railcar flow number of time-space arc a;
y_p	Loaded railcar flow number of time-space path p;
z_p^{f1}	Loaded railcar flow number of feasible path p of initial loaded railcar flow $f1$;
v_a^{f2}	Empty railcar flow number of feasible arc a of initial empty railcar flow $f2$.
d_s^g	Number of railcars at station p in period g;

3.3.2. Objective Function

The objective is to compress the transit time of the loaded railcars at the station, reduce the running time of the empty railcars, and minimize the number of off-axle railcars. The expression is as follows:

$$
minZ = \sum_{p\in P}\left(t_p - \sum_{a\in A_p|b_a=1} t_a\right)\times y_p + \sum_{a\in A|(h_a\neq 1, b_a\neq 7, b_a\neq 10)} t_a \times x_a
$$
$$
+ \sum_{a\in A|b_a=1}\left(m - \sum_{p\in P|a\in A_p} y_p - x_a\right).
\tag{1}
$$

3.3.3. Constraint Conditions

Initial loaded railcar flow constraints: For each loaded railcar flow, the sum of the number of flows allocated on the feasible path is equal to the number of initial loaded railcar flows.

$$
\sum_{p\in P_{f1}} z_p^{f1} = q_{f1}, \quad \forall f_1 \in F_1
\tag{2}
$$

Initial empty railcar flow constraints: For each empty railcar flow, the sum of the number of flows allocated on the feasible arc is equal to the number of initial empty railcar flow.

$$
\sum_{a\in A_{f1}} v_a^{f2} = q_{f2}, \quad \forall f_2 \in F_2
\tag{3}
$$

Consistent constraints of decision variables: For each IP or DP, the sum of the number of initial loaded railcar flows allocated on that path is equal to the number of flows of that path; for each IA or DA, the sum of the number of initial empty railcar flows allocated on that arc is equal to the number of flows of that arc.

$$
\sum_{f_1\in F_1|p\in P_{f1}} z_p^{f1} = y_p, \quad \forall p \in P\big|(b_p = 1 \ or \ b_p = 4)
\tag{4}
$$

$$
\sum_{f_2\in F_2|a\in A_{f2}} v_a^{f2} = x_a, \quad \forall a \in A\big|(b_a = 8 \ or \ b_a = 10), h_a = 2
\tag{5}
$$

Time-space arc capability constraints: For each time-space arc, the sum of the number of loaded railcar flows on that arc and the number of empty railcar flows on that arc cannot exceed the capacity of that arc.

$$
\sum_{p\in P|a\in A_p} y_p + x_a \leq c_a, \quad \forall a \in A
\tag{6}
$$

Consistent constraints of exchange of LRL and ERL: For each LRSEN, the loaded railcar flow number of the IP ending at that node is equal to the sum of the loaded railcar flow number of the SOP starting from that node and the empty railcar flow number of the ERAA starting from that node. For each ERSEN, the empty railcar flow number of the LOA ending at that node is equal to the sum of the loaded railcar flow numbers of the ETLP starting from that node.

$$
\sum_{p\in P_{ne}|b_p=1} y_p = \sum_{p\in P_{ns}|b_p=2} y_p + \sum_{a\in A_{ns}|b_a=5} x_a, \quad \forall n \in N|b_n = 3
\tag{7}
$$

$$
\sum_{a\in A_{ne}|b_a=2 \ or \ b_a=8} x_a = \sum_{p\in P_{ns}|b_p=3} y_p, \quad \forall n \in N|b_n = 4
\tag{8}
$$

Flow conservation constraints of node in ERL: For each time-space node in ERL or PL, the sum of the empty railcar flow number of the time-space arc ending at that node in ERL or PL is equal to the sum of the empty railcar flow number of the time-space arc starting from that node in ERL or PL.

$$\sum_{a\in A_{ne}|h_a\neq 1} x_a = \sum_{a\in A_{ns}|h_a\neq 1} x_a, \quad \forall n \in N|h_n \neq 1 \tag{9}$$

Loading plan constraints: For each loading plan, the sum of the loaded railcar flow number of the time-space paths in line with the loading plan cannot exceed the planned loading number for one day of the plan, nor can it be lower than the completed loading number of the plan when an emergency occurs.

$$r_l \leq \sum_{p\in P_l} y_p \leq q_l, \quad \forall l \in L \tag{10}$$

Emptying plan constraints: For each emptying plan, the sum of the empty railcar flow number of time-space arcs in line with the emptying plan cannot exceed the planned emptying number for one day of the plan, nor can it be lower than the completed emptying number of the plan when an emergency occurs.

$$r_e \leq \sum_{a\in A_e} x_a \leq q_e, \quad \forall e \in E \tag{11}$$

Train operation restrictions constraints in an emergency: For each affected TA, the sum of the number of loaded railcar flow on that arc and the number of empty railcar flow on that arc is equal to zero.

$$\sum_{p\in P|a\in A_p} y_p + x_a = 0, \quad \forall a \in A_T \tag{12}$$

3.3.4. Model Synthesis

By combining constraints (1) to (12), the Dynamic Railcar Flow Distribution Model in an emergency (DRFDMUE) can be obtained.

DRFDMUE:

$$minZ = \sum_{p\in P}(t_p - \sum_{a\in A_p|b_a=1} t_a) \times y_p + \sum_{a\in A|(h_a\neq 1,b_a\neq 7,b_a\neq 10)} t_a \times x_a$$
$$+ \sum_{a\in A|b_a=1}(m - \sum_{p\in P|a\in A_p} y_p - x_a),$$

s.t.

$$\sum_{p\in P_{f1}} z_p^{f1} = q_{f1}, \quad \forall f_1 \in F_1$$

$$\sum_{a\in A_{f1}} v_a^{f2} = q_{f2}, \quad \forall f_2 \in F_2$$

$$\sum_{f_1\in F_1|p\in P_{f1}} z_p^{f1} = y_p, \quad \forall p \in P|(b_p = 1 \ or \ b_p = 4)$$

$$\sum_{f_2\in F_2|a\in A_{f2}} v_a^{f2} = x_a, \quad \forall a \in A|(b_a = 8 \ or \ b_a = 10), h_a = 2$$

$$\sum_{p\in P|a\in A_p} y_p + x_a \leq c_a, \quad \forall a \in A$$

$$\sum_{p\in P_{ne}|b_p=1} y_p = \sum_{p\in P_{ns}|b_p=2} y_p + \sum_{a\in A_{ns}|b_a=5} x_a, \quad \forall n \in N|b_n = 3$$

$$\sum_{a\in A_{ne}|b_a=2 \text{ or } b_a=8} x_a = \sum_{p\in P_{ns}|b_p=3} y_p, \quad \forall n \in N|b_n = 4$$

$$\sum_{a\in A_{ne}|h_a\neq 1} x_a = \sum_{a\in A_{ns}|h_a\neq 1} x_a, \quad \forall n \in N|h_n \neq 1$$

$$r_l \leq \sum_{p\in P_l} y_p \leq q_l, \quad \forall l \in L$$

$$r_e \leq \sum_{a\in A_e} x_a \leq q_e, \quad \forall e \in E$$

$$\sum_{p\in P|a\in A_p} y_p + x_a = 0, \quad \forall a \in A_T$$

Change constraint (10) and constraint (11) into constraint (13) and constraint (14) as follows:

$$\sum_{p\in P_l} y_p \leq q_l, \quad \forall l \in L \tag{13}$$

$$\sum_{a\in A_e} x_a \leq q_e, \quad \forall e \in E \tag{14}$$

Combining constraints (1) to (9), together with constraint (13) and constraint (14), the Dynamic Railcar Flow Distribution Model under Normality (DRFDMUN) can be obtained.

DRFDMUN:

$$minZ = \sum_{p\in P}\left(t_p - \sum_{a\in A_p|b_a=1} t_a\right) \times y_p + \sum_{a\in A|(h_a\neq 1, b_a\neq 7, b_a\neq 10)} t_a \times x_a$$

$$+ \sum_{a\in A|b_a=1}\left(m - \sum_{p\in P|a\in A_p} y_p - x_a\right),$$

s.t.

$$\sum_{p\in P_{f1}} z_p^{f1} = q_{f1}, \quad \forall f_1 \in F_1$$

$$\sum_{a\in A_{f1}} v_a^{f2} = q_{f2}, \quad \forall f_2 \in F_2$$

$$\sum_{f_1\in F_1|p\in P_{f1}} z_p^{f1} = y_p, \quad \forall p \in P|(b_p = 1 \text{ or } b_p = 4)$$

$$\sum_{f_2\in F_2|a\in A_{f2}} v_a^{f2} = x_a, \quad \forall a \in A|(b_a = 8 \text{ or } b_a = 10), h_a = 2$$

$$\sum_{p\in P|a\in A_p} y_p + x_a \leq c_a, \quad \forall a \in A$$

$$\sum_{p\in P_{ne}|b_p=1} y_p = \sum_{p\in P_{ns}|b_p=2} y_p + \sum_{a\in A_{ns}|b_a=5} x_a, \quad \forall n \in N|b_n = 3$$

$$\sum_{a\in A_{ne}|b_a=2 \text{ or } b_a=8} x_a = \sum_{p\in P_{ns}|b_p=3} y_p, \quad \forall n \in N|b_n = 4$$

$$\sum_{a\in A_{ne}|h_a\neq 1} x_a = \sum_{a\in A_{ns}|h_a\neq 1} x_a, \quad \forall n \in N|h_n \neq 1$$

$$\sum_{p\in P_l} y_p \leq q_l, \quad \forall l \in L$$

$$\sum_{a\in A_e} x_a \leq q_e, \quad \forall e \in E$$

3.3.5. Solution Method

The model is a pure integer linear programming, so it can be solved by a branch and bound algorithm [36]. We implement the algorithm by c# programming. We use branch and bound algorithm to search the solution space tree of the problem and each node can become an extension node at most once. Its search strategy is as follows:

Step 1: Generate all children of the current extension node;

Step 2: In the generated child nodes, the nodes that cannot generate a feasible solution (or an optimal solution) are dropped.

Step 3: Add the remaining children to the list of open nodes;

Step 4: Select the next node from the list of open nodes as the new extension node;

Step 5: This loop continues until either the feasible solution (optimal solution) or the open node list is left empty.

3.4. The Method of Calculating the Real-Time Number of Railcars at Each Station in Each Period

We set the decision cycle as one day and divide the day into 24 periods (i.e., one hour is one period). We mainly analyze the real-time number of railcars at each station at the end of each period (i.e., the real-time number of railcars at station A at 18:00 represents the number of railcars at the station during the period from 17:00 to 18:00). The algorithm of calculating the number of railcars in period g of station s is as follows:

Step 1: Search out the space-time arc set A_s^g based on the time-space network. Traverse the time-space arc set A, for any $a \in A$, if $s_{n_{as}} = s_{n_{ae}} = s$ and $st_a < et_g \leq et_a$, add the arc a to the set A_s^g;

Step 2: Calculate

$$d_s^g = \sum_{a\in A_s^g} \sum_{p\in P|a\in A_p} y_p + x_a$$

4. Numerical Studies

4.1. Data Input

Set up a small network, where A, B, C, and D are fulcrum stations and D is the boundary station. Set the arrival time, disintegration time, assembly time, grouping time and departure time of each fulcrum station as 30 min, loading time and unloading time of each fulcrum station as 180 min, and pick-up time and delivery time of each fulcrum station as 40 min. Set the capacity at each fulcrum station as 1300, 800, 1200, 900 in turn. The maximum number of marshaling railcars in a train is set to 55. Suppose station C stopped operation at 23:00 that day due to natural disasters. The train diagram is shown in Table S1 in Supplementary Materials. The initial railcar flows and their routes are shown in Table 7, where the generation moment is the conversion minutes (the difference between the time and 18:00, for example, 60 refers to 19:00). The loading plan is shown in Table 8 and the emptying plan is shown in Table 9.

Table 7. Initial railcar flows and their routes.

O	D	Railcar Flow Number (Generation Moment in minutes)	Type	Railcar Flow Route
A	B	100(0), 200(180), 150(250)	Loaded	A–B
A	C	50(0), 60(120), 90(160)	Loaded	A–C
A	D	30(0), 30(160), 40(200)	Loaded	A–C–D
B	C	20(100), 50(300)	Loaded	B–C
B	D	150(0)	Loaded	B–D
C	D	60(0), 50(100), 80(160), 100(230), 50(350)	Loaded	C–D
A		200(100), 20(170), 100(230), 80(310), 30(350)	Empty	
B		60(0), 100(160), 60(200), 100(270), 60(360)	Empty	
C		100(120), 80(250), 160(300), 150(360), 150(450)	Empty	

Table 8. Loading plan.

Loading Station	Destination Station	Planned Loading Number	Completed Loading Number
A	B	120	0
A	C	160	0
B	C	100	0
B	D	100	0
C	D	240	0

Table 9. Emptying plan.

Boundary Station	Planned Emptying Number	Completed Emptying Number
D	300	0

4.2. Results and Discussion

C# programming was adopted to realize the construction of a time-space network, and 307 time-space nodes and 584 time-space arcs were generated. A feasible path set generation algorithm based on an improved A* algorithm was used to generate the feasible path set between any pair of OD demands. The generation results of the feasible path sets between OD demands are shown in Figure 6.

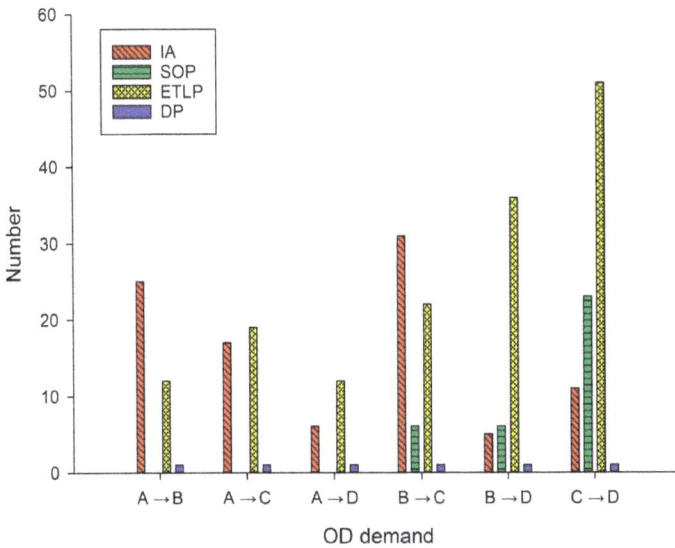

Figure 6. The generation results of feasible path sets.

Since this model was a linear programming model, the distribution results of railcar flow in the time-space network can be obtained by using the optimization solution software CPLEX. The loaded railcar flows were loaded onto the time-space arc, and the flow distribution results of time-space arcs are shown in Table S2 in Supplementary Materials. The railcar flow number of the TA indexed 2 and 36 in the table was 0, indicating that some trains stop running after the emergency occurred at station C. According to the distribution results of the time-space arc, the real-time number of railcars at the station can be obtained. The ratio of the real-time number of railcars at the station to the storage capacity of the station was taken as the indicator to measure whether the station became congested, that is, the Station Capacity Utilization (SCU). When SCU was greater than 1, the station was in congestion at that moment. SCU time series curves of each station are shown in Figure 7. It can be concluded from the figure that station B was in congestion from 16:00 to 18:00 of the next day, and other stations were in the normal state.

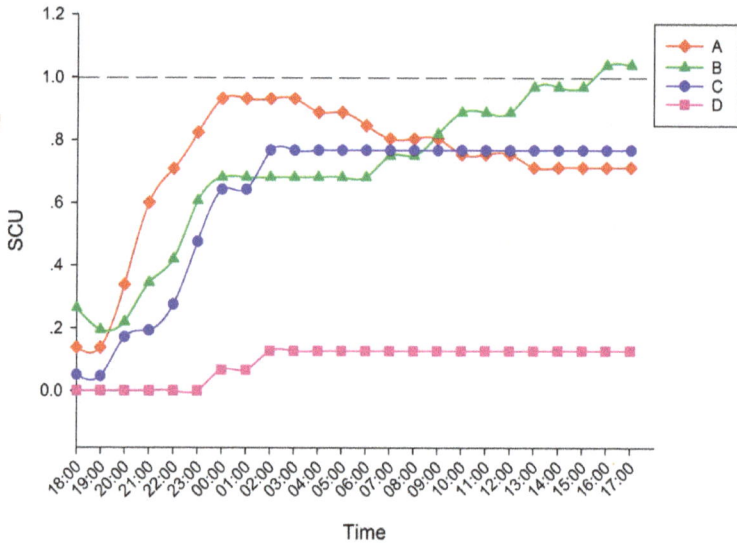

Figure 7. Station Capacity Utilization (SCU) time series curves corresponding to different stations in an emergency.

The solution of DRFDMUN was used to represent the dynamic railcar flow operation process under normality, and it was compared with the solution of DRFDMUE. Comparison of Total Loss (TL), comparison of Average Loading Plan Complete Rate (ALPCR) and comparison of Average Emptying Plan Complete Rate (AEPCR) under two different conditions are shown in Figure 8. As can be seen from the figure, when an emergency occurred at station C at 23:00, the TL of the transport system would increase by 16.6%, the ALPCR would be reduced from 100% to 25%, and the AEPCR would be reduced from 58% to 19%.

Aiming at the overstock phenomenon of railcar flow in station B, a restriction loading strategy can be adopted to solve the congestion of the station. In order to obtain the appropriate loading number, the loading number of the loading plan whose destination station was station B can be continuously reduced, and then the SCU time series curves of station B can be obtained by using the solution of DRFDMUE. In this paper, the gradient was set as 20, and the loading number of the loading plan whose destination station was station B was set as 100, 80, 60, 40, and 20 successively. The SCU time series curves of station B corresponding to different loading numbers of the loading plan are shown in Figure 9. In order to facilitate observation, the SCU time series curves from 13:00 to 18:00 of the next day at station B were mainly intercepted. As can be seen from Figure 9, when the loading number of the loading plan was constantly reduced, the real-time SCU of station B constantly decreased. When the loading number of the loading plan was 80, the real-time SCU of station B during this period was less than 1, indicating that the congestion of station B had been solved. Although the loading number of the loading plan could be reduced, and the SCU of station B would be lower, the economic benefits of the transportation system would be reduced. Therefore, it is relatively reasonable to adjust the loading number of the loading plan to 80. Of course, in order to get a more ideal loading number for the loading plan, the gradient can be set smaller.

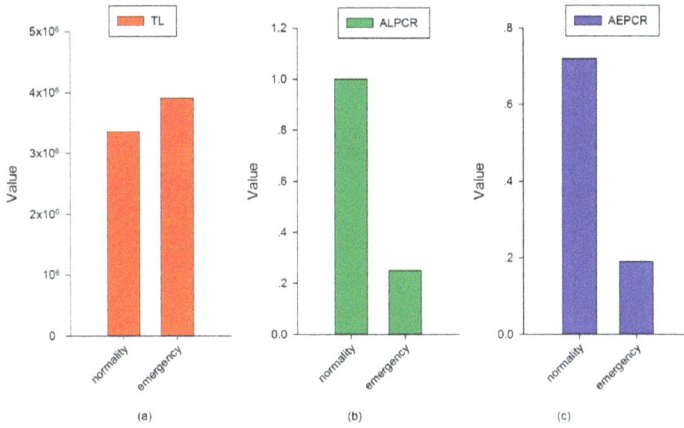

Figure 8. Indicator comparison in an emergency and normality.

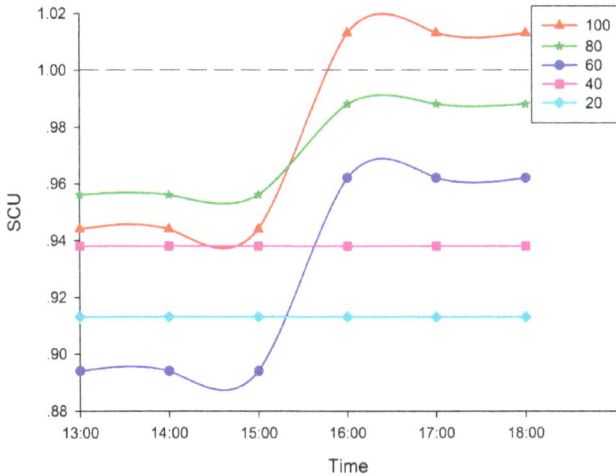

Figure 9. SCU time series curves of station B corresponding to different loading numbers.

5. Conclusions

In this paper, we analyzed strong timeliness, time-space, and the strongly controllable and multiple highly dynamic constraints of traffic congestion prediction and control in an emergency, which demands accuracy and efficiency in terms of a solution methodology. Then, we analyzed the turnover process of the railcar and the transition of the railcar state in detail and abstracted the railcar management model into a multi-commodity flow problem. Furthermore, we disassembled the railcar management problem into two sub-problems. One of the sub-problems was the construction of a time-space network based on the turnover process of the railcar and the transition of the railcar state, the other sub-problem was the construction of a dynamic railcar flow distribution model in an emergency. The innovation points of the proposed model are as follows: (1) Based on the train diagram, we constructed a time-space network for the first time by considering the transition of the railcar state; (2) an improved A* algorithm based on the railcar flow route was used to generate a feasible path set of any pair of transportation demands; (3) we obtained the real-time number of railcars at each station basing the results to predict the traffic congestion in the station.

The numerical studies demonstrated that: (1) this proposed model can be used to predict which station will become congested first and when the station begins to become congested (i.e., station B was in congestion from 16:00 to 18:00 of the next day); (2) the emergency can lead to a dramatic increase in the overall loss of the transport system (i.e., compared with the solution of DRFDMUN, DRFDMUE showed that an emergency caused the TL to increase by 16.6%, the ALPCR to drop from 100% to 25%, and the AEPCR to drop 58% to 19%); (3) we can control the traffic congestion of the station by adjusting the loading plan (i.e., the reasonable loading number to solve the traffic congestion of station B is 80).

Even though the relatively comprehensive features of traffic congestion prediction and control in an emergency are considered in this paper, stochastic and large-scale computing problems are lacking. For example, the arrival and departure times of trains do not follow the train diagram exactly. When the calculation scale is too large, the calculation efficiency will be affected. In the future, the stochastic of train arrival and departure will be considered and we intend to choose a large-scale domain search algorithm as the solution method.

Supplementary Materials: The following are available online at http://www.mdpi.com/2073-8994/11/6/780/s1, Table S1: Train diagram, Table S2: The flow distribution results of time-space arcs.

Author Contributions: Methodology, Z.Q.; Resources, S.H.; Data Curating, Z.Q.; Writing—Original Draft Preparation, Z.Q.; Writing—Review & Editing, Z.Q.; Project Administration, S.H.; Funding Acquisition, S.H.

Funding: This paper was supported by the National Key R&D Program of China (2018YFB1201402).

Acknowledgments: The authors wish to acknowledge the anonymous reviewers for their insightful comments.

Conflicts of Interest: The authors declare no conflict of interest.

Abbreviations

All the abbreviations are listed as follows:

AN	Arrival Node
DN	Departure Node
LRSEN	Loaded Railcar State End Node
ERSEN	Empty Railcar State End Node
SOCN	Secondary Operation Completion Node
IN	Initial Node
EN	End Node
TA	Train Arc
LOA	Local Operation Arc
SOA	Secondary Operation Arc
FADA	Formation And Departure Arc
ERAA	Empty Railcar Allocation Arc
TOA	Transfer Operation Arc
SA	Stay Arc
IA	Initial Arc
EA	End Arc
DA	Detention Arc
LRL	Loaded Railcar Layer

ERL	Empty Railcar Layer
PL	Public Layer
IP	Initial Path
SOP	Secondary Operation Path
ETLP	Empty To Loaded Path
DP	Detention Path
DRFDMUE	Dynamic Railcar Flow Distribution Model Under Emergency
DRFDMUN	Dynamic Railcar Flow Distribution Model Under Normality
SCU	Station Capacity Utilization
TL	Total Loss
ALPCR	Average Loading Plan Complete Rate
AEPCR	Average Emptying Plan Complete Rate

References

1. Backfrieder, C.; Ostermayer, G.; Mecklenbrauker, C.F. Increased traffic flow through node-based congestion prediction and v2x communication. *IEEE Trans. Intell. Transp. Syst.* **2017**, *18*, 349–363. [CrossRef]
2. Ros, B.G.; Knoop, V.L.; Shiomi, Y.; Takahashi, T.; Arem, B.V.; Hoogendoorn, S.P. Modeling traffic at sags. *Int. J. Intell. Transp. Syst. Res.* **2016**, *14*, 64–74.
3. Lee, W.H.; Tseng, S.S.; Shieh, J.L.; Chen, H.H. Discovering traffic congestion in an urban network by spatiotemporal data mining on location-based services. *IEEE Trans. Intell. Transp. Syst.* **2011**, *12*, 1047–1056. [CrossRef]
4. Arnott, R.; Palma, A.D.; Lindsey, R. A structural model of peak-period congestion: A traffic congestion with elastic demand. *Am. Econ. Rev.* **1993**, *83*, 161–179.
5. Davis, L.C. Mitigation of congestion at a traffic congestion with diversion and lane restrictions. *Phys. A* **2012**, *391*, 1679–1691. [CrossRef]
6. Sun, L.; Wei, L.; Yao, L.; Shi, Q.; Jian, R. A comparative study of funnel shape congestion in subway stations. *Transp. Res. Part A Policy Pract.* **2017**, *98*, 14–27. [CrossRef]
7. Xu, X.; Liu, J.; Li, H.; Hu, J.Q. Analysis of subway station capacity with the use of queueing theory. *Transp. Res. Part C Emerg. Technol.* **2014**, *38*, 28–43. [CrossRef]
8. Ma, J.; Xu, S.M.; Li, T.; Mu, H.L.; Wen, C.; Song, W.G.; Lo, S.M. Method of congestion identification and evaluation during crowd evacuation process. *Procedia Eng.* **2014**, *71*, 454–461. [CrossRef]
9. Liu, J.; Shengxue, H.E.; Zhang, H.; School, B. Optimization analysis for serial congestion system of urban rail transit station. *J. Comput. Appl.* **2016**, *36*, 271–274.
10. Yi-Fan, L.I.; Chen, J.M.; Jie, J.I.; Ying, Z.; Sun, J.H. Analysis of crowded degree of emergency evacuation at "congestion" position in subway station based on stairway level of service. *Procedia Eng.* **2011**, *11*, 242–251. [CrossRef]
11. Sun, Y.; Li, X.; Liang, X.; Zhang, C. A bi-objective fuzzy credibilistic chance-constrained programming approach for the hazardous materials road-rail multimodal route problem under uncertainty and sustainability. *Sustainability* **2019**, *11*, 2577. [CrossRef]
12. Jiang, J.; Li, Q.; Wu, L.; Tu, W. Multi-objective emergency material vehicle dispatching and route under dynamic constraints in an earthquake disaster environment. *ISPRS Int. J. Geo-Inf.* **2017**, *6*, 142. [CrossRef]
13. Lu, Q.-C.; Lin, S. Vulnerability analysis of urban rail transit network within multi-modal public transport networks. *Sustainability* **2019**, *11*, 2109. [CrossRef]
14. Deng, Y.; Ru, X.; Dou, Z.; Liang, G. Design of bus bridging routes in response to disruption of urban rail transit. *Sustainability* **2018**, *10*, 4427. [CrossRef]
15. White, W.; Bomberault, A. A network algorithm for empty freight car allocation. *IBM Syst. J.* **2010**, *8*, 147–169. [CrossRef]
16. Allman, W.P. Application series ‖ an optimization approach to freight car allocation under time-mileage per diem rental rates. *Manag. Sci.* **1972**, *18*, 567–574. [CrossRef]
17. Spieckermann, S.; Voß, S. A case study in empty railcar distribution. *Eur. J. Oper. Res.* **1995**, *87*, 586–598. [CrossRef]
18. Holmberg, K.; Joborn, M.; Lundgren, J.T. Improved empty freight car distribution. *Transp. Sci.* **1998**, *32*, 163–173. [CrossRef]

19. Narisetty, A.K.; Richard, J.P.P.; Ramcharan, D.; Murphy, D.; Minks, G.; Fuller, J. An optimization model for empty freight car assignment at union pacific railroad. *Interfaces* **2008**, *38*, 89–102. [CrossRef]
20. Gorman, M.F.; Crook, K.; Sellers, D. North American freight rail industry real-time optimized equipment distribution systems: State of the practice. *Transp. Res. Part C Emerg. Technol.* **2011**, *19*, 103–114. [CrossRef]
21. Heydari, R.; Melachrinoudis, E. A path-based capacitated network flow model for empty railcar distribution. *Ann. Oper. Res.* **2017**, *253*, 773–798. [CrossRef]
22. Jiang, Y.; Zhou, X.; Xu, Q. Scenario Analysis–Based Decision and Coordination in Supply Chain Management with Production and Transportation Scheduling. *Symmetry* **2019**, *11*, 160. [CrossRef]
23. Sun, Y.; Liang, X.; Li, X.; Zhang, C. A Fuzzy Programming Method for Modeling Demand Uncertainty in the Capacitated Road–Rail Multimodal Route Problem with Time Windows. *Symmetry* **2019**, *11*, 91. [CrossRef]
24. Lin, B.; Wu, J.; Wang, J.; Duan, J.; Zhao, Y. A Bi-Level Programming Model for the Railway Express Cargo Service Network Design Problem. *Symmetry* **2018**, *10*, 227. [CrossRef]
25. Lawley, M.; Parmeshwaran, V.; Richard, J.P.; Turkcan, A.; Dalal, M.; Ramcharan, D. A time-space scheduling model for optimizing recurring bulk railcar deliveries. *Transp. Res. Part B Methodol.* **2008**, *42*, 438–454. [CrossRef]
26. Sayarshad, H.R.; Tavakkoli-Moghaddam, R. Solving a multi periodic stochastic model of the rail–car fleet sizing by two-stage optimization formulation. *Appl. Math. Model.* **2010**, *34*, 1164–1174. [CrossRef]
27. Liang, D.; Lin, B. Research on the model of optimizing strategically dynamic railcar operation. *Syst. Eng. Theory Pract.* **2007**, *27*, 77–84.
28. Kwon, O.K.; Martland, C.D.; Sussman, J.M. Routing and scheduling temporal and heterogeneous freight car traffic on rail networks. *Transp. Res. Part E Logist. Transp. Rev.* **1998**, *34*, 101–115. [CrossRef]
29. Tian, Y.; Lin, B.; Ji, L. Railway car flow distribution node-arc and arc-path models based on multi-commodity and virtual arc. *J. China Railw. Soc.* **2011**, *33*, 7–12.
30. Wang, L.; Ma, J.; Lin, B. Optimal route choice model for loaded and empty car flows in railway network. *J. Beijing Jiaotong Univ.* **2014**, *38*, 12–18.
31. Guo, P.; Lin, B. Section center optimization method for distribution of empty railcars over large scale road network. *China Railw. Sci.* **2001**, *22*, 122–128.
32. Alfaki, M.; Haugland, D. A multi-commodity flow formulation for the generalized pooling problem. *J. Glob. Optim.* **2013**, *56*, 917–937. [CrossRef]
33. Chabini, I.; Shan, L. Adaptations of the a* algorithm for the computation of fastest paths in deterministic discrete-time dynamic networks. *IEEE Trans. Intell. Transp. Syst.* **2002**, *3*, 60–74. [CrossRef]
34. Zhan, W.; Wang, W.; Chen, N.; Wang, C. Path planning strategies for UAV based on improved a* algorithm. *Geomat. Inf. Sci. Wuhan Univ.* **2015**, *40*, 315–320.
35. Qian, H.; Wenfeng, G.E.; Zhong, M.; Ming, G.E. Application of improved a* algorithm based on hierarchy for route planning. *Comput. Eng. Appl.* **2014**, *50*, 225–229.
36. D'Ariano, A.; Pacciarelli, D.; Pranzo, M. A branch and bound algorithm for scheduling trains in a railway network. *Eur. J. Oper. Res.* **2007**, *183*, 643–657. [CrossRef]

symmetry

MDPI

Article

Study on a Novel Fault Diagnosis Method Based on VMD and BLM

Jianjie Zheng [1], Yu Yuan [1], Li Zou [1], Wu Deng [1,2,3], Chen Guo [4] and Huimin Zhao [1,2,3,*]

[1]   Software Institute, Dalian Jiaotong University, Dalian 116028, China; zheng853796151@126.com (J.Z.);
      yuanyu_knife@163.com (Y.Y.); lizou@djtu.edu.cn (L.Z.); dw7689@djtu.edu.cn (W.D.)
[2]   The State Key Laboratory of Mechanical Transmissions, Chongqing University, Chongqing 400044, China
[3]   Traction Power State Key Laboratory of Southwest Jiaotong University, Chengdu 610031, China
[4]   College of Marine Electrical Engineering, Dalian Maritime University, Dalian 116026, China;
      guoc@dlmu.edu.cn
*   Correspondence: hm_zhao1977@126.com; Tel.: +86-411-8410-5386

Received: 24 April 2019; Accepted: 30 May 2019; Published: 2 June 2019

Abstract: The bearing system of an alternating current (AC) motor is a nonlinear dynamics system. The working state of rolling bearings directly determines whether the machine is in reliable operation. Therefore, it is very meaningful to study the fault diagnosis and prediction of rolling bearings. In this paper, a new fault diagnosis method based on variational mode decomposition (VMD), Hilbert transform (HT), and broad learning model (BLM), called VHBLFD is proposed for rolling bearings. In the VHBLFD method, the VMD is used to decompose the vibration signals to obtain intrinsic mode functions (IMFs). The HT is used to process the IMFs to obtain Hilbert envelope spectra, which are transformed into the mapped features and the enhancement nodes of BLM according to the complexity of the modeling tasks, and the nonlinear transformation mean according to the characteristics of input data. The BLM is used to classify faults of the rolling bearings of the AC motor. Next, the pseudo-inverse operation is used to obtain the fault diagnosis results. Finally, the VHBLFD is validated by actual vibration data. The experiment results show that the BLM can quickly and accurately be trained. The VHBLFD method can achieve higher identification accuracy for multi-states of rolling bearings and takes on fast operation speed and strong generalization ability.

Keywords: rolling bearings; fault diagnosis; broad learning model; variational mode decomposition; Hilbert transform

---

1. Introduction

The working state of rolling bearings directly determines the reliable operation of a machine [1,2]. However, the occurrence probability of fault is always higher due to the influences of the load of rolling bearings of AC motor [3–6]. Thus, it is very important to improve the operation reliability and accurately diagnose faults for rolling bearings in time [7–9].

Fault diagnosis methods of rolling bearings are used to essentially recognize the working states [10–13]. To effectively recognize the working state of rolling bearings, many signal processing methods have been proposed in recent years, such as short time Fourier transform (STFT) [14,15], wavelet transform (WT) [16], Hilbert–Huang transform (HHT) [17–19], empirical mode decomposition (EMD) [20–22], entropy [23–25], support vector machine (SVM) [26], artificial intelligence methods [27–29], and other processing methods [30,31]. In addition, some new methods have also been applied in the field of signal analysis and fault diagnosis [32,33]. Gao et al. [34] proposed a fault identification method based on time-frequency distribution (TFD) for rolling bearings. Zhang et al. [35] proposed a flexible wavelet transform to obtain weak fault feature. Kabla et al. [36] applied HHT and marginal spectrum to analyze the signals of the stator current. Yuan et al. [37] proposed an ensemble noise-reconstructed

EMD method. The SVM is widely applied for fault diagnosis. Du et al. [38] proposed a stochastic fault diagnosis method using EMD and principal component analysis (PCA). To solve the classification ability of SVM, Fei et al. [39] proposed a power transformer fault diagnosis model using a rough set and SVM. Gao et al. [40] proposed a matrix factorization method to represent and identify the bearing faults. Cheng et al. [41] proposed a fault diagnosis model using a band decomposition method. Huang et al. proposed a fault diagnosis method for rolling bearings [42,43].

Deep learning is a new area of machine learning research that uses multilayer artificial neural networks to provide the most advanced accuracy in speech recognition, object detection, and so on. It can automatically study representations from text, images, or video data. The flexible structure can directly study from more raw data and improve forecasting accuracy. Due to these advantages of deep learning, it has been applied in fault diagnosis. In recent years, a lot of fault diagnosis methods based on deep learning have been proposed, and good diagnostic results have been obtained. Van Tung et al. [44] proposed a deep belief networks (DBN)-fault diagnosis method in reciprocating compressors. Guo et al. [45] proposed an adaptive deep convolutional neural networks (DCNN) to classify and diagnose mechanical faults. Qi et al. [46] proposed a stacked sparse auto-encoder-fault diagnosis method. Shao et al. [47] proposed an adaptive DBN to identify the faults. Li et al. [48] proposed a novel new fault diagnosis model for rolling bearings. Shao et al. [49] proposed an improved convolutional deep belief networks (CDBN) for rolling bearing fault diagnosis. Sun et al. [50] proposed a sparse deep learning method. Zhang et al. [51] proposed a DCNN for bearing fault diagnosis under different loads and noisy environments. Shao et al. [52] proposed a fault diagnosis model for electric locomotive bearings. Wang et al. [53] proposed a fault diagnosis method for rolling bearings. Wang et al. [54] proposed a DBN with RBM based on a data indicator for multiple faults. Liu et al. [55] proposed a deep neural networks(DNN)-unsupervised fault diagnosis model. Zhao and Jia [56] proposed a deep fuzzy clustering neural network to realize the fault recognition of rotating machinery. Hu and Jiang [57] proposed a new fault diagnosis model using modified DNN with incremental imbalance.

However, the structure of a deep learning network is complex and has many parameters, which results in an extremely time consuming training process. In order to obtain higher diagnosis accuracy, the deep learning network has to continuously increase the number of network layers or adjust the parameters using optimization algorithms. The fault diagnosis requires the rapidity and high accuracy to ensure safe and smooth operation. Therefore, it is necessary to use a new deep network model and further study a corresponding combination with other methods. The broad learning model (BLM) is an effective incremental learning system model. It could realize competitive results in various applications. At the same time, if the network needs to be extended, the model can be efficiently reconstructed through incremental learning. Therefore, it is significant to deeply research the new fault diagnosis model for rolling bearings.

The key of fault diagnosis is to choose proper methods to diagnose the fault type, the position and the severity. It is easier to diagnose the fault type, but the fault development is a gradual process. When the fault degree is different, the vibration signal also shows different features. To effectively diagnose the fault and reveal the development and evolution of faults, VMD, HT and BLM are introduced into the fault diagnosis to deeply study new fault diagnosis model for AC motor rolling bearings.

2. Basic Method

2.1. VMD

The VMD is a completely non-recursive signal decomposition method. Its essence is multiple Wiener filter banks. The VMD can decompose a signal into a number of discrete sparse sub-signals. Therefore, the VMD is applied in fault diagnosis. Li et al. [58] proposed an independence-oriented VMD via correlation analysis to adaptively obtain weak fault features. Jiang et al. [59] proposed an initial center frequency-guided VMD to accurately extract weak damage features. Li et al. [60]

proposed an adaptive VMD for extracting periodic impulses. Wang et al. [61] proposed an adaptive parameter optimized VMD. The other signal decomposition methods have also been proposed in recent years [62–76].

Assuming that each mode u_k has a center frequency $\omega(k)$ and a limited bandwidth, the constraint condition is that the sum of each mode is equal to the input signal, and the sum of the estimated mode bandwidth is the minimum. The $\omega(k)$ and the bandwidth of each mode are updated continuously during the iterative process of solving the variational model. Finally, the adaptive decomposition for signal is realized.

The signal is decomposed at scale K and the variational problem is constructed with the minimum of the sum of the estimated bandwidths of the IMF components.

$$\min_{\{u_k\},\{\omega_k\}}\left\{\sum_k \left\|\partial_t\left[\left(\delta(t)+\frac{j}{\pi t}\right)*u_k(t)\right]e^{-j\omega_k t}\right\|_2^2\right\}. \tag{1}$$

where $\{u_k\}=\{u_1,u_2,\ldots,u_K\}$ represents each modal function and $\{\omega_k\}=\{\omega_1,\omega_2,\ldots,\omega_K\}$ represents the central frequencies of each modal function. $\delta(t)$ is the Dirichlet distribution function; * is the convolution.

The quadratic penalty factor is used to guarantee the fidelity of the reconstructed signal, and the Lagrange multiplier is used to guarantee the strictness of the constraint. The extended Lagrange expression is as follows:

$$\begin{aligned}L(\{u_k\},\{\omega_k\},\lambda) &= \alpha\sum_k\left\|\partial_t\left[\left(\delta(t)+\frac{j}{\pi t}\right)*u_k(t)\right]e^{-j\omega_k t}\right\|_2^2 + \left\|f(t)-\sum_k u_k(t)\right\|_2^2 \\ &+ <\lambda(t),f(t)-\sum_k u_k(t)>.\end{aligned} \tag{2}$$

where, α represents a penalty factor; λ represents a Lagrangian multiplier.

In the VMD, the multiplicative operator alternating direction method is used to solve thee variational problems. By alternately updating u_k^{n+1}, ω_k^{n+1} and λ, we seek the "saddle point" of the extended Lagrangian expression. The component u_k and the center frequency ω_k are described as follows.

$$\hat{u}_k^{n+1}(\omega) = \left(\hat{f}(\omega)-\sum_{i\neq k}\hat{u}_i(\omega)+\frac{\hat{\lambda}(\omega)}{2}\right)\frac{1}{1+2\alpha(\omega-\omega_k)^2} \tag{3}$$

$$\omega_k^{n+1} = \frac{\int_0^\infty \omega|\hat{u}_k(\omega)|^2 d\omega}{\int_0^\infty |\hat{u}_k(\omega)|^2 d\omega} \tag{4}$$

where $\hat{u}_k^{n+1}(\omega)$ is equivalent to the Wiener filtering of the current residual $\hat{f}(\omega)-\sum_{i\neq k}\hat{u}_i(\omega)$ and the real part of $\hat{u}_k(\omega)$ after inverse Fourier transform is $u_k(t)$.

2.2. Deep Belief Network

Deep learning can learn the discriminative features from data [77]. The basic models of deep learning can be divided into a multi-layer model, a deep neural network model and a recursive neural network model. The Deep belief network (DBN) is a generating graphical model, composed of multilayer hidden units. The DBN can generate training data according to the maximum probability in the whole neural network by training the weights of its neurons. The Deep Boltzmann machine (DBM) can learn input data probability distributions by latent or hidden variables. The RBM is an undirected graphical model $v=\{0,1\}^F$ and hidden units $h=\{0,1\}^D$. The structure of RBM is shown in Figure 1.

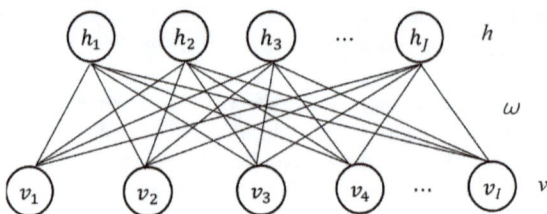

Figure 1. Structure of RBM.

For a given set (v, h), it can be defined as follows:

$$E(v,h) = - \sum_{i \in visible} a_i v_i - \sum_{j \in hidden} b_j h_j - \sum_{i,j} v_i h_j w_{ij} \tag{5}$$

The joint distribution over v and h is defined as follows:

$$P(v,h) = \frac{1}{Z} \exp(-E(v,h)) \tag{6}$$

The unbiased sample can be obtained:

$$p(v_i = 1|h) = \sigma(a_i + \sum_j h_j w_{ij}) \tag{7}$$

$$p(h_j = 1|v) = \sigma(b_j + \sum_i v_i w_{ij}) \tag{8}$$

where w_{ij} is the connection weight, a_i and b_j are bias coefficients of the ith neuron and the jth neuron, v is the input vector and the h is output vector.

2.3. Broad Learning Model

The Broad learning model (BLM) is an effective incremental learning system [78]. It is essentially designed for various applications. The mapping feature nodes can efficiently extract features. At the same time, the random connection from mapping features to enhancing nodes can compensate for the non-linearity of mapping feature nodes and improve the speed of the model. The BLM can achieve competitive results with state-of-art methods on various applications. The BLM is represented in Figure 2.

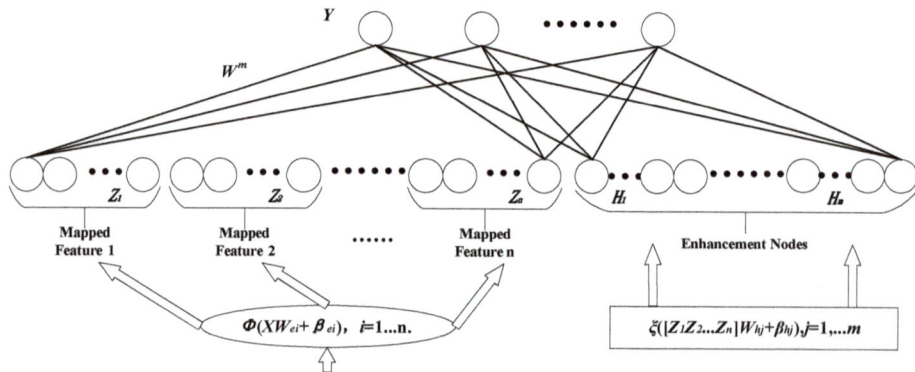

Figure 2. The broad learning model.

Assume that input data X and project data are presented using $\varphi_i(XW_{ei} + \beta_{ei})$ in order to obtain the ith mapping features Z_i. $Z^i = [Z_1, \ldots, Z_i]$ is the concatenation of all mapping features of the first i groups. Likely, the enhancement nodes of the jth group is $\xi_j(Z_iW_{hj} + \beta_{hj})$, which can be regarded as H_j. $H^j = [H_1, \ldots, H_j]$ is concatenation of all enhancement node of the first j groups. In addition, φ_i and φ_k are different functions when there is i ≠ k. Likely, ξ_j and ξ_r are also different functions when there is j ≠ r.

In the BLM, the linear inverse problem is used, and the initial W_{ei} is fine tuned in order to obtain better features. Next, the BLM is described in detail. Assume the input data set X with N samples, each sample is M dimensions. The output matrix is Y.

$$Z_i = \varphi(XW_{ei} + \beta_{ei}), i = 1, \ldots, n \tag{9}$$

where W_{ei} and β_{ei} are generated randomly, and $Z^n = [Z_1, \ldots, Z_n]$. The enhancement nodes of mth group are described.

$$H_m = \xi(Z^n W_{hm} + \beta_{hm}) \tag{10}$$

Hence, the BLM can be represented.

$$
\begin{aligned}
Y &= [Z_1, \ldots, Z_n | \xi(Z^n W_{h1} + \beta_{h1}), \ldots, \xi(Z^n W_{hm} + \beta_{hm})] W^m \\
&= [Z_1, \ldots, Z_n | H_1, \ldots, H_m] W^m \\
&= [Z^n | H^m] W^m
\end{aligned} \tag{11}
$$

where the $W^m = [Z^n | H^m]^+ Y$.

W^m are the connecting weights of BLM.

$$\underset{\hat{W}}{\operatorname{argmin}} : \|Z\hat{W} - X\|_2^2 + \lambda\|\hat{W}\|_1 \tag{12}$$

3. A New Fault Diagnosis Method Based on VMD, HT and BLM

3.1. The Idea of the VHBLFD Method

Many researchers have deeply researched fault bearing diagnosis; some results have been achieved, and some signal analysis methods have been proposed successively in recent years. The time domain features are easy to be calculated, but the anti-jamming ability for fault vibration data is poor. The frequency domain features are based on the global transformation of signals, which cannot effectively analyze non-stationary signals. The VMD has the advantages of effectively reducing pseudo components and modal aliasing. Hilbert transform (HT) is applied to obtain accurate time-frequency distributions of signal energy and further construct the corresponding marginal spectrum. The Hilbert marginal spectrum can accurately reflect the change rule of signal amplitude with frequency. Compared with the existing signal feature extraction methods, HT has better noise robustness. The deep learning can better solve the problems of feature learning, feature extraction, and deep network training, but there exists many parameters to be optimized, which usually requires a great deal of time and machine resources. The BLM provides an alternative method. It wadesigned by expanding the broad features nd enhancement nodes. Therefore, the BLM with fast calculation speed and strong generalization ability could be used to a new fault diagnosis (VHBLFD) method. The VHBLFD determines the numbers of enhancement nodes and mapped features and according to the complexity of the modeling tasks, as well as the nonlinear transformation mean according to the features of input data. Then, the vibration signals are decomposed using the VMD, and the HT is used to process the IMFs to obtain Hilbert envelope spectra, which are transformed into the mapped features and enhancement nodes. The BLM is used to realize the fault diagnosis, and the pseudo-inverse operation is used to obtain the fault diagnosis results.

3.2. The Fault Diagnosis Model and Steps

The model of the proposed the VHBLFD method using the VMD, HT and BLM is shown in Figure 3.

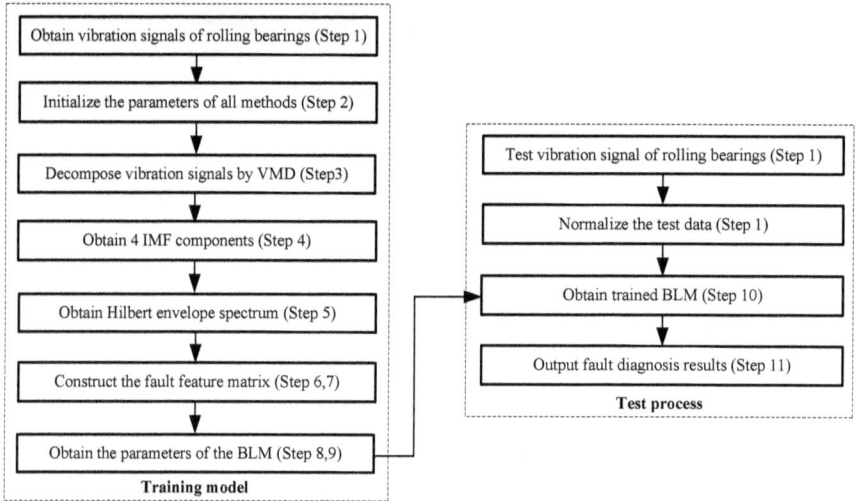

```
┌─────────────────────────────────────────────┐
│  Obtain vibration signals of rolling bearings (Step 1)  │
│                      ↓                         │
│  Initialize the parameters of all methods (Step 2)      │
│                      ↓                         │
│  Decompose vibration signals by VMD (Step3)             │
│                      ↓                         │
│  Obtain 4 IMF components (Step 4)                        │
│                      ↓                         │
│  Obtain Hilbert envelope spectrum (Step 5)              │
│                      ↓                         │
│  Construct the fault feature matrix (Step 6,7)          │
│                      ↓                         │
│  Obtain the parameters of the BLM (Step 8,9)            │
│              Training model                    │
└─────────────────────────────────────────────┘

┌─────────────────────────────────────────────┐
│  Test vibration signal of rolling bearings (Step 1)     │
│                      ↓                         │
│  Normalize the test data (Step 1)                       │
│                      ↓                         │
│  Obtain trained BLM (Step 10)                           │
│                      ↓                         │
│  Output fault diagnosis results (Step 11)               │
│              Test process                      │
└─────────────────────────────────────────────┘
```

Figure 3. The flow of fault diagnosis method.

3.3. The Steps of the Fault Diagnosis Method

The steps of the proposed VHBLFD method for rolling bearings of the AC motor are described in detail.

Step 1: The acceleration sensors are used to collect vibration acceleration signals of rolling bearings of the AC motor.

Step 2: Initialize these parameters of the proposed VHBLFD method using VMD, HT and BLM. These parameters mainly include the number of decompositions of VMD, the number of feature nodes per window, the windows and the enhancement nodes of BLM, and so on.

Step 3: The VMD is used to decompose the vibration acceleration signals into a series of IMFs.

Step 4: According to the number of decompositions of the VMD method, four IMF components are determined.

Step 5: The HT is used to process the four IMF components to obtain the Hilbert envelope spectrum for obtaining fault features.

Step 6: The Hilbert envelope spectrums of four IMF components are connected by the beginning and the end to construct the feature matrix.

Step 7: The fault features are proportionally divided into the training feature samples and the test feature samples.

Step 8: Calculate the feature nodes of the BLM according to Formula (8) and the enhancement nodes of the BLM according to Formula (9).

Step 9: Calculate the output of the BLM based on the feature nodes and the enhancement nodes using the pseudo inverse operation.

Step 10: Input the training feature samples to train BLM in order to obtain the trained BLM for realizing the fault diagnosis.

Step 11: Test feature samples are used to validate the effectiveness of the proposed VHBLFD to obtain diagnosis results. Analyze and verify the effectiveness and the rapidity of the VHBLFD method.

4. Validation and Analysis of the VHBLFD Method

4.1. Experiment Data and Environment

The vibration data are selected to validate the VHBLFD method here in [79]. The platform is shown in Figure 4. The vibration data are obtained under 0 HP at 1730 r/min. The different faults of outer race, inner race and rolling element are given. These fault diameters are 0.1778 mm, 0.3556 mm, and 0.5334 mm. There are 10 kinds of vibration data. The vibration data is sampled at 12,000 Hz frequency. Each sample consists of 2048 data points in Table 1.

Figure 4. The experiment platform.

Table 1. The sample data.

No.	Inner Race	Outer Race	Rolling Element
1	−0.0830	0.0085	−0.0028
2	−0.1957	0.4235	−0.0963
3	0.2334	0.0130	0.1137
4	0.1040	−0.2652	0.2573
5	−0.1811	0.2372	−0.0583
6	0.0556	0.5909	−0.1260
7	0.1738	−0.0930	0.2074
8	−0.0469	−0.4069	0.1727
9	−0.1119	0.2794	−0.2199
10	0.0596	0.4370	−0.1561
11	0	−0.3529	0.2240
⋯	⋯	⋯	⋯
2041	0.2305	0.0309	0.2375
2042	0.0461	0.1186	−0.0271
2043	−0.5122	−0.0061	−0.1327
2044	0.1481	−0.0979	0.0929
2045	0.6280	0.0914	0.1106
2046	−0.2043	0.1494	−0.1499
2047	−0.2640	−0.2355	−0.1108
2048	0.4662	−0.3224	0.1467

The experiment scheme is divided into two schemes. The first experiment scheme is to determine the fault types. The second experiment scheme not only determines the fault types, but also determines the severity of the fault. Each experiment scheme contains four data sets under four different working loads; 2072 data under no-load (0HP) are taken as training sets and 540 data sets under other working loads are taken as test sets.

4.2. Feature Extraction

The mode number of VMD decomposition is selected as four according to the empirical value. The VMD is applied to decompose the inner race vibration signal into four IMF components (fault diameter is 0.3556 mm) in Figure 5.

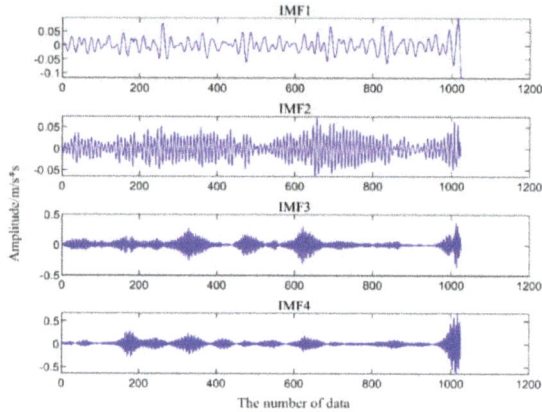

Figure 5. Decomposition of inner race fault signal.

The instantaneous frequency and the amplitude of the vibration signal can be obtained from each IMF, and the Hilbert envelope spectra are also obtained. The four Hilbert envelope spectra are connected from the head to the end in order to arrange a row. The length of the connected data is 4096, that is, the input dimensions are 4096. The VMD-Hilbert envelope spectra of normal data, inner race fault data (the fault diameters are 0.1778 mm, 0.3556 mm, and 0.5334 mm), outer race fault data (the fault diameters are 0.1778 mm, 0.3556 mm, and 0.5334 mm) and rolling element fault data (the fault diameters are 0.1778 mm, 0.3556 mm, and 0.5334 mm) are shown in Figure 6.

4.3. Fault Diagnosis Results

The VHBLFD method is used to recognize the fault of rolling bearings. The parameters of the BLM mainly include the number of feature nodes, the window number of the feature nodes, the number of enhancement nodes, the regularization parameter C, and the reduction rate s of the enhancement node. Because the parameters of the VHBLFD method are critical to classify, it is very important to select reasonable values of parameters. The size and the characteristics of experiment data are analyzed, and the selecting methods of parameters of BLS in the original paper are studied herein. The values of the parameters are determined. The training times are 80. The learning rate is 0.01. The weight penalty coefficient is 0.0002. The initial momentums are 0.5 and 0.9. For the VHBLFD1 method, the feature nodes are 100, the window number is five, and the enhanced nodes are 1000. For the VHBLFD2 method, the feature nodes are 100, the window number is 15, and the enhanced nodes are 17,000.

The 4096-dimension Hilbert envelope spectra of 10 kinds of state data are applied to construct the feature matrix for an input of the BLM. The 5300 data sets are regarded as training sets, and 1300 data sets are regarded as test sets in the experiment. The diagnosis results and the test times of the proposed VHBLFD method are shown in Table 2.

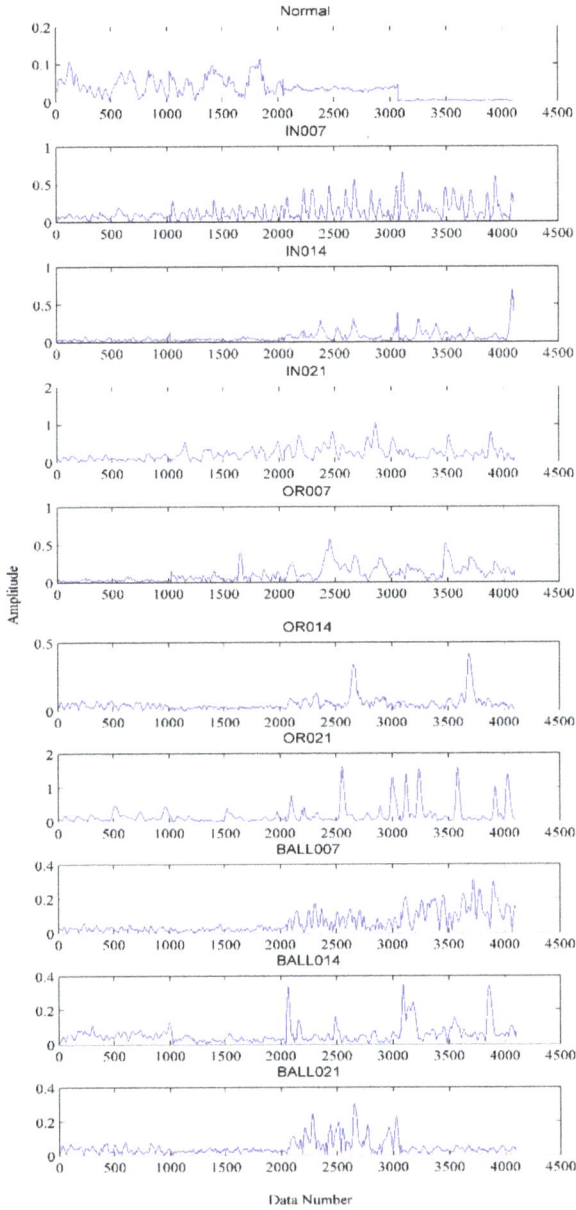

Figure 6. The VMD-Hilbert envelope spectra for different fault data.

Table 2. The diagnosis results and test times of the proposed VHBLFD method.

Fault Diagnosis Method	Diagnostic Accuracy (%)	Test Time (s)
VHBLFD1 (100,5,1000)	95.99	6.45
VHBLFD2 (100,15,17000)	97.74	22.29

As can be seen from Table 2, the diagnosis accuracy and the test time of the VHBLFD1 (100,5,1000) are 95.99% and 6.45 s. The diagnostic accuracy and test time of the VHBLFD2 (100,15,17000) are 97.74% and 22.29 s. The experimental results show that the BLM can construct a fault diagnosis model with better diagnosis efficiency and faster diagnosis speed. The proposed VHBLFD method can obtain higher diagnostic accuracy and takes less test time.

4.4. Comparision and Analysis for Diagnosis Results

In order to test and verify the effectiveness of the proposed VHSMFD method for rolling bearings of the AC motor, the VHSMFD method based on VMD, HT, and SVM, the EHDNFD method based on EMD, HT, and DBN, the EEHDNFD based on EEMD, HT and DBN, the VHDNFD method based on VMD, HT and DBN are compared with the proposed VHSMFD method. The four limited Boltzmann machines are used in this paper according to many experiments, that is, the five layer DBN, which allows for the shorter training time and obtain good diagnosis results. The number of nodes of the DBN (50-50-200) and the BLM (100-15-17,000) are set in this experiment. The initial values of parameters for SVM are described as c = 380, g = 0.4710, and p = 0.010375. The values of other parameters are the same as those in Section 4.3. The diagnosis results and the test times of different methods are shown in Table 3, Figure 7, and Figure 8.

Table 3. Comparison of diagnosis results and test time.

Diagnosis Methods	Diagnostic Accuracy (%)	Test Time (s)
VHSMFD	40.46	274.71
EHDNFD	95.02	664.57
EEHDNFD	96.55	630.37
VHDNFD	97.68	459.21
VHBLFD	97.74	22.29

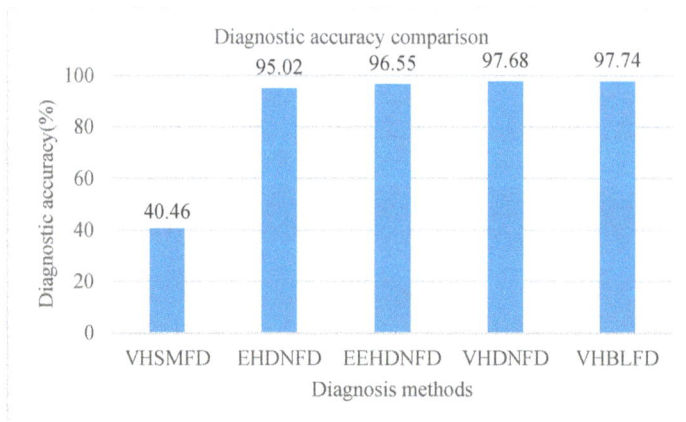

Figure 7. The comparison results of diagnostic accuracy.

From Table 3, Figure 7, and Figure 8, the diagnosis accuracies of VHSMFD, EHDNFD, EEHDNFD, VHDNFD, and VHBLFD are 40.46%, 95.02%, 96.55%, 97.68%, and 97.74%, respectively. For the VHSMFD method based on VMD, HT and SVM, the fault diagnosis accuracy is only 40.46%, and the diagnosis effect is the worst of the five diagnosis methods. The results show that the SVM cannot construct a fault diagnosis model with strong generalization ability for difference data. For the VHBLFD method, the diagnosis accuracy is 97.74%, and the diagnosis effect is better than that of the VHSMFD, EHDNFD, EEHDNFD and VHDNFD methods. The results show that the BLM can construct a fault diagnosis model with strong generalization ability and the higher accuracy for difference data. The test time of the VHSMFD, EHDNFD, EEHDNFD, VHDNFD and VHBLFD are 274.71 s, 664.57 s, 630.37, 459.21 s, and 22.29 s, respectively. The test times of EEHDNFD method is 664.57 s, and he fault diagnosis efficiency is the lowest of these diagnosis methods. The test time of the VHBLFD is 22.29 s, and the fault diagnosis efficiency is the highest of these fault diagnosis methods. The results show that the BLM can construct a fault diagnosis model with better diagnosis efficiency and faster diagnosis speed. Therefore, the VHBLFD takes on the higher diagnosis accuracy and better diagnosis efficiency.

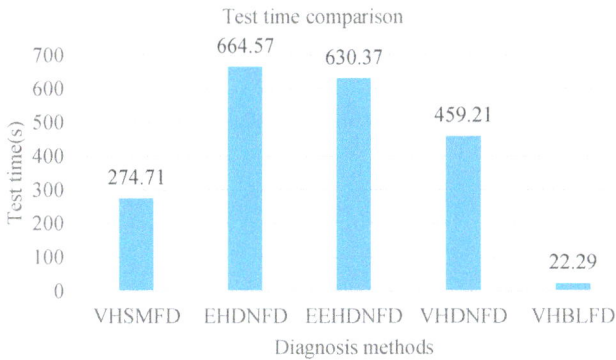

Figure 8. The comparison results of test time.

4.5. The Influences of Parameters in BLM for Diagnosis Accuracy

4.5.1. The Influences of the Number of Feature Nodes for Diagnosis Accuracy

In this section, when the number of feature node windows and the number of enhancement nodes are unchanged, the number of feature nodes is changed, and ten different states of the rolling bearings are identified. The experiment under each parameter is carried out ten times, and the test results and the running times are averaged for 10 times. The test results and running time are shown in Table 4, Figure 9, and Figure 10. The regularization parameter C is 2×10^{-30}, and the enhance node reduction ratio s is 0.8 in Table 1. N11 is the number of feature nodes. N2 is the number of feature node windows. N33 is the number of enhancement nodes.

Table 4. Test results for different feature nodes. (N11 is the number of feature nodes, N2 is the number of feature node windows, and N33 is the number of enhancement nodes).

(N11, N2, N33)	Test Accuracy (%)	Total Average Time (s)
40, 15, 3000	96.9902	4.8618
50, 15, 3000	96.9601	5.2248
60, 15, 3000	96.3506	5.6163
70, 15, 3000	96.2904	6.0630
80, 15, 3000	96.2302	6.5115
90, 15, 3000	95.8239	7.0634
100, 15, 3000	95.5982	7.5683
200, 15, 3000	92.0692	15.0772
300, 15, 3000	89.7968	21.7082

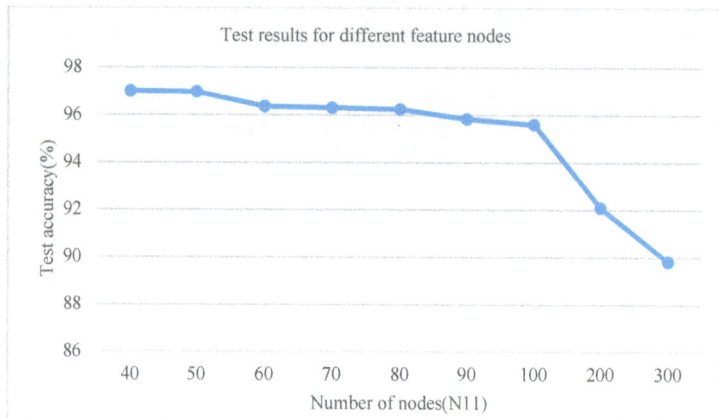

Figure 9. The test accuracy of different numbers of nodes (N11).

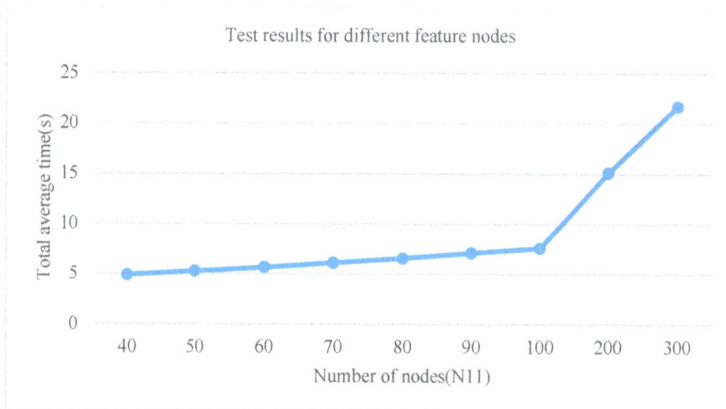

Figure 10. The total average times of different numbers of nodes (N11).

From Table 4, Figures 9 and 10, the total test time increases from 4.8618 s to 21.7082 s in the process of increasing feature nodes from 40 to 300, but the test accuracy takes on a decreasing trend from 96.9902% to 89.7968%. The experiment results show that the number of feature nodes can be reasonably selected according to the size of the input data in order to obtain the best diagnosis result in practical applications.

4.5.2. The Influences of the Number of Feature Node Windows for Diagnosis Accuracy

In this section, when the number of enhancement nodes and feature nodes is unchanged, the number of feature node windows is changed, and ten different states of the rolling bearings are identified. The experiment under each parameter is carried out ten times, and the test results and are running times are averaged for 10 times. The results and the running times are shown in Table 4, Figure 11, and Figure 12. The regularization parameter C is 2×10^{-30}, and the enhance node reduction ratio s is 0.8 in Table 5.

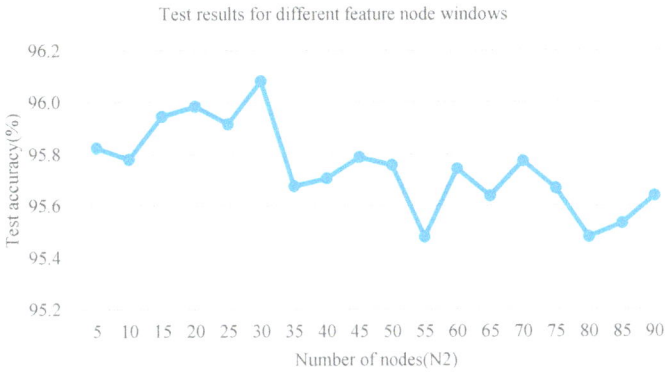

Figure 11. The test accuracy of different numbers of nodes (N2).

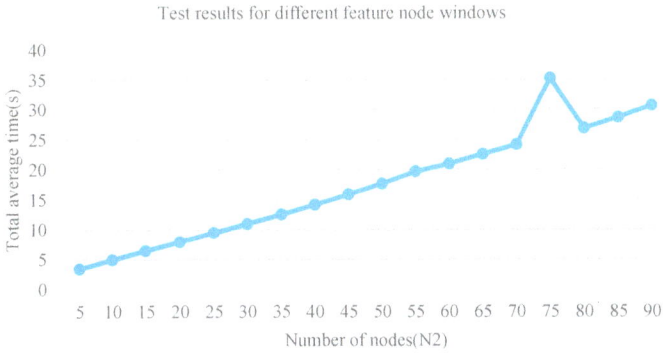

Figure 12. The total average time of different numbers of nodes (N2).

Table 5. Test results for different feature node windows.

Number of Nodes (N11, N2, N33)	Test Accuracy (%)	Total Average Time (s)
100, 5, 1000	95.8239	3.4090
100, 10, 1000	95.7787	4.9404
100, 15, 1000	95.9443	6.4226
100, 20, 1000	95.9819	7.9057
100, 25, 1000	95.9142	9.4303
100, 30, 1000	96.0797	10.9200
100, 35, 1000	95.6734	12.4819
100, 40, 1000	95.7035	14.1060
100, 45, 1000	95.7863	15.8277
100, 50, 1000	95.7562	17.6226
100, 55, 1000	95.4778	19.6315
100, 60, 1000	95.7411	20.9249
100, 65, 1000	95.6358	22.5161
100, 70, 1000	95.7712	24.0979
100, 75, 1000	95.6659	35.2004
100, 80, 1000	95.4778	26.8084
100, 85, 1000	95.5304	28.5762
100, 90, 1000	95.6358	30.5728

As can be seen from Table 5, Figure 11, and Figure 12, in the case of only changing the number of feature node windows, the total test time increases from 3.4090 s to 35.2004 s in the process of increasing feature windows from 5 to 90, and the test accuracy almost remains unchanged. The experimental results show that the training time of BLM is affected, and the test accuracy is less affected by the number of feature node windows. Therefore, the number of feature node windows only affects the training time of BLM.

4.5.3. The Influences of the Number of Enhancement Nodes for Diagnosis Accuracy

In this section, when the number of feature nodes and feature node windows is unchanged, the number of enhancement nodes is changed, and ten different states of the bearing are identified. The experiment under each parameter is carried out ten times, and the test results and the running times are averaged for 10 times. The results and running time are shown in Table 6, Figure 13, and Figure 14. The regularization parameter C is 2×10^{-30}, and the enhance node reduction ratio s is 0.8 in Table 6.

Table 6. Test results for different enhancement nodes.

Number of Nodes (N11, N2, N33)	Test Accuracy (%)	Total Average Time (s)
100, 15, 1000	95.9970	6.4500
100, 15, 2000	96.5613	7.1869
100, 15, 3000	95.5982	7.5683
100, 15, 4000	90.0000	8.0937
100, 15, 5000	80.7374	8.9812
100, 15, 6000	93.4989	9.2012
100, 15, 7000	95.6810	9.9691
100, 15, 8000	96.5162	10.7368
100, 15, 9000	97.0880	11.6575
100, 15, 10000	97.1257	12.7143
100, 15, 11000	97.2611	13.6536
100, 15, 12000	97.3589	14.8280
100, 15, 13000	97.4643	16.0800
100, 15, 14000	97.6072	17.3710
100, 15, 15000	97.5320	18.6380
100, 15, 16000	97.7200	20.8649
100, 15, 17000	97.7351	22.2932

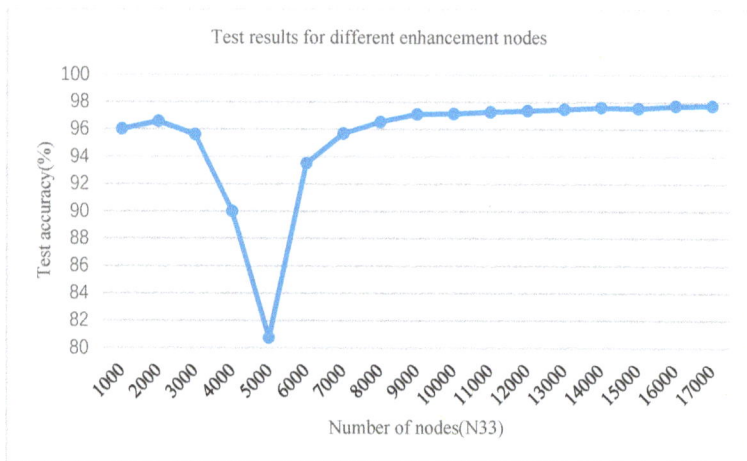

Figure 13. The test accuracy of different numbers of nodes (N33).

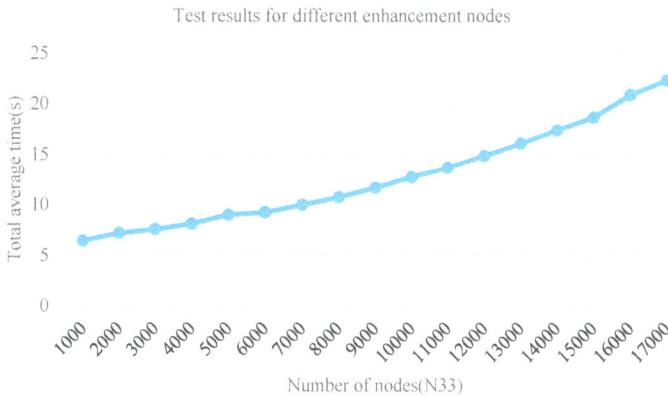

Figure 14. The total average times of different numbers of nodes (N33).

As can be seen from Table 6, Figure 13, and Figure 14, in the case of changing only the number of the enhancement nodes, the training time of the BLM increases with the increase enhancement nodes. Under the influence of eliminating some special nodes, the test accuracy will also increase with the increasing of the enhancement nodes. The experiment results show that the number of enhancement nodes can be flexibly selected according to the different requirements of the BLM for training time and test accuracy. Therefore, the number of enhancement nodes not only affects the training time but also affects the test accuracy.

To sum up, the feature nodes and enhancement nodes can influence the accuracy of fault diagnosis, and the number of feature node windows cannot influence the accuracy using the BLM for fault diagnosis. Using the BLM for classifying the fault provide better diagnosis efficiency and faster diagnosis speed.

5. Conclusions

To effectively diagnose the faults of rolling bearings of the AC motor, a new fault diagnosis (VHBLFD) method based on VMD, HT, and BLM is proposed. VMD and HT are used to obtain Hilbert envelope spectra. The BLM is an effective incremental learning model to achieve competitive results of the state-of-art method. The BLM with fast calculation speed and strong generalization ability is used to realize the fault classification. The actual vibration data are used to validate the effectiveness of the VHBLFD. The fault diagnosis accuracy of the VHBLFD is 97.74%, and the test time of the VHBLFD is 22.29s. The results show that the BLM can construct the VHBLFD model with a higher diagnosis accuracy and a better diagnosis efficiency. Compared with the DBN and the SVM, the BLM is more sensitive for fault features and has faster diagnosis speed and better robustness. Therefore, the VHBLFD method provides higher diagnosis accuracy and better diagnosis efficiency.

In order to solve the practical engineering problem, we plan to design a new experiment platform to obtain a new dataset containing much more data. We will use the new dataset to study the fault diagnosis method in future work.

Author Contributions: Methodology, H.Z. and J.Z.; validation, L.Z.; resources, C.G.; data curation, W.D. and Y.Y.; writing—original draft preparation, J.Z. and Y.Y.; writing—review and editing, W.D.; visualization, L.Z.; funding acquisition, H.Z., W.D. and C.G.

Funding: This research was funded by the National Natural Science Foundation of China grant number 51605068, grant number 61771087, grant number 51879027, grant number 51579024, the Open Project Program of State Key Laboratory of Mechanical Transmissions of Chongqing University grant number SKLMT-KFKT-201803,

Open Project Program of the Traction Power State Key Laboratory of Southwest Jiaotong University under Grant TPL1803, and Liaoning BaiQianWan Talents Program.

Conflicts of Interest: The authors declare no conflict of interest.

References

1. Lu, S.L.; He, Q.B.; Zhang, H.B.; Kong, F.R. Rotating machine fault diagnosis through enhancement stochastic resonance by full-wave signal construction. *Mech. Syst. Signal Process.* **2017**, *85*, 82–97. [CrossRef]
2. Yang, Y.; Yu, D.J.; Cheng, J.S. A fault diagnosis approach for roller bearing based on IMF envelope spectrum and SVM. *Measurement* **2007**, *40*, 943–950. [CrossRef]
3. Deng, W.; Zhang, S.J.; Zhao, H.M.; Yang, X.H. A novel fault diagnosis method based on integrating empirical wavelet transform and fuzzy entropy for motor bearing. *IEEE Access* **2018**, *6*, 35042–35056. [CrossRef]
4. Guo, S.K.; Chen, R.; Wei, M.M.; Li, H.; Liu, Y.Q. Ensemble data reduction techniques and Multi-RSMOTE via fuzzy integral for bug report classification. *IEEE Access* **2018**, *6*, 45934–45950. [CrossRef]
5. Yu, J.; Ding, B.; He, Y.J. Rolling bearing fault diagnosis based on mean multigranulation decision-theoretic rough set and non-naive Bayesian classifier. *J. Mech. Sci. Technol.* **2018**, *32*, 5201–5211. [CrossRef]
6. Liu, Y.Q.; Yi, X.K.; Chen, R.; Zhai, Z.G.; Gu, J.X. Feature extraction based on information gain and sequential pattern for English question classification. *IET Softw.* **2018**, *12*, 520–526. [CrossRef]
7. Deng, W.; Yao, R.; Zhao, H.M.; Yang, X.H.; Li, G.Y. A novel intelligent diagnosis method using optimal LS-SVM with improved PSO algorithm. *Soft Comput.* **2019**, *23*, 2445–2462. [CrossRef]
8. Ren, Z.R.; Skjetne, R.; Jiang, Z.Y.; Gao, Z.; Verma, A.S. Integrated GNSS/IMU hub motion estimator for offshore wind turbine blade installation. *Mech. Syst. Signal Process.* **2019**, *123*, 222–243. [CrossRef]
9. Chen, H.L.; Xu, Y.T.; Wang, M.J.; Zhao, X.H. Chaos-induced and Mutation-driven algorithm for constrained engineering design problems. *Appl. Math. Model.* **2019**, *71*, 45–59. [CrossRef]
10. Lu, S.L.; Zhou, P.; Wang, X.X.; Liu, B.Y.; Liu, F.; Zhao, J.W. Condition monitoring and fault diagnosis of motor bearings using undersampled vibration signals from a wireless sensor network. *J. Sound Vib.* **2018**, *414*, 81–96. [CrossRef]
11. Xie, H.M.; Yang, K.; Li, S.W.; Yin, S.; Peng, J.; Zhu, F.; Li, H.; Zhang, L. Microwave heating-assisted pyrolysis of mercury from sludge. *Mater. Res. Express* **2019**, *6*, 015507. [CrossRef]
12. Deng, W.; Xu, J.J.; Zhao, H.M. An improved ant colony optimization algorithm based on hybrid strategies for scheduling problem. *IEEE Access* **2019**, *7*, 20281–20292. [CrossRef]
13. Yu, J.; He, Y.J. Planetary gearbox fault diagnosis based on data-driven valued characteristic multigranulation model with incomplete diagnostic information. *J. Sound Vib.* **2018**, *429*, 63–77. [CrossRef]
14. Osman, S.; Wang, W. A normalized Hilbert-Huang transform technique for bearing fault detection. *J. Vib. Control* **2014**, *22*, 2771–2787. [CrossRef]
15. Wang, L.H.; Zhao, X.P.; Wu, J.X.; Xie, Y.Y.; Zhang, Y.H. Motor fault diagnosis based on short-time Fourier transform and convolutional neural network. *Chin. J. Mech. Eng.* **2017**, *30*, 1357–1368. [CrossRef]
16. Pan, Y.N.; Chen, J. The changes of complexity in the performance degradation process of rolling element bearing. *J. Vib. Control* **2016**, *22*, 344–357. [CrossRef]
17. Li, T.; Hu, Z.; Jia, Y.; Wu, J.; Zhou, Y. Forecasting crude oil prices using ensemble empirical mode decomposition and sparse Bayesian learning. *Energies* **2018**, *11*, 1882. [CrossRef]
18. Ren, Z.; Skjetne, R.; Gao, Z. A crane overload protection controller for blade lifting operation based on model predictive control. *Energies* **2019**, *12*, 50. [CrossRef]
19. Liu, G.; Chen, B.; Jiang, S.; Fu, H.; Wang, L.; Jiang, W. Double entropy joint distribution function and its application in calculation of design wave height. *Entropy* **2019**, *21*, 64. [CrossRef]
20. Zhao, H.M.; Sun, M.; Deng, W.; Yang, X.H. A new feature extraction method based on EEMD and multi-scale fuzzy entropy for motor bearing. *Entropy* **2017**, *19*, 14. [CrossRef]
21. Chen, R.; Guo, S.K.; Wang, X.Z.; Zhang, T.L. Fusion of multi-RSMOTE with fuzzy integral to classify bug reports with an imbalanced distribution. *IEEE Trans. Fuzzy Syst.* **2019**. [CrossRef]
22. Zhao, H.M.; Yao, R.; Xu, L.; Yuan, Y.; Li, G.Y.; Deng, W. Study on a novel fault damage degree identification method using high-order differential mathematical morphology gradient spectrum entropy. *Entropy* **2018**, *20*, 682. [CrossRef]

23. Sun, F.R.; Yao, Y.D.; Li, X.F. The heat and mass transfer characteristics of superheated steam coupled with non-condensing gases in horizontal wells with multi-point injection technique. *Energy* **2018**, *143*, 995–1005. [CrossRef]

24. Guo, S.K.; Chen, R.; Li, H.; Zhang, T.L.; Liu, Y.Q. Identify severity bug report with distribution imbalance by CR-SMOTE and ELM. *Int. J. Softw. Eng. Knowl. Eng.* **2019**, *29*, 139–175. [CrossRef]

25. Zhou, Y.R.; Li, T.Y.; Shi, J.Y.; Qian, Z.J. A CEEMDAN and XGBOOST–based approach to forecast crude oil prices. *Complexity* **2019**. [CrossRef]

26. Deng, W.; Zhao, H.M.; Yang, X.H.; Xiong, J.X.; Sun, M.; Li, B. Study on an improved adaptive PSO algorithm for solving multi-objective gate assignment. *Appl. Soft Comput.* **2017**, *59*, 288–302. [CrossRef]

27. Guo, J.H.; Mu, Y.; Xiong, M.D.; Liu, Y.Q.; Gu, J.X. Activity feature solving based on TF-IDF for activity recognition in smart homes. *Complexity* **2019**. [CrossRef]

28. Fu, H.; Li, Z.; Liu, Z.; Wang, Z. Research on big data digging of hot topics about recycled water use on micro-blog based on particle swarm optimization. *Sustainability* **2018**, *10*, 2488. [CrossRef]

29. Tang, G.; Zhang, Y.; Wang, H.Q. Multivariable LS-SVM with moving window over time slices for the prediction of bearing performance degradation. *J. Intell. Fuzzy Syst.* **2018**, *34*, 3747–3757. [CrossRef]

30. Kim, K.J.; Cho, S.B. Ensemble Bayesian networks evolved with speciation for high-performance prediction in data mining. *Soft Comput.* **2017**, *21*, 1065–1080. [CrossRef]

31. Deng, W.; Zhao, H.M.; Zou, L.; Li, G.Y.; Yang, X.H.; Wu, D.Q. A novel collaborative optimization algorithm in solving complex optimization problems. *Soft Comput.* **2017**, *21*, 4387–4398. [CrossRef]

32. Wang, H.C.; Chen, J. Performance degradation assessment of rolling bearing based on bispectrum and support vector data description. *J. Vib. Control* **2014**, *20*, 2032–2041. [CrossRef]

33. Yu, J.; Bai, M.Y.; Wang, G.N.; Shi, X.J. Fault diagnosis of planetary gearbox with incomplete information using assignment reduction and flexible naive Bayesian classifier. *J. Mech. Sci. Technol.* **2018**, *32*, 37–47. [CrossRef]

34. Gao, H.Z.; Liang, L.; Chen, X.G.; Xu, G.H. Feature extraction and recognition for rolling element bearing fault utilizing short-time Fourier transform and non-negative matrix factorization. *Chin. J. Mech. Eng.* **2015**, *28*, 96–105. [CrossRef]

35. Zhang, C.L.; Li, B.; Chen, B.Q.; Cao, H.; Zi, Y.; He, Z. Weak fault signature extraction of rotating machinery using flexible analytic wavelet transform. *Mech. Syst. Signal. Process.* **2015**, *64*, 162–187. [CrossRef]

36. Kabla, A.; Mokrani, K. Bearing fault diagnosis using Hilbert-Huang transform (HHT) and support vector machine (SVM). *Mech. Ind.* **2016**, *17*, 308. [CrossRef]

37. Yuan, J.; Ji, F.; Gao, Y.; Zhu, J.; Wei, C.; Zhou, Y. Integrated ensemble noise-reconstructed empirical mode decomposition for mechanical fault detection. *Mech. Syst. Signal Process.* **2018**, *104*, 323–346. [CrossRef]

38. Du, Y.C.; Du, D.P. Fault detection and diagnosis using empirical mode decomposition based principal component analysis. *Comput. Chem. Eng.* **2018**, *115*, 1–21. [CrossRef]

39. Fei, S.W.; He, Y. A multi-layer KMC-RS-SVM classifier and DGA for fault diagnosis of power transformer. *Recent Pat. Comput. Sci.* **2012**, *5*, 238–243. [CrossRef]

40. Kang, L.; Zhao, L.; Yao, S.; Duan, C.X. A new architecture of super-hydrophilic β-SiAlON/graphene oxide ceramic membrane for enhanced anti-fouling and separation of water/oil emulsion. *Ceram. Int.* **2019**. [CrossRef]

41. Cheng, J.S.; Peng, Y.F.; Yang, Y.; Wu, Z. Adaptive sparsest narrow-band decomposition method and its applications to rolling element bearing fault diagnosis. *Mech. Syst. Signal Process.* **2017**, *85*, 947–962. [CrossRef]

42. Huang, W.Y.; Cheng, J.S.; Yang, Y. Rolling bearing performance degradation assessment based on convolutional sparse combination learning. *IEEE Access* **2019**, *7*, 17834–17846. [CrossRef]

43. Huang, W.Y.; Cheng, J.S.; Yang, Y. Rolling bearing fault diagnosis and performance degradation assessment under variable operation conditions based on nuisance attribute projection. *Mech. Syst. Signal Process.* **2019**, *114*, 165–188. [CrossRef]

44. Van Tung, T.; AlThobiani, F.; Ball, A. An approach to fault diagnosis of reciprocating compressor valves using Teager-Kaiser energy operator and deep belief networks. *Expert Syst. Appl.* **2014**, *41*, 4113–4122.

45. Guo, X.J.; Chen, L.; Shen, C.Q. Hierarchical adaptive deep convolution neural network and its application to bearing fault diagnosis. *Measurement* **2016**, *93*, 490–502. [CrossRef]

46. Qi, Y.M.; Shen, C.Q.; Wang, D.; Shi, J.; Jiang, X.; Zhu, Z. Stacked sparse autoencoder-based deep network for fault diagnosis of rotating machinery. *IEEE Access* **2017**, *5*, 15066–15079. [CrossRef]

47. Shao, H.D.; Jiang, H.K.; Wang, F.A.; Wang, Y. Rolling bearing fault diagnosis using adaptive deep belief network with dual-tree complex wavelet packet. *ISA Trans.* **2017**, *69*, 187–201. [CrossRef]

48. Li, S.B.; Liu, G.K.; Tang, X.H.; Lu, J.; Hu, J. An ensemble deep convolutional neural network model with improved d-s evidence fusion for bearing fault diagnosis. *Sensors* **2017**, *17*, 1729. [CrossRef]

49. Shao, H.D.; Jiang, H.K.; Zhang, H.Z.; Duan, W.; Liang, T.; Wu, S. Rolling bearing fault feature learning using improved convolutional deep belief network with compressed sensing. *Mech. Syst. Signal Process.* **2018**, *100*, 743–765. [CrossRef]

50. Sun, C.; Ma, M.; Zhao, Z.B.; Chen, X.F. Sparse deep stacking network for fault diagnosis of motor. *IEEE Trans. Ind. Inform.* **2018**, *14*, 3261–3270. [CrossRef]

51. Zhang, W.; Li, C.H.; Peng, G.L.; Chen, Y.; Zhang, Z. A deep convolutional neural network with new training methods for bearing fault diagnosis under noisy environment and different working load. *Mech. Syst. Signal Process.* **2018**, *100*, 439–453. [CrossRef]

52. Shao, H.D.; Jiang, H.K.; Zhang, H.Z.; Liang, T. Electric locomotive bearing fault diagnosis using a novel convolutional deep belief network. *IEEE Trans. Ind. Electron.* **2018**, *65*, 2727–2736. [CrossRef]

53. Wang, Z.R.; Wang, J.; Wang, Y.R. An intelligent diagnosis scheme based on generative adversarial learning deep neural networks and its application to planetary gearbox fault pattern recognition. *Neurocomputing* **2018**, *310*, 213–222. [CrossRef]

54. Wang, S.H.; Xiang, J.W.; Zhong, Y.T.; Tang, H. A data indicator-based deep belief networks to detect multiple faults in axial piston pumps. *Mech. Syst. Signal Process.* **2018**, *112*, 154–170. [CrossRef]

55. Liu, H.; Zhou, J.Z.; Xu, Y.H.; Zheng, Y.; Peng, X.; Jiang, W. Unsupervised fault diagnosis of rolling bearings using a deep neural network based on generative adversarial networks. *Neurocomputing* **2018**, *315*, 412–424. [CrossRef]

56. Zhao, X.L.; Jia, M.P. A novel deep fuzzy clustering neural network model and its application in rolling bearing fault recognition. *Meas. Sci. Technol.* **2018**, *29*, 125005. [CrossRef]

57. Hu, Z.X.; Jiang, P. An imbalance modified deep neural network with dynamical incremental learning for chemical fault diagnosis. *IEEE Trans. Ind. Electron.* **2019**, *66*, 540–550. [CrossRef]

58. Li, Z.P.; Chen, J.L.; Zi, Y.Y.; Pan, J. Independence-oriented VMD to identify fault feature for wheel set bearing fault diagnosis of high speed locomotive. *Mech. Syst. Signal Process.* **2017**, *85*, 512–529. [CrossRef]

59. Jiang, X.X.; Shen, C.Q.; Shi, J.J.; Zhu, Z. Initial center frequency-guided VMD for fault diagnosis of rotating machines. *J. Sound Vib.* **2018**, *435*, 36–55. [CrossRef]

60. Li, J.M.; Yao, X.F.; Wang, H.; Zhang, J. Periodic impulses extraction based on improved adaptive VMD and sparse code shrinkage denoising and its application in rotating machinery fault diagnosis. *Mech. Syst. Signal Process.* **2019**, *126*, 568–589. [CrossRef]

61. Wang, C.G.; Li, H.K.; Huan, G.J.; Ou, J. Early fault diagnosis for planetary gearbox based on adaptive parameter optimized VMD and singular kurtosis difference spectrum. *IEEE Access* **2019**, *7*, 31501–31516. [CrossRef]

62. Zhang, F.Q.; Lei, T.S.; Li, J.H.; Cai, X.; Shao, X.; Chang, J.; Tian, F. Real-time calibration and registration method for indoor scene with joint depth and color camera. *J. Pattern Recognit. Artif. Intell.* **2018**, *32*, 1854021. [CrossRef]

63. Guo, S.K.; Liu, Y.Q.; Chen, R.; Sun, X.; Wang, X.X. Using an improved SMOTE algorithm to deal imbalanced activity classes in smart home. *Neural Process. Lett.* **2018**. [CrossRef]

64. Wen, J.; Zhong, Z.F.; Zhang, Z.; Fei, L.K.; Lai, Z.H.; Chen, R.Z. Adaptive locality preserving regression. *IEEE Trsns. Circ. Syst. Vid.* **2018**. [CrossRef]

65. Liu, Y.Q.; Wang, X.X.; Zhai, Z.G.; Chen, R.; Zhang, B.; Jiang, Y. Timely daily activity recognition from headmost sensor events. *ISA Trans.* **2019**. [CrossRef] [PubMed]

66. Zhang, F.Q.; Wang, Z.W.; Chang, J.; Zhang, J.; Tian, F. A fast framework construction and visualization method for particle-based fluid. *EUPASIP J. Image Video Process.* **2017**, *2017*, 79. [CrossRef]

67. Huang, F.; Yao, C.; Liu, W.; Li, Y.; Liu, X. Landslide susceptibility assessment in the nantian area of china: A comparison of frequency ratio model and support vector machine. *Geomat. Nat. Haz. Risk* **2018**, *9*, 919–938. [CrossRef]

68. Zhang, H.; Fang, Y. Temperature dependent photoluminescence of surfactant assisted electrochemically synthesized ZnSe nanostructures. *J. Alloy Compd.* **2019**, *781*, 201–208. [CrossRef]

69. Liu, G.; Chen, B.; Gao, Z.; Fu, H.; Jiang, S.; Wang, L.; Yi, K. Calculation of joint return period for connected edge data. *Water* **2019**, *11*, 300. [CrossRef]
70. Zhou, J.; Du, Z.; Yang, Z.; Xu, Z. Dynamic parameters optimization of straddle-type monorail vehicles based multiobjective collaborative optimization algorithm. *Vehicle Syst. Dyn.* **2019**. [CrossRef]
71. Lin, J.; Yuan, J.S. Analysis and simulation of capacitor-less ReRAM-based stochastic neurons for the in-memory spiking neural network. *IEEE Trans. Biomed Circ. Syst.* **2018**, *12*, 1004–1017. [CrossRef] [PubMed]
72. Wen, J.; Fang, X.Z.; Cui, J.R.; Fei, L.K.; Yan, K.; Chen, Y.; Xu, Y. Robust sparse linear discriminant analysis. *IEEE Trans. Circ. Syst. Vid.* **2019**, *29*, 390–403. [CrossRef]
73. Chen, H.L.; Jiao, S.; Heidarib, A.A.; Wang, M.J.; Chen, X.; Zhao, X.H. An opposition-based sine cosine approach with local search for parameter estimation of photovoltaic models. *Energ. Convers. Manage.* **2019**, *195*, 927–942. [CrossRef]
74. Yu, W.J.; Zeng, Z.; Peng, B.; Yan, S.; Huang, Y.H.; Jiang, H.; Li, X.B.; Fan, T. Multi-objective optimum design of high-speed backplane connector using particle swarm optimization. *IEEE Access* **2018**, *6*, 35182–35193. [CrossRef]
75. Luo, J.; Chen, H.L.; Zhang, Q.; Xu, Y.T.; Huang, H.; Zhao, X.H. An improved grasshopper optimization algorithm with application to financial stress prediction. *Appl. Math. Model.* **2018**, *64*, 654–668. [CrossRef]
76. Liu, G.; Gao, Z.; Chen, B.; Fu, H.; Jiang, S.; Wang, L.; Koi, Y. Study on Threshold selection methods in calculation of ocean environmental design parameters. *IEEE ACCESS* **2019**, *7*, 39515–39527. [CrossRef]
77. Hinton, G.E.; Osindero, S.; Teh, Y.W. A fast learning algorithm for deep belief nets. *Neural Comput.* **2006**, *18*, 1527–1554. [CrossRef] [PubMed]
78. Chen, C.P.; Liu, Z.L. Broad learning system: An effective and efficient incremental learning system without the need for deep architecture. *IEEE Trans. Neural Netw. Learn. Syst.* **2018**, *29*, 10–24. [CrossRef]
79. Bearing Data Center. Available online: http://csegroups.case.edu/bearingdatacenter/home (accessed on 13 July 2017).

symmetry

MDPI

Article

Rampant Arch and Its Optimum Geometrical Generation

Cristina Velilla [1], Alfredo Alcayde [2], Carlos San-Antonio-Gómez [1], Francisco G. Montoya [2], Ignacio Zavala [1] and Francisco Manzano-Agugliaro [2],*

[1]   Department of Ingeniería Agroforestal, Universidad Politécnica de Madrid, Av. De Puerta de Hierro, 28040 Madrid, Spain; cristina.velilla@upm.es (C.V.); c.sanantonio@upm.es (C.S.-A.-G.); ignacio.zavala.morencos@upm.es (I.Z.)
[2]   Department of Engineering, University of Almeria, ceiA3, 04120 Almeria, Spain; aalcayde@ual.es (A.A.); pagilm@ual.es (F.G.M.)
*   Correspondence: fmanzano@ual.es; Tel.: +34-950-015396; Fax: +34-950-015491

Received: 15 April 2019; Accepted: 28 April 2019; Published: 3 May 2019

Abstract: Gothic art was developed in western Europe from the second half of the 12th century to the end of the 15th century. The most characteristic Gothic building is the cathedral. Gothic architecture uses well-carved stone ashlars, and its essential elements include the arch. The thrust is transferred by means of external arches (flying buttresses) to external buttresses that end in pinnacles, which accentuates the verticality. The evolution of the flying buttresses should not only be considered as an aesthetic consideration, but also from a constructive point of view as an element of transmission of forces or loads. Thus, one evolves from a beam-type buttress to a simple arch, and finally to a rampant arch. In this work, we study the geometry of the rampant arch to determine which is the optimum from the constructive point of view. The optimum rampant arch obtained is the one with the common tangent to the two arches parallel to the slope line. A computer program was created to determine this optimal rampant arch by means of a numerical or graphical input. It was applied to several well-known and representative cases of Gothic art in France (church of Saint Urbain de Troyes) and Spain (Cathedral of Palma de Mallorca), establishing if they were designs of optimal rampant arches or not.

Keywords: rampant arch; geometry; optimum; flying buttresses; cathedral

---

1. Introduction

The rampant arch is an arch whose starts, in walls or buttresses, are located at different levels, often with a considerable difference in height [1]. It was used extensively in Gothic architecture to shape the buttress, and its function was to transmit the thrust of the vaults to the buttresses and these to the foundations [2]. In addition to this structural function, the buttress serves to carry rainwater from the vaults [3] to the exterior through the pinnacles.

The most common rampant arch used in the flying buttresses of most well-known Gothic cathedrals is the one formed by a single circumferential arch [4]. According to Viollet-le-Duc (1854) [5] there are two types: flying buttresses in which the center of the circumference is in the wall (Figure 1A), and flying buttresses in which its center is displaced toward the interior of the building (Figure 1B). The first are the oldest flying buttresses, while the second are the most used because, from a mechanical point of view, they perform better than the first [6].

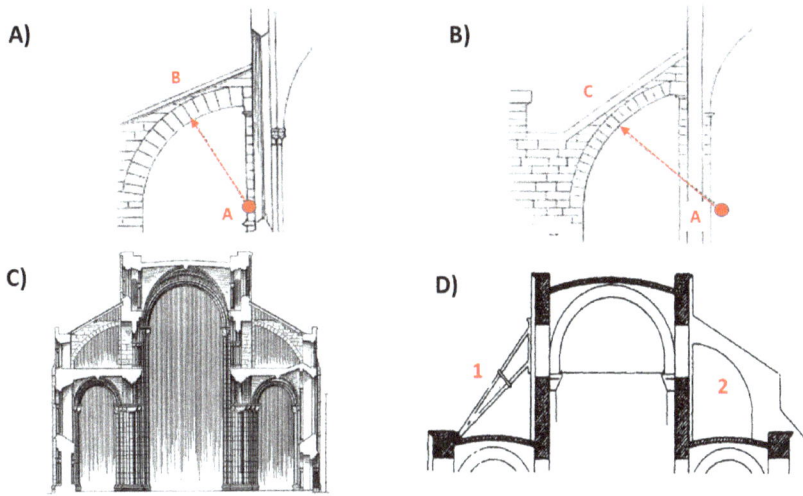

Figure 1. (**A**,**B**) The two types of flying buttresses according to Viollet-le-Duc [5]. (**C**) Section of Durham Cathedral [7]. (**D**) Wooden props from the vault of the church of Vézelay (1) replaced by stone flying buttresses (2) [8].

The flying buttresses appeared for the first time in Durham Cathedral around the year 1100 as an evolution of the hidden buttress in the triforium, where an opening is made that allows a longitudinal route parallel to the main nave [9]. In Durham, the buttress did not yet have the function of balancing the lateral thrusts of the vaults; its mission was to support the roof that covers the triforium (Figure 1C).

The flying buttresses of a single circumferential arch could also have been the result of the evolution of the framework of auxiliary wooden constructions that, from a certain moment on, appeared to support the Romanesque barrel vaults that were beginning to open up. Over time, these frameworks would begin to be built in stone (Figure 1D). This moment perhaps occurred in France at the Abbey of Vézelay [8], where the buttress would not be an "invention ex novo" but the evolution of those wooden structures known throughout France [10].

The rampant arch of a single circumferential arch played an essential role in the structural and formal conception of Gothic architecture, creating in the external volume of the cathedrals a succession of dematerialized planes composed of one, two, and even three rows of buttresses, contributing to the spatial richness of the building [11]. The successive rhythm of the flying buttresses and buttresses with their pinnacles of reinforcement makes it possible to open large windows with stained-glass windows that give great luminosity to the interior of the cathedral and give symbolic and immaterial value to the Gothic architecture, which contrasts with the heaviness of the Romanesque architecture it replaced (Figure 1D) [12].

Although the single-arch flying buttresses are usually used in large cathedrals, there are also flying buttresses made up of two circumferential arches (Figure 2A). In this case, the rampant arch results from the union of two circumferential arches, tangent in the key, whose centers are at different levels, so that their outcrops in the walls or buttresses are at different heights. Examples that could be cited include, in France, the flying buttresses of the Church of Saint Urbain de Troyes [13] (Figure 2B), where the red points (A and C) are the starting points of the two arches, and between which the unevenness of the flying buttress must be bridged, and those of the Cathedral of Beauvais, which alternate a first line of ramps of a single arch of circumference with another of two arches (Figure 3A) [14]. In Spain, for example, there are the original flying buttresses of the cathedral of Palma de Mallorca, made up of two rampant arches of one and two circumferential arches, integrated into a single flying buttress (Figure 3B) [15]. The rampant arches consisting of two circumferential arches are used in Gothic architecture to shape the flying buttresses, but also to build stairs. In Spain, there are many examples such as the Monastery of Uclés in Cuenca (Figure 3C), in the palaces of the Generalitat de Barcelona and Valencia, and, in the latter city, the Lonja staircase (Figure 3D).

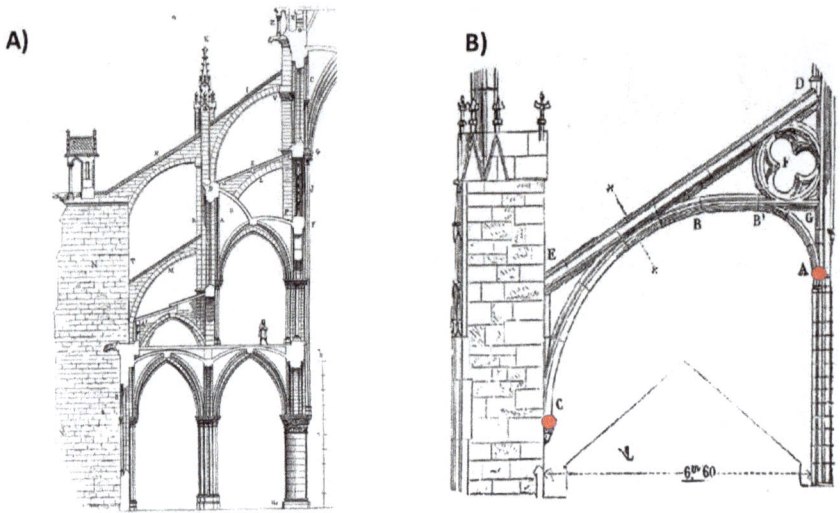

Figure 2. (**A**) Flying buttresses of Notre Dame (Paris) [5]. (**B**) Flying buttress of the church of Saint Urbain de Troyes [8].

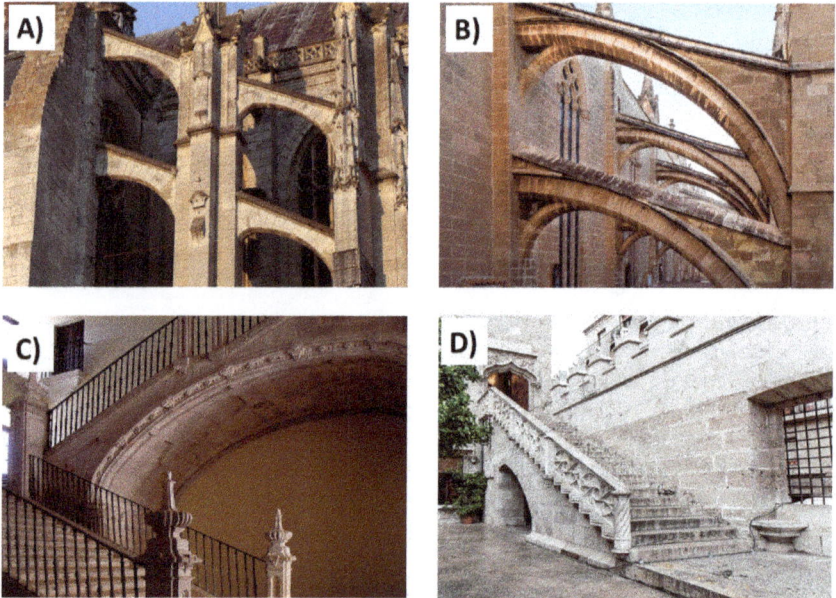

Figure 3. (**A**) One- and two-arch circumference flying buttresses of Beauvais Cathedral (France). (**B**) Flying buttresses of the Cathedral of Palma de Mallorca. (**C**) Rampant arch in the staircase of the Monastery of Uclés (Cuenca). (**D**) Rampant arch in the staircase of La Lonja (Valencia).

The rampant arches were used profusely again at the end of the 19th century with the appearance of neo-Gothic architecture. Many churches were built throughout Europe, especially in France. In America, for example, cathedrals such as New York or Washington were built in the United States (Figure 4). In the same way, in modernism, these arches were also used.

Figure 4. Washington cathedral and its flying buttresses.

In general, the design of a rampant arch of a single arch does not present any difficulty, but the most difficult are the rampant arches of two arches, which throughout history were calculated graphically, and there is no consensus on which the optimal rampant arch of two arches is. The objectives of this work are to review the state of the art in the execution of rampant arches of two arches, to define which is the optimal one, to calculate the numerical solution, and to program it for its calculation and verification of rampant arches of already existing arches. As an additional objective, this work seeks to determine the starting points of the arches in the construction of the rampant arch.

2. Rampant Arches of Two Arches Graphically Constructed

The Gothic buildings raised the question of what methods were used by the builders in the design of their structures. The answer is that the Gothic builders had no method, only a great structural intuition nourished by the experience of successive collapses of structures. In classical texts, there are "structural rules" for dimensioning walls, pillars, and abutments of Gothic churches, all based on the proportionality of the elements. These rules were applied or calculated using geometric or graphical constructions; see, for instance, treaties of Derand [16] or the later one of Blondel [17].

Thus, the usual for the execution of rampant arches was graphical construction, in which two cases were distinguished depending on which parameters were known (see Figure 5).

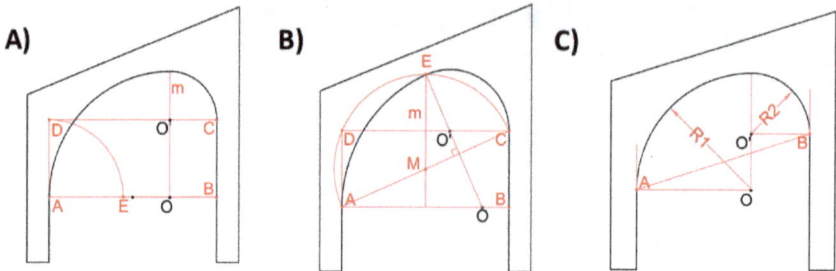

Figure 5. Rampant arches of two arches graphically constructed: (**A**) known A and C (method 1); (**B**) known A and C (method 2); (**C**) known radii of the two arcs (R1 and R2).

2.1. Rampant Arch 1: Known A and C (Starting of the Two Arches)

2.1.1. Method 1

In Figure 5A, the graphical drawing of the rampant arch is described, with the initial (A) and final (C) points known, where the tracing can be summarized as follows:

a.   The ABCD rectangle is drawn;
b.   With center in A, circumference of $R = AD$, obtaining E;
c.   Mediatrix m of BE is drawn;
d.   The intersection of m with the rectangle determines centers O and O'.

2.1.2. Method 2

In Figure 5B, the graphical drawing of the rampant arch is described, with the initial (A) and final (C) points known. Note that the centers are no longer parallel to the support pillars, where the tracing can be summarized as follows:

a.   The ABCD rectangle is drawn;
b.   Mediatrix m of AB determines M;
c.   With center in M, circumference of diameter AC is drawn;
d.   The intersection of this circumference with m is E;
e.   The perpendicular by E to AC determines the centers O and O'.

2.2. Rampant Arch 2: The Radii of the Two Arcs (R1 and R2) Are Known

In this other method, the radii of the two arcs (R1 and R2) are known, which add up the horizontal distance between both points, and it is a question of determining the centers of both.

1. $R1 + R2 =$ horizontal distance between both points (span);
2.   In A and B, perpendicular to the pillars where they are taken, R1 and R2 are used to obtain the centers O and O', respectively, of the arches.

3. The Optimal Rampant Arch

In the literature, there are different points of view for the optimal rampant arch. For example, there are studies that define it from the aesthetic point of view, or from a more harmonious view, in which the curvatures of the different arches that compose it were more similar; here, the optimal rampant arch corresponds to the one in which the ratio between the curvatures of the lateral circumference and the central circumference is closer to one [18].

However, the rampant arch, apart from its aesthetic consideration, as explained in the introduction, has a structural load transmission function, as it replaces the function of a beam [19]. In addition, however, loads can only be transmitted by compression, as this is how the masonry resists bending moments and, therefore, sets the limits of the resistance capacity of the arches [20]. Figure 6 shows examples of classical studies concerned with this issue, i.e., the possible breaking points of rampant arches [21,22]. In this sense, classical research attempted to find the minimum value of "h" (the rise) for which masonry is able to withstand bending moments without any tensile strength [23], estimated to be between 3 and 5 tons of passive horizontal thrust. The optimum rampant arch, from the point of view of the support function or the transmission of the constructive loads, is the one in which the tangent common to the two circles that compose the arch is parallel to the straight unevenness between the points.

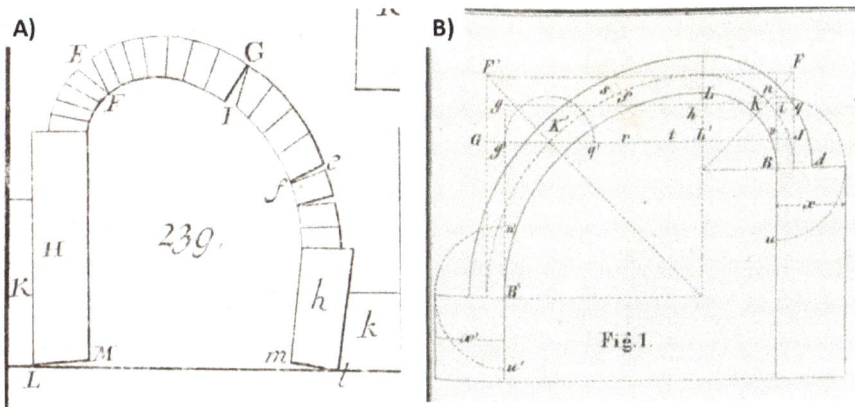

Figure 6. Classical studies of the rupture of rampant arches: (**A**) breakage of masonry by Frezier (1737) [21]; (**B**) calculation of the trust line by Breymann (1881) [22].

Note that there are examples where an inadequate intervention in the structure caused the fall of the dome or the vault, e.g., at Edinburgh in 1768, where the church of Holyrood Abbey collapsed after an inappropriate intervention in the vaults, where the essential buttressing system was rampant arches abutting the nave vault and their pier buttresses were capped with stabilizing pinnacles [24].

3.1. Graphical Design

This section describes the graphic drawing of an optimum rampant arch, i.e., the one whose tangent common to the two arches is parallel to the straight unevenness. The main ideas are as follows:

1. Once the gradient "d" is defined, which indicates the slope of a flying buttress or, where appropriate, of a staircase, the start of the arches is determined by points A and B, which are on a line parallel to "d".
2. Assuming the problem is solved, centers O and O′ are on the straight perpendicular to the gradient line "d". In addition, O and O′ are on the horizontal straight lines from A and B.
3. If from B, the line BC is drawn perpendicular to "d", where logically BC = OO′.
4. At this point, the question is reduced to place a segment OO′ on the bisectors of the angles in M and N.

Thus, the graphical construction of Figure 7 was as follows:

a.   Any PQ line was drawn parallel to "d".
b.   The bisectors b_1 and b_2 of the angles were in P and Q.
c.   BC was traced perpendicular to "d".
d.   BC was moved (by parallelism) until the TS = BC segment was obtained on the bisectors.
e.   TS was moved (by parallelism on the perpendicular to AC) to points O and O', which are the centers of the arches.

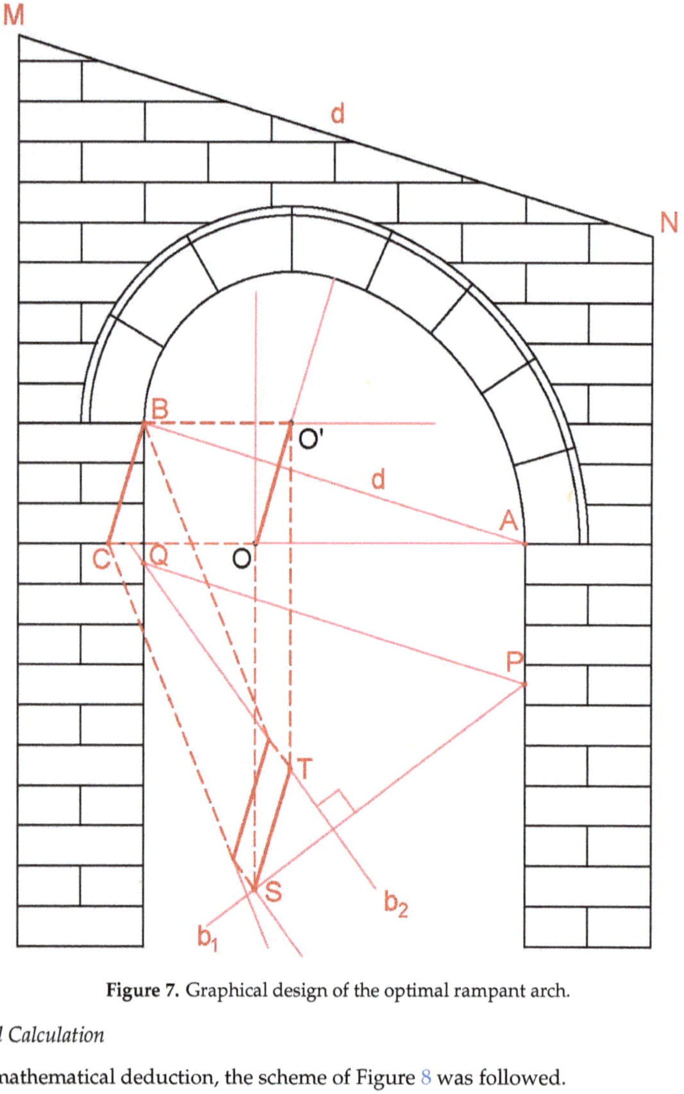

Figure 7. Graphical design of the optimal rampant arch.

3.2. Analytical Calculation

For the mathematical deduction, the scheme of Figure 8 was followed.

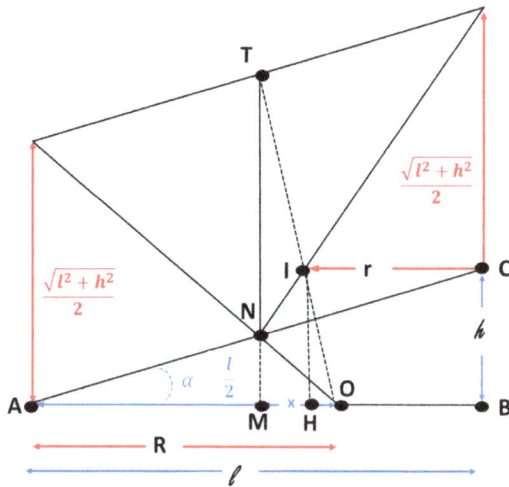

Figure 8. Geometric elements to define the optimal rampant arch.

This scheme traces a rampant arch any ATC, of span l and rise h where the following relationships exist:

$$AB = l,$$

$$BC = HI = h,$$

$$OA = OT = R,$$

$$IC = HB = r,$$

$$OI = R - r.$$

In the triangle OMN,

$$\tan \alpha = \frac{\frac{h}{2}}{R - \frac{l}{2}} = \frac{h}{2R - l}.$$

In the triangle IHO,

$$\sin 2\alpha = \frac{h}{R - r};$$

therefore, it is

$$2R - l = \frac{h \cos \alpha}{\sin \alpha},$$

$$2R - 2r = \frac{h}{\sin \alpha \cos \alpha},$$

where

$$2r - l = \frac{h(\cos^2 \alpha - 1)}{\sin \alpha \cos \alpha} = -\frac{h \sin \alpha}{\cos \alpha}.$$

Then, we get

$$2R = l + \frac{h \cos \alpha}{\sin \alpha},$$

$$2r = l - \frac{h \sin \alpha}{\cos \alpha},$$

$$\frac{R}{r} = \frac{\frac{l \sin \alpha + h \cos \alpha}{\sin \alpha}}{\frac{l \cos \alpha - h \sin \alpha}{\cos \alpha}} = \frac{l \sin \alpha \cos \alpha + h \cos^2 \alpha}{l \sin \alpha \cos \alpha - h \sin^2 \alpha \propto} = \frac{\frac{l \sin 2\alpha}{2} + \frac{h(1 + \cos 2\alpha)}{2}}{\frac{l \sin 2\alpha}{2} - \frac{h(1 - \cos 2\alpha)}{2}} = \frac{l \sin 2 \propto + h \cos 2\alpha + h}{l \sin 2\alpha + h \cos 2\alpha - h}.$$

That is to say, in short,

$$\frac{R}{r} = 1 + \frac{2h}{l\sin 2\propto + h\cos 2\propto - h'}$$

deriving

$$\left(\frac{R}{r}\right)' = \frac{2h\,(2l\cos 2\,\alpha - 2h\sin 2\,\alpha)}{(l\sin 2\,\alpha + h\cos 2\,\alpha - h)^2} = 0,$$

$$\cot 2\,\alpha = \frac{h}{l}.$$

In other words, the optimum rampant arch is the one with the tangent parallel to the uneven line.

The relationship between the radii of curvature in a rampant arch will be found when the tangent is parallel to the straight line of unevenness. Let l the distance between walls (span) and h be the rise. One has, by resemblance of triangles,

$$R = \frac{l}{2} + x,$$

$$R.r = \frac{l^2 + h^2}{4},$$

$$\frac{\sqrt{l^2 + h^2}}{2} = \frac{\frac{h}{2}}{x},$$

where

$$x = \frac{lh}{2} \frac{1}{\left(\sqrt{l^2 + h^2}\right) - h},$$

and

$$R = \frac{l}{2}\left|1 + \frac{h}{\left(\sqrt{l^2 + h^2}\right) - h}\right| = \frac{l^2 + h^2 + h\sqrt{l^2 + h^2}}{2l};$$

Therefore,

$$r = \frac{l^2 + h^2}{4R} = \frac{\frac{l^2 + h^2}{2(l^2 + h^2 + h\sqrt{l^2 + h^2})}}{l} = \frac{\left(\sqrt{l^2 + h^2}\right)l}{2\left(\sqrt{l^2 + h^2} + h\right)},$$

where

$$\frac{R}{r} = \frac{\left(\sqrt{l^2 + h^2} + h\right)^2}{l^2}.$$

Dividing by l^2 numerator and denominator and calling $\frac{h}{l} = p$; then $\frac{R}{r} = \left(\sqrt{1 + p^2} + p\right)^2$. If we call $\frac{R}{r} = k$, it results in $p = \frac{k-1}{2\sqrt{k}}$.

3.3. Computer Software

The above calculus was programmed in MATLAB. Figure 9 shows the programming diagram, with the calculations described in the previous section. The program can calculate the arcs either analytically, by entering the coordinates of the starting and ending points, or graphically, on a scaled image. In Figure 10, several examples of rampant arch calculations are shown, where the coordinates of the start and end points were entered. Note that they are shown in blue on the graph obtained (Figure 10A). In the first example, the coordinates were for the initial point (1, 1), and for the final point (3, 3), so that the horizontal separation or span (*l*) between both was 2 m and the vertical or rise (*h*) was also 2 m, obtaining the coordinates of the center of the major arch (5, 1) and those of the minor arch (2.33, 3). In Figure 10B, an example of calculation of rampant arches in stairs is shown, where the two circumferences and their radii are highlighted.

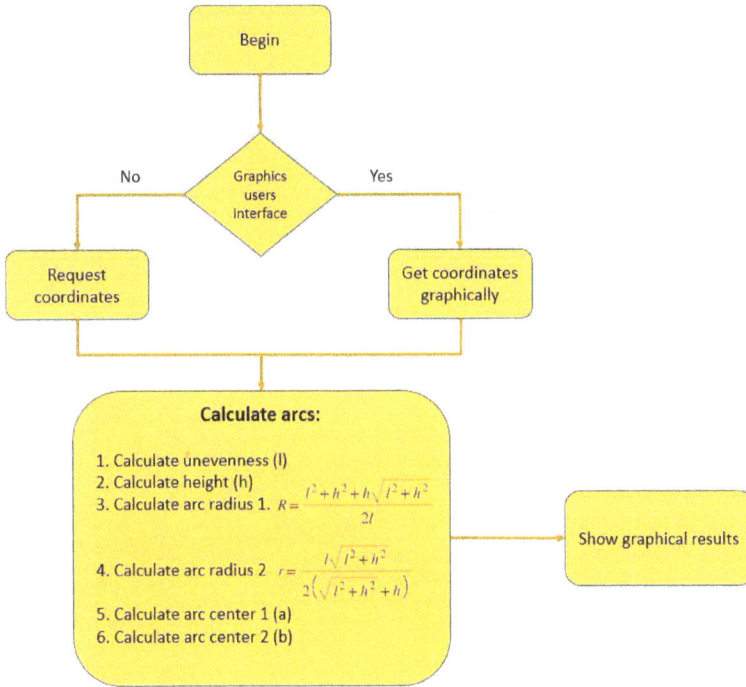

Figure 9. Software calculation diagram for optimal rampant arch calculation.

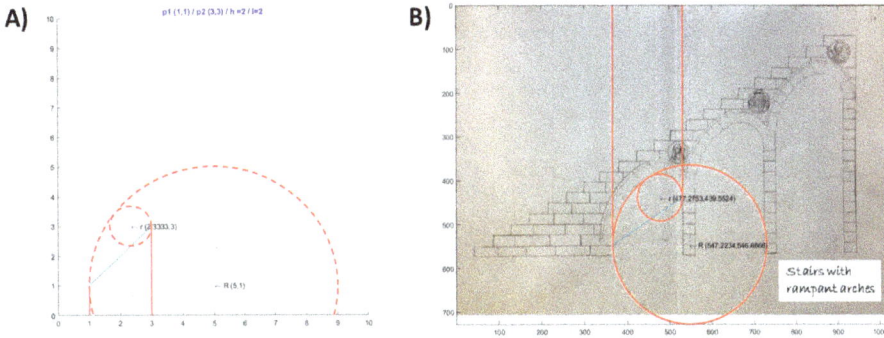

Figure 10. Rampant arches with different geometries: (**A**) numerical data input; (**B**) graphical data input.

4. Case Studies

In this section, two case studies of real rampant arches were selected, representative cases of Gothic art in France (church of Saint Urban de Troyes) and Spain (Cathedral of Palma de Mallorca), where one comes out as an optimal rampant arch and the other does not.

4.1. Church of Saint Urbain de Troyes (France)

The Saint Urban Church (Église Saint-Urbain) is a large medieval French church erected in the city of Troyes, now the capital of the department of Aube. It was a collegiate church, endowed in 1262 by Pope Urban IV, and it is a remarkable example of late Gothic architecture. Construction of the building probably began in 1263, where the main section of the church was not completed until the

16th century, and the tower was not completed until 1630. One of its highlights is the collegiate choir, long acclaimed as a masterpiece of French Gothic architecture, and demonstrates the advanced degree of sophistication of French architecture in the third quarter of the 13th century [25,26]. This church was the object of numerous architectural studies as an example of the French Gothic style [13]. Figure 11 shows the calculation of the optimum rampant arch calculated with the software overlapped over the original arch of the Church of Saint Urbain de Troyes. The starting (A) and ending (C) points are marked in the section of the rampant arch. Note that the actual arch is far from the optimal rampant arch calculated with the software.

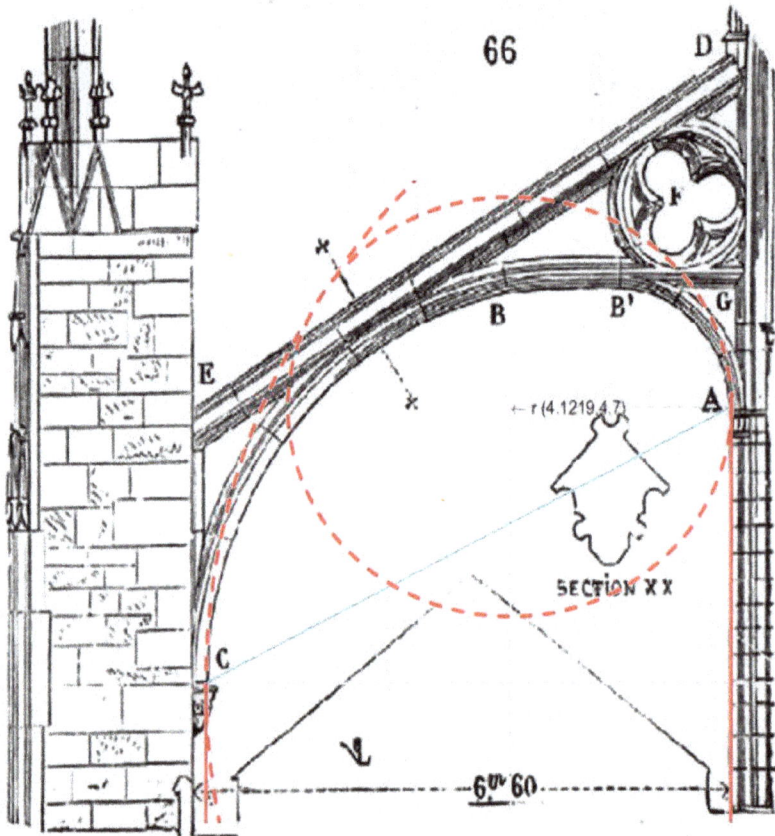

Figure 11. Calculation of the optimal rampant arch (in red) over a rampant arch of the flying buttress of the church of Saint Urbain de Troyes (France).

4.2. Cathedral of Palma de Mallorca (Spain)

The Cathedral-Basilica of Santa María de Palma de Mallorca, also known as the Cathedral of Mallorca, is one of the most significant monuments of Spanish Gothic architecture [27]. Its construction started around the year 1300 and was completed around the year 1600 [28]. Its vault reaches a height of 44 m, only surpassed by the Cathedral of Beauvais (48 m), the highest Gothic cathedral in the world, and slightly lower than the height of the Cathedral of Milan (45 m). Moreover, it surpasses the Cathedral of Cologne 43 m, and it is, therefore, considered as one of those with a nave of greater height among the cathedrals of European Gothic style [29]. In Figure 12, the optimal rampant arch was calculated on one of the arches of the façade of the cathedral of Palma de Mallorca. Here, you can see how the arch calculated with the software and the real arch match. Indeed, a recent study on the structure of the cathedral of Palma de Mallorca showed the horizontal displacements of the façade [27], where it was observed that the rampant arches do not have any displacement, as a result of its proper construction and design.

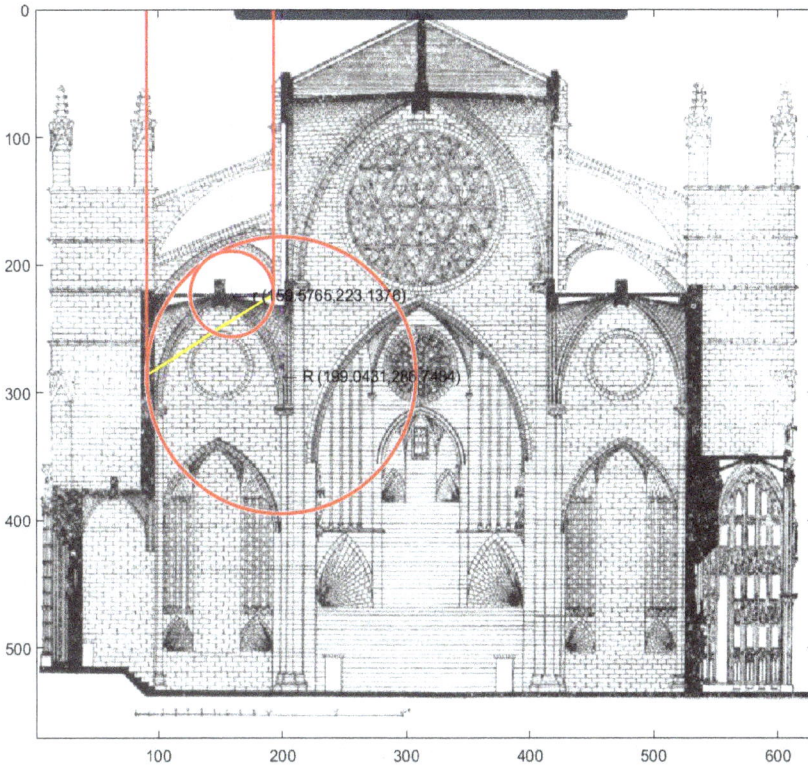

Figure 12. Calculation of the optimal rampant arch over a flying buttress (in red) of the Cathedral of Palma de Mallorca (Spain). Note that the yellow line highlights the initial and final points of the rampant arch.

5. Discussion

In Gothic constructions, the flying buttresses of the first constructions were generally circular, where the shape of the line of thrust did not match, normally, with the geometric forms of traditional construction formed by circumferential arches [4]. In addition, when the line of thrust reaches or touches the edge of the arch, an articulation is formed that can cause the arch to collapse [21]. Enough arch thickness will easily accommodate an infinite number of thrust lines. By progressively reducing the thickness of the arch, the moment will come when there will be only one line of thrusts and the thickness of the arch cannot be reduced without it exiting the line of the arch. Note that load-bearing structures in Gothic cathedrals account for 85% to 90% of the building material [30]. Therefore, the design of the rampant arches of two arches tried lightening the structure while maintaining or improving the thrust resistance of the flying buttresses.

In general, the only action to be considered in constructions made with ashlars, such as Gothic cathedrals, is the position of the line of thrusts by its own weight. Scientific calculation, therefore, justifies the use of geometric or graphical patterns. Therefore, the ancient builders understood the essence of building with ashlars: the proper placement of weights, the balance of thrusts, and counterbalances. This knowledge of the decisive importance of geometry is perhaps the secret of Gothic constructions. In this research, it was shown how the rampant arches of cathedrals like the Palma de Mallorca are optimum from the geometric point of view.

This work may have limitations from a practical point of view. Firstly, if it is analyzed numerically, the precision of the calculation of centers and radii will depend on the precision of the coordinates of the starting and finishing points of the arc. Secondly, from the point of view of analyzing existing constructions, historical plans were analyzed to scale, which are not always available, and can sometimes have distortions or errors [31–33].

In order to analyze real existing cases, a centimetric precision survey would be needed. Although, currently, with the technology of unmanned aerial vehicles, it could be solved in a relatively economic way by realizing a photogrammetric survey [34]. In the literature, several examples can be found [35,36].

On the other hand, if the rampant arches to be analyzed are close to the ground, such as those commented on for staircases or inside buildings, other techniques can be used, such as close-range photogrammetry [37–39] or terrestrial laser scanning [40].

6. Conclusions

In this work, the geometry of the rampant arch was studied from different approaches, the graphical point of view known in classical literature and the current mathematical point of view, with its subsequent programming. An optimal geometrical solution was defined from the constructive point of view and was calculated and programmed. The optimum rampant arch obtained is the one with the common tangent to the two arches parallel to the slope line. Additionally, knowing the radii of the two arcs, the software allows calculating the starting points of both arches on the pillars or walls. It was applied to several well-known and representative cases of Gothic art in France (church of Saint Urbain de Troyes) and Spain (Cathedral of Palma de Mallorca), establishing that the first was not an optimal design while the second was. The conception and study of certain architectural elements, such as the rampant arch, not only help identify and describe existing aspects of our traditional architecture, but can also serve as a starting point for the verification of existing structures, which, despite standing for many years, may require intervention. This research opens new perspectives for the geometric study of buttresses of cathedrals in general and of rampant arches in particular. It was possible to analyze the geometries used in the design of cathedrals, and specifically through the proposed case studies. It was verified how the constructors reached the optimum rampant arch from a structural point of view, defined as the one whose tangent between the two arches is parallel to the straight unevenness between the initial and final points of the rampant arch.

Author Contributions: C.V., A.A., C.S.-A.-G., F.G.M., I.Z., and F.M.-A. conceived the study, designed the method, and wrote the manuscript.

Funding: This research received no external funding.

Acknowledgments: The authors would like to thank CIAIMBITAL (University of Almeria, CeiA3) for its support.

Conflicts of Interest: The authors declare no conflict of interest.

References

1.   Warland, E.G. *Modern Practical Masonry*; Routledge: Abingdon-on-Thames, UK, 2006. [CrossRef]
2.   Theodossopoulos, D. Structural scheme of the Cathedral of Burgos. *Struct. Anal. Hist. Constr.* **2004**, *4*, 643–652.
3.   Sánchez-Beitia, S. On-site Stress Measurements of a Flying Buttress in the Palma de Mallorca (Spain) Cathedral. *Rock Mech. Rock Eng.* **2016**, *49*, 315–319. [CrossRef]
4.   Llopis-Pulido, V.; Durá, A.A.; Fenollosa, E.; Martínez, A. Analysis of the Structural Behavior of the Historical Constructions: Seismic Evaluation of the Cathedral of Valencia (Spain). *Int. J. Archit. Herit.* **2019**, *13*, 205–214. [CrossRef]

5. Viollet le Duc, E. Dictionnaire Raisonné de L'architecture Française du XIe au XVIe Siècle. B. Bance, París. 1854. Available online: https://fr.wikisource.org/wiki/Livre:Viollet-le-Duc_-_Dictionnaire_raisonn%C3%A9_de_l%E2%80%99architecture_fran%C3%A7aise_du_XIe_au_XVIe_si%C3%A8cle,_1854-1868,_tome_1.djvu (accessed on 15 April 2019).

6. Tarrío, I. Los arbotantes en el sistema de contrarresto de construcciones medievales: Teorías sobre su comportamiento estructural. In *Actas del IX Congreso Nacional y I Internacional Hispanoamericano de Historia de la Construcción*; Instituto Juan de Herrera: Madrid, Spain, 2015; Volume 3, p. 1677.

7. Deio, G. Durham Cathedral Plans. Available online: https://www.medart.pitt.edu/image/england/Durham/Cathedral/Plans/dc149dur-b.jpg (accessed on 15 April 2019).

8. Choisy, A. *Histoire de L'architecture Tome II*; Béránger, G., Ed.; Gauthier-Villars: Paris, France, 1899.

9. Courtenay, L.T. *The Engineering of Medieval Cathedrals*; Routledge: Abingdon-on-Thames, UK, 2016. [CrossRef]

10. Moya, D. *El Origen de los Arbotantes en la Historia de la Arquitectura, X Certamen Arquímedes de Introducción a la Investigación Científica*; Ministerio de Educación y CSIC: Madrid, Spain, 2011; p. 2.

11. Tempesta, G.; Galassi, S. Safety evaluation of masonry arches. A numerical procedure based on the thrust line closest to the geometrical axis. *Int. J. Mech. Sci.* **2019**, *155*, 206–221. [CrossRef]

12. Panofsky, E. *Gothic and Architecture and Scholasticism*; Archabbey Press: New York, NY, USA, 1957; ISBN 0529020920.

13. Davis, M.T.; Neagley, L.E. Mechanics and Meaning: Plan Design at Saint-Urbain, Troyes and Saint-Ouen, Rouen. *Gesta* **2000**, *39*, 161–182. [CrossRef]

14. Heyman, J. Beauvais cathedral. *Trans. Newcom. Soc.* **1967**, *40*, 15–35. [CrossRef]

15. Pelà, L.; Bourgeois, J.; Roca, P.; Cervera, M.; Chiumenti, M. Analysis of the effect of provisional ties on the construction and current deformation of Mallorca Cathedral. *Int. J. Archit. Herit.* **2016**, *10*, 418–437. [CrossRef]

16. Derand, F. L'Architecture des Voutes, ou l'art, traits et coupes des voutes; traite tres-utile, méme nécessaire à tous les architectes, maitres-maçons, appareilleurs, tailleurs de pierres, et généralement à tous ceux qui se mélent de l'Architecture, meme militaire, París: S. Cramoysi, 1643. Available online: https://gallica.bnf.fr/ark:/12148/btv1b8626566f/f7.image (accessed on 24 April 2019).

17. Blondel, F.N. Cours d' Architecture enseigné dans l'Académie royale d' architecture. Ed. P. Auboin et F. Clouzier (Paris). 1675. Available online: https://gallica.bnf.fr/ark:/12148/bpt6k85661p.r=cours%20architecture%20cours%20architecture?rk=21459;2 (accessed on 24 April 2019).

18. Casado, E.A.; Sánchez, J.M.D. Geometría del arco carpanel. *Suma: Revista sobre Enseñanza y Aprendizaje de las Matemáticas* **2015**, *79*, 17–25.

19. Quintas, V. Structural analysis of flying buttresses. *Eur. J. Environ. Civ. Eng.* **2017**, *21*, 471–507. [CrossRef]

20. Portioli, F.; Cascini, L.; Casapulla, C.; D'Aniello, M. Limit analysis of masonry walls by rigid block modelling with cracking units and cohesive joints using linear programming. *Eng. Struct.* **2013**, *57*, 232–247. [CrossRef]

21. Frezier, A.F. La Théorie et la Pratique de la Coupe des Pierres et des Bois pour la Construction des Voûtes et Autres Parties des Bâtiments Civils & Militaires, ou Traité de Stéréotomie, à L'usage de L'architecture. Tome 3. Strasbourg and Paris (1737-9). Available online: https://gallica.bnf.fr/ark:/12148/bpt6k1040142z/f491.image (accessed on 15 April 2019).

22. Breymann, G.U. Bau-Constructions-Lehre. Berlag von Gustav Beise. 1881. Available online: https://www.e-rara.ch/zut/content/pageview/8642945 (accessed on 15 April 2019).

23. Heyman, J. The stone skeleton. *Int. J. Solids Struct.* **1966**, *2*, 249–279. [CrossRef]

24. Theodossopoulos, D.; Sinha, B.P.; Usmani, A.S. Case study of the failure of a cross vault: Church of Holyrood Abbey. *J. Archit. Eng.* **2003**, *9*, 109–117. [CrossRef]

25. Bruzelius, C. The Second Campaign at Saint-Urbain at Troyes. *Speculum* **1987**, *62*, 635–640. [CrossRef]

26. Davis, M.T. On the Threshold of the Flamboyant: The Second Campaign of Construction of Saint-Urbain, Troyes. *Speculum* **1984**, *59*, 847–884. [CrossRef]

27. Roca, P.; Cervera, M.; Pelà, L.; Clemente, R.; Chiumenti, M. Continuum FE models for the analysis of Mallorca Cathedral. *Eng. Struct.* **2013**, *46*, 653–670. [CrossRef]

28. Elyamani, A.; Roca, P.; Caselles, O.; Clapes, J. Seismic safety assessment of historical structures using updated numerical models: The case of Mallorca cathedral in Spain. *Eng. Fail. Anal.* **2017**, *74*, 54–79. [CrossRef]

29. Pérez-Gracia, V.; Caselles, J.O.; Clapés, J.; Martinez, G.; Osorio, R. Non-destructive analysis in cultural heritage buildings: Evaluating the Mallorca cathedral supporting structures. *NDT E Int.* **2013**, *59*, 40–47. [CrossRef]

30. Fitchen, J. *The Construction of Gothic Cathedrals: A Study of Medieval Vault Erection*; University of Chicago Press: Chicago, IL, USA, 1981.

31. San-Antonio-Gómez, C.; Velilla, C.; Manzano-Agugliaro, F. Photogrammetric techniques and surveying applied to historical map analysis. *Surv. Rev.* **2015**, *47*, 115–128. [CrossRef]

32. Manzano-Agugliaro, F.; Montoya, F.G.; San-Antonio-Gómez, C.; López-Márquez, S.; Aguilera, M.J.; Gil, C. The assessment of evolutionary algorithms for analyzing the positional accuracy and uncertainty of maps. *Expert Syst. Appl.* **2014**, *41*, 6346–6360. [CrossRef]

33. San-Antonio-Gómez, C.; Velilla, C.; Manzano-Agugliaro, F. Urban and landscape changes through historical maps: The Real Sitio of Aranjuez (1775–2005), a case study. *Comput. Environ. Urban Syst.* **2014**, *44*, 47–58. [CrossRef]

34. Perea-Moreno, A.J.; Aguilera-Ureña, M.J.; Larriva, M.D.; Manzano-Agugliaro, F. Assessment of the potential of UAV video image analysis for planning irrigation needs of golf courses. *Water* **2016**, *8*, 584. [CrossRef]

35. Achille, C.; Adami, A.; Chiarini, S.; Cremonesi, S.; Fassi, F.; Fregonese, L.; Taffurelli, L. UAV-based photogrammetry and integrated technologies for architectural applications—Methodological strategies for the after-quake survey of vertical structures in Mantua (Italy). *Sensors* **2015**, *15*, 15520–15539. [CrossRef] [PubMed]

36. Karachaliou, E.; Georgiou, E.; Psaltis, D.; Stylianidis, E. Uav for Mapping Historic Buildings: From 3d Modelling to Bim. In *International Archives of the Photogrammetry, Remote Sensing and Spatial Information Sciences, Proceedings of the 8th International Workshop 3D-ARCH "3D Virtual Reconstruction and Visualization of Complex Architectures", Bergamo, Italy, 6–8 February 2019*; International Society of Photogrammetry and Remote Sensing (ISPRS): Hannover, Germany, 2019.

37. Yilmaz, H.M.; Yakar, M.; Gulec, S.A.; Dulgerler, O.N. Importance of digital close-range photogrammetry in documentation of cultural heritage. *J. Cult. Herit.* **2007**, *8*, 428–433. [CrossRef]

38. Jiang, R.; Jáuregui, D.V.; White, K.R. Close-range photogrammetry applications in bridge measurement: Literature review. *Measurement* **2008**, *41*, 823–834. [CrossRef]

39. Grussenmeyer, P.; Landes, T.; Voegtle, T.; Ringle, K. Comparison methods of terrestrial laser scanning, photogrammetry and tacheometry data for recording of cultural heritage buildings. *Int. Arch. Photogramm. Remote Sens. Spat. Inf. Sci.* **2008**, *37*, 213–218.

40. Lerma, J.L.; Navarro, S.; Cabrelles, M.; Villaverde, V. Terrestrial laser scanning and close range photogrammetry for 3D archaeological documentation: The Upper Palaeolithic Cave of Parpalló as a case study. *J. Archaeol. Sci.* **2010**, *37*, 499–507. [CrossRef]

![symmetry logo] *symmetry*

MDPI

Article

The Hay Inclined Plane in Coalbrookdale (Shropshire, England): Geometric Modeling and Virtual Reconstruction

José Ignacio Rojas-Sola [1],* and Eduardo De la Morena-De la Fuente [2]

[1] Department of Engineering Graphics, Design and Projects, University of Jaén, Campus de las Lagunillas, s/n, 23071 Jaén, Spain

[2] Research Group 'Engineering Graphics and Industrial Archaeology', University of Jaén, Campus de las Lagunillas, s/n, 23071 Jaén, Spain; edumorena@gmail.com

* Correspondence: jirojas@ujaen.es; Tel.: +34-953-212452

Received: 9 April 2019; Accepted: 22 April 2019; Published: 24 April 2019

Abstract: This article shows the geometric modeling and virtual reconstruction of the inclined plane of Coalbrookdale (Shropshire, England) that was in operation from 1792 to 1894. This historical invention, work of the Englishman William Reynolds, allowed the transportation of boats through channels located at different levels. Autodesk Inventor Professional software has been used to obtain the 3D CAD model of this historical invention and its geometric documentation. The material for the research is available on the website of the Betancourt Project of the Canary Orotava Foundation for the History of Science. Also, because the single sheet does not have a scale, it has been necessary to adopt a graphic scale so that the dimensions of the different elements are coherent. Furthermore, it has been necessary to establish some dimensional, geometric, and movement restrictions (degrees of freedom) so that the set will work properly. One of the main conclusions is that William Reynolds designed a mechanism seeking a longitudinal symmetry so that, from a single continuous movement, the mechanism allows two vessels to ascend and descend simultaneously. This engineering solution facilitated a doubling of the working capacity of the device, as well as a reduction of the energy needs of the system.

Keywords: inclined plane; Coalbrookdale (Shropshire); Agustín de Betancourt; geometric modeling; virtual reconstruction; industrial heritage; industrial archaeology; symmetry

---

1. Introduction

The present study is part of the research line of industrial archeology whose purpose is the systematic study of the industrial memory of an era. Addressing industrial archaeology from the point of view of engineering provides a necessary vision for the correct understanding of industrial heritage, since in many occasions this study is carried out in an unscientific way, and the written record of said heritage being incomplete and with the consequent risk of loss.

The work presented in this article has been developed within a research project on the work of an outstanding Spanish Enlightenment engineer, Agustín de Betancourt, and Molina [1,2], analyzing his best known inventions from an engineering graphics standpoint in order to obtain his geometric modeling [3–5].

As is well known, a first step for the recovery and study of industrial heritage is the creation of realistic 3D CAD models. The article shows the 3D digital restitution of the inclined plane for the transport of vessels that operated in Coalbrookdale (Shropshire, England) at the end of the 18th century. This 3D model has been obtained with CAD (computer-aided design) techniques on which we carry out subsequent CAE (computer-aided engineering) studies, following the objectives established in the document Principles of Seville [6] on virtual archaeology that cites the London Charter [7] regarding the computer-based visualization of cultural heritage.

The Shropshire canal was created in 1790 and closed to river traffic in 1944. Its construction allowed the channeling of one of the mining regions of the County of Shropshire through several minor canals that connected the Union Shropshire Canal to the north with the population of Coalbrookdale to the south, along the River Severn [8,9].

Initially, the project consisted of the creation of a fluvial transport network in a difficult region to channel due to its geography, but with a high economic interest due to the presence of factories and coal mines. His main promoter was William Reynolds who already had some experience personally planning the Ketley channel. To solve the problem of level differences, Reynolds used a novel solution for the channels of the time, the use of dry inclined planes by which to transport the boats through channels at different levels [10]. The first channel in England in which this technique was used was that of Ketley (1788), and based on this first experience, Reynolds himself designed the inclined plane of Coalbrookdale, popularly known as *'The Hay inclined plane'* [11].

This inclined plane was launched in 1792, saved a drop of 207 feet (63.1 m), and had a length of 350 yards (320 m), which supposed a ramp of 11.15 sexagesimal degrees (slope of 19.71%). The inclined plane of Coalbrookdale differs from that of Ketley in that at the top, it does not end in a lock. Also, it was a bidirectional channel that allowed for ascending and descending boats at the same time. The upper part of the inclined plane rose a few meters above the level of the upper channel, and from this point descended another much shorter inclined plane that connected with the upper channel [12].

The inclined plane was a considerable benefit for the coal mines and smithies of the area, enabling the trade of their products through the River Severn to countless destinations, and the river port Coalport became an industrial hub of the first magnitude [13].

During his stay in England (1793–1796), Agustín de Betancourt devoted an important part of his time to the study of navigation channels [1]. This is what is recorded in his work 'Mémoire sur un nouveau système de navigation intérieure' written in Paris many years later (1807) [14]. One advantage of this stay was the acquisition of knowledge of the inclined plane of Coalbrookdale, which allowed saving a huge drop without loss of water in the process of the ascent and descent of boats. Specifically, Agustín de Betancourt copied in detail the mechanism as it had been built by businessman William Reynolds and engineer John Lowden, drawing a colored sheet of the inclined plane with much detail and a handwritten memory of three pages explaining the parts of the plane and its functioning. These are the only documents that arrived in Paris in reference to the inclined plane of Coalbrookdale [15]. A few months after the development of this memory, the French engineer François de Recicourt wrote a short report explaining the use that the English gave of the inclined plane to overcome considerable level differences in their channels. In this memory, he mentions the drawings of the inclined plane donated by Robert Fulton [16] and other American authors to the French Academy of Sciences, and in the final part of the document, there is an explanatory sheet signed by Betancourt. This sheet is a scheme based on his own drawing, which includes explanations for the understanding of its operation [17].

Almost three decades later, in 1807, Betancourt wrote his work on a new interior navigation system [14]. This report proposes a channel navigation system for France very similar to the English navigation system: Shallow channels and an advanced system of locks that avoids the loss of water in the ascent and descent of the boats. The report proposes a new lock system consisting of a plunger that pushes water from a reservoir and locally increases the water level of the channel, allowing the ascent and descent of these vessels. In that memory, Betancourt applies his plunger lock to the inclined plane and mentions Robert Fulton's previous work [16], indicating that his inclined plane is designed according to the Reynolds procedure. To overcome differences of over 5 m, Betancourt proposes using a series of locks or the use of an inclined plane. This is where he refers again to the inclined plane of Coalbrookdale, although modifying some elements of the same and adding to the system his plunger lock, in order to flood the boat zone of the upper channel. The inclined plane of Betancourt has no descent ramp, unlike that of Reynolds, so the ascent ramp ends at the level of the upper channel. Of this modification of the inclined plane of Coalbrookdale there are two color plates (and a more modern version based on the original one), as well as the explanatory texts of the memory that Betancourt presented to the Academy of Sciences of Paris [14].

At present, of the Coalbrookdale inclined plane, only the ruins remain and some vintage photographs taken from a distance and kept at the Coalbrookdale Museum of Iron located in Shropshire (England) [18], which still show the inclined plane and the remains of the tracks, although it is also possible to distinguish the upper cargo basin and the remains of the brick factory of the building that housed the steam engine. Some photographs taken of the Hay inclined plane show its condition in various years from 1879 to 2012 [19].

The main objective of this research is to obtain a reliable 3D CAD model of this historical invention that allows us to know, in detail, this outstanding engineering work that significantly influenced the socioeconomic development of the region, and which will enable future studies of computer-aided engineering. The originality and novelty of this research is that there is no 3D CAD model of this historical invention with this degree of detail, helping in an outstanding way in the detailed understanding of its operation. An educational goal is also pursued, through its exhibition on the websites of the foundations that have supported this research (Fundación Canaria Orotava de Historia de la Ciencia [20] and Fundación Agustín de Betancourt [21]), as well as in other museums of the history of technology.

The remainder of the paper is structured as follows: Section 2 presents the materials and methods used in this investigation; then, Section 3 includes the main results in the process of geometric modeling and their discussion to explain the operation of this device (both in the ascending and descending movement), and Section 4 shows the main conclusions.

2. Materials and Methods

The material used in this research was obtained from two files on the Betancourt Digital Project website of the Canary Orotava Foundation for the History of Science [22], assigned for its digitalization by the National School of Bridges and Roads of ParisTech University. The first of these is manuscript 1558 (MS 1558), consisting of a descriptive memory of 3 pages and a sheet (Figure 1) made by Betancourt [15], and the second is the manuscript 2812 (MS 2812) of Recicourt that contains several documents, one of which is an explanatory sheet of the Hay inclined plane (Figure 2) [17].

The graphic legacy that exists on the inclined plane of Reynolds is valuable but scarce, and does not provide the information that today would be expected from a descriptive plane of a mechanism. Specifically, the sheet printed in A3 format from which the dimensions of the elements have been taken by direct measurement does not have a graphic scale (Figure 1), but only the dimensions of the ramps that make up the gap to be saved. Therefore, there is no direct reference that can be used to obtain an accurate graphic scale of the sheet, nor does it exist in the descriptive memory. As in other documents of the time, they are not plans made with the level of detail of a contemporary plan, but rather a descriptive model that seeks only to facilitate understanding of the operation of the mechanism.

To alleviate in part the lack of information, the research is based on the industrial remains of the current inclined plane, such as the lower access channel to the inclined plane, the tracks on the inclined plane, the stone factory on which the wooden structure sat, the short upper plane, and the upper channel. There are also the ruins of a building that housed a steam engine with a chimney. Therefore, these industrial remains [19] have been used to check the actual dimensions of the different elements of the inclined plane in order to find a real scale of the Agustín de Betancourt plate (Figure 1), and in turn, they also serve to confirm that all the elements that appear in this sheet correspond to the signals that can be seen in the industrial remains.

In order to obtain a 3D reconstruction that represents a functional set, it is very important to understand the workings of the mechanism and what each of its elements are for. As happens with a great number of inventions of the time, the two dihedral projections of the sheet (front view and top view) are insufficient to obtain information of all the elements, which has forced us to establish different conjectures and/or proposals that have had to be validated when compared with other mechanisms of the time. In addition, it has been necessary to incorporate knowledge of mechanical engineering in order to geometrically model some elements, since each one of them had to have its specific mission in the set. This task has been one of the most complicated, since there is no information in the file of many of the elements.

Figure 1. Hay inclined plane drawn by Agustín de Betancourt [15] (Courtesy of Fundación Canaria Orotava de Historia de la Ciencia).

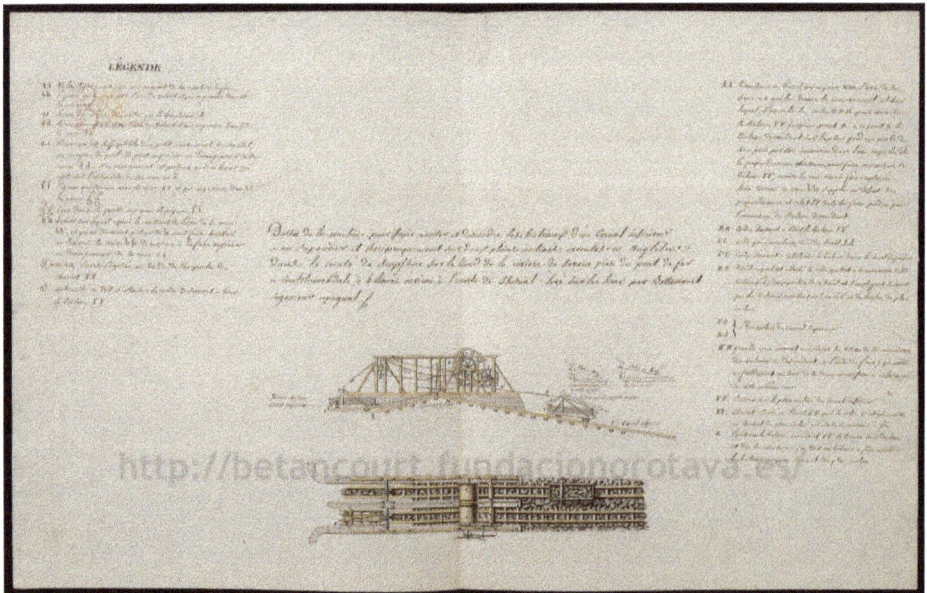

Figure 2. Sheet drawn by Agustín de Betancourt in the Recicourt memory [17] (Courtesy of Fundación Canaria Orotava de Historia de la Ciencia).

The methodology followed in the 3D modeling of the device of William Reynolds consists of the following steps:

1. Transcription and translation of the notes of Agustín de Betancourt

In order to fully understand the functioning of William Reynolds' device, it is not enough to analyze the only sheet with the graphic representation that Agustin de Betancourt made of the inclined plane [15], but it is necessary to translate the three-page handwritten memory in French and dated in 1796 where the different parts of it and its operation are explained. Here, the main elements are detailed, accompanied by an alphabetic reference and the function that each element performs.

2. Proportional impression of the existing sheet depicting the inclined plane of Coalbrookdale

Once the file MS 1558 (descriptive memory of 3 pages and an explanatory sheet (Figure 1)) of the website of the Betancourt Digital Project [22] had been downloaded, the sheet was printed in color, keeping the proportionality of the image on a format of A3 paper, in order to be able to directly measure the dimensions of all the elements.

3. Determination of the general scale of the sheet and verification with the existing remains

Given that the sheet does not have a graphic scale, and that there is no related information in the descriptive memory, it is necessary to determine the scale thereof for the correct 3D CAD modeling of the invention. It is therefore essential to determine a known reference dimension, which in this investigation has been the width of the trailer as this was standardized throughout the county of Shropshire.

It is known that Shropshire was a county where mining had a special relevance, and Coalbrookdale specifically was the epicenter of the industrial revolution, with the cars for the extraction of coal being what determined the width of the roads. According to the English engineer George Stephenson, before 1832, the track width was 4 feet and 8 inches (1.422 m). Later, this engineer extended the width of the tracks by half an inch more for the construction of the Liverpool-Manchester railway, and since then this measure has been the most widely used in most of the railways of the world [23].

Therefore, the measure of reference adopted in this investigation to determine the scale of the sheet has been the track width with a value of 1.422 m. Thus, if the width of the A3 printed sheet is measured, it results in a value of 12.5 mm, so that the approximate scale of said sheet turns out to be 11: 1250; that is, a value of around 1: 113, which, as a first approximation, is enough to give some idea of the proportions of the different elements.

However, it is necessary to check dimensionally that the general scale we have determined is correct. Agustín de Betancourt detailed in his sheet particularities of the foundation of the structure, which allows us to compare the current dimensions of the ruins with the 3D model.

The walls on which the wooden structure rests on the Betancourt plate from which the dimensions have been taken correspond, without any doubt, to the walls that currently exist [19]. The successive reforms that the mechanism had to undergo, especially in its materials, did not affect the stone factory of the canal that has remained to this day. Thus, the width of the channel between the walls is 7 feet and 10 inches (approximately 2.387 m), but in the model the width is 2.370 m, so the difference between one and another is 17 mm.

Taking into account that the measures have been taken with a caliber with an error of 0.1 mm and applying the scale 11: 1250, the error attributable to the caliber is +/- 11.36 mm. Therefore, it was not necessary to adjust the scale for the precision that can be requested at the level of detail of the plane, which confirms that the scale chosen is correct.

Undoubtedly, the existence of industrial remains (called 'material culture' in the discipline of industrial archeology) has been a great advantage in the process of confirmation and determination of the actual graphic scale of the sheet.

4. Identification of the different systems by their functionality

In this section, the different elements or systems have been defined by their function in the inclined plane based on the determination of their dihedral projections (orthographic views) thanks to the knowledge of the descriptive geometry. These have been:

- Stone factory and canal. This system comprises the immobile part of the inclined plane: The grading, sleepers, tracks, and walls of the upper part that delimit the two accesses to the inclined plane. Its materials are stone, oak wood, and cast iron.
- Support structure. This is the set of elements that serve as structural support to the mechanism for raising and lowering the boats. The frames are all made of oak and on them, the axes will be supported by supports fixed to the structure.
- Traction system (transmission shaft). The traction system consists of the drive axles and pulleys, specifically referring to the four axes that come into play in the process of ascent and descent: The axle with its inertia wheel in contact with the steam engine, an intermediate shaft that serves to change the direction of rotation, and two main axes that are responsible for movement in the long plane and the short plane. In addition to these four axes are also included pulleys that facilitate movement. The axle of the inertia wheel and the intermediate one are made of cast iron, like the pulleys, while the main axes have an iron core, like their gears, but their drums are made of wood like their wheels.
- Braking system. This refers to the brakes that affect the wheels of the two main axes to stop the transmission system. A very functional friction brake system made of wood, driven by bars made of cast iron, and fixed in the frame by metal supports of the same material.
- Trailers-boats. The trailer is the frame of oak wood suspended on six cast iron metal wheels on which the wooden boat rests. Also, in order to provide the boat with stability during the climb, this frame is equipped with several hooks and metal chains.
- Finally, there are the cords that serve as a link between the different systems. The ropes made of hemp fiber are used to join the traction axles with the trailers, and both are traction ropes since they serve for both the ascent and the descent of the boats.

5. Determination of the dimensions and geometry of each element

Once the different systems of the whole of the invention have been identified, the dimensions and geometry of each element defined by its dihedral projections are determined. This step is essential in order to make a model faithful to the original plane. For this, it is important to combine collaborative work with CAD with a consistent and symmetric geometric modeling process [24].

As previously indicated for the direct measurement of each element on the sheet (Figure 1), the aforementioned scale (11: 1250) is applied and modeled with the aid of the parametric software Autodesk Inventor Professional [25], generating a file with extension (.ipt). Similarly, in the design process, each element must be assigned a material with certain physical properties that are essential for the model to be real and not just a geometric approximation.

However, it has been necessary to define a series of elements since, from their position on the sheet (Figure 1), it has not been possible to determine their geometry, perhaps because Betancourt did not want to draw them to give greater clarity to others.

The dihedral projections drawn by Betancourt are partially sectioned, and in addition, there are some elements drawn with some transparency in order to be able to observe the elements that they cover. In the case of the friction braking system, all the brake elements have not been drawn. The braking bars and the inertia wheels of the axes are perfectly observed, but the brake support of the other end at the point on which it is supported is not distinguishable and does not appear in any of the dihedral projections. Since the brake worked by friction, it is understood that this end had to rest on the frame so that it was very close to the inertia wheel of the shaft that was to be braked. Thus, in the case of both brakes, it has been necessary to make an assumption of the geometry and dimensions of this support.

On the other hand, the axle that transmits the movement from the steam engine to the rest of the transmission shaft has the ability to change position and disengage its wheels from the gears it moves.

For this, there is a bar on which it rests that can be moved. This bar and its function are described in the descriptive memory, but the illustration does not allow us to appreciate it clearly. Therefore, it has also been necessary to define this bar and its support dimensionally so that it could perform the function for which it was conceived.

6. Assembly of all the elements and obtaining of the 3D CAD model

It is necessary to highlight the importance of the assembly of all the elements being carried out correctly, applying the necessary restrictions (geometric, dimensional, and movement) so that there is a single assembly allowing us to obtain a very complete model that perfectly reflects the mechanical and structural characteristics of the mechanism.

In order to assemble the elements, we have to define the relationship between them, by means of restrictions and unions. The restrictions used allow adjacent elements to define the degrees of freedom they have in their movement, while in the unions, there is no freedom. The correct execution of this process is crucial in order for the model to behave in a realistic manner.

The software used (Autodesk Inventor Professional) has its limitations when it comes to reproducing the relationship between elements, for example the relationship between links in the chain, which is very complicated. The toroidal geometry makes it very difficult to define correctly between surfaces in the contact between two links, and therefore when modeling the link, a pair of points have been defined in the internal region of the link (at the ends of their breasts). Thus, two contiguous links define their relationship by opposing these points, but this definition gives rise to an unreal relationship that can give problems when using the model in a static analysis to obtain the Von Mises stresses, for example.

The relationship of the brake with the axle of inertia of the shaft is also problematic. Depending on the position of the brake rod, it will embrace the inertia wheel to a greater or lesser extent, exerting an unequal frictional force on it. However, the changing behavior of this brake cannot be modeled although there is a direct relationship with the position of the bar. Also, there is no data available on the coefficient of friction of the surfaces that come into contact, so both static and dynamic simulations of the mechanism should take this factor into account.

The rest of the elements (saving those that behave dynamically like the ropes) have been correctly assembled without major difficulty. Thus, once the model is assembled, it could be useful for studying simulations, both dynamic and static, which will allow us to know more about the invention.

7. Preparing 3D CAD model visualization

Obtaining a realistic image of the inclined plane provides valuable information on how the device was during the first years of operation. To obtain these images, the software allows for the use of a digital image editing tool named Autodesk Inventor Studio, which is incorporated as a menu of the main interface. This tool allows us to define a camera from a few simple parameters: Position, framing, angle, and focal length. Once all the parameters of the camera have been defined, the software allows choosing the type of perspective to be taken on the object: Conical (more realistic) or axonometric.

On the other hand, the software defines a specific appearance and texture for each material that can be modified by defining each element. These default values are those obtained from the library of materials, but each user can define new textures and aspects from their own images.

The next step is to define the lighting of the element and the possible shadows it generates. Autodesk Inventor Studio allows us to define lighting in two ways: Generating a lighting environment or defining the position and intensity of lights or spotlights. In the present case, it has been decided to take a simple lighting environment taken from those defined by the same library of the software. Shadows, dependent on the lighting definition, can also be configured through the software. Specifically, the software establishes the ones that the object throws on the ground, the shadows that projects on itself, and those due to the lights defined by the user. This set of parameters help to give realism to the modeled object.

Finally, once the camera, materials, lighting, and shadows have been defined, the parameters of the rendering engine must be defined: The sizes of the output image, the type of file and the most important, the number of iteration time for obtaining the image. The rendering engine develops, from the 3D geometry of the model, the simulation of the physical, and luminous conditions of the invention, obtaining a very realistic image of the model through a complex calculation.

The entire process described above is shown as summative scheme in Figure 3.

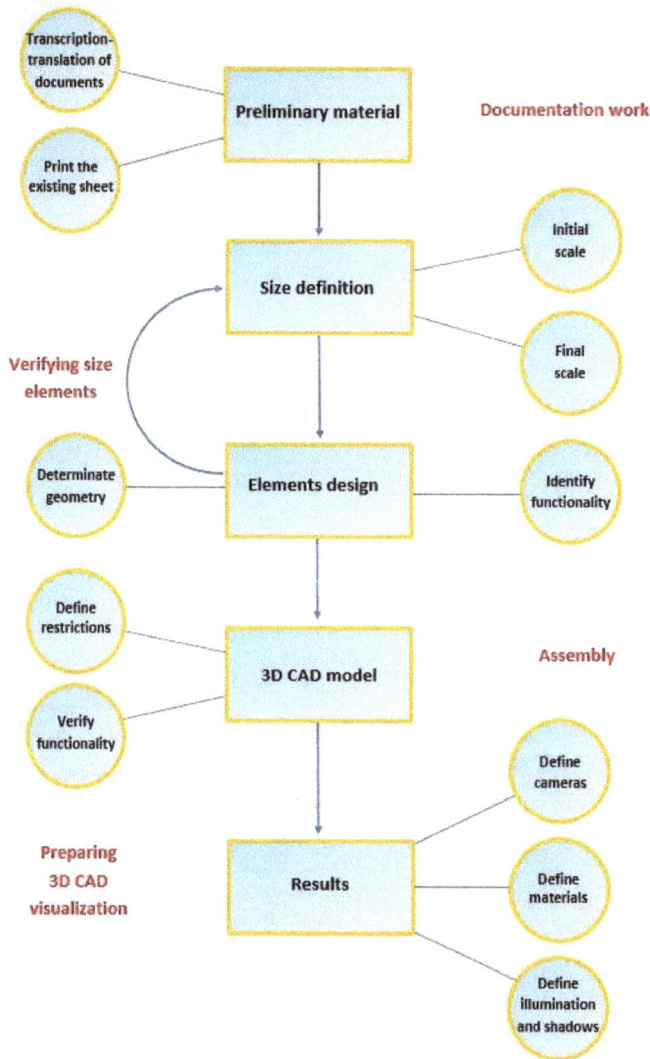

Figure 3. Summative scheme of the methodology followed in the 3D modeling process.

3. Results and Discussion

Figure 4 shows a rendered image of the 3D CAD model of the historical invention from one viewing direction and from its opposite, and Figure 5 shows an exploded view of the whole of the device in order to give a better understanding of how the different elements are assembled.

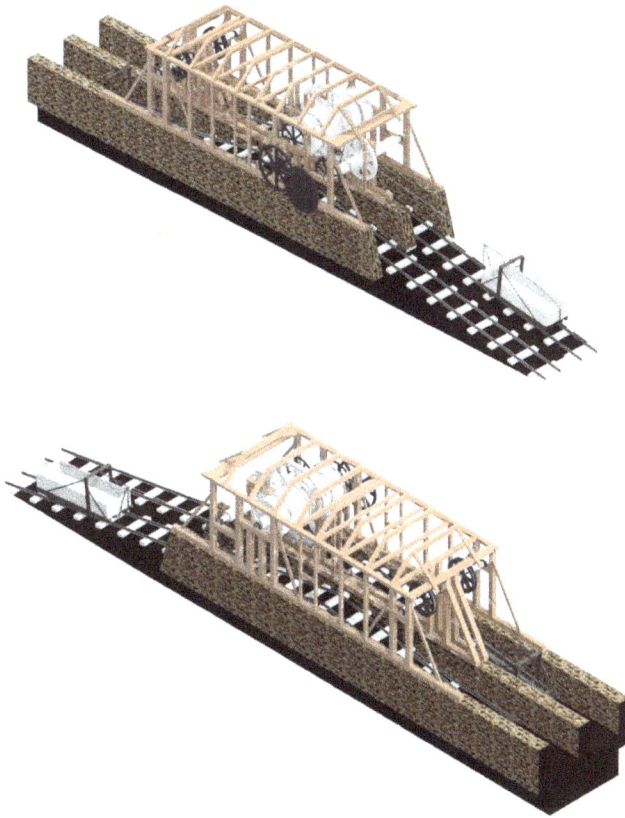

Figure 4. Rendered isometric view of the 3D CAD model (above) and opposite (below).

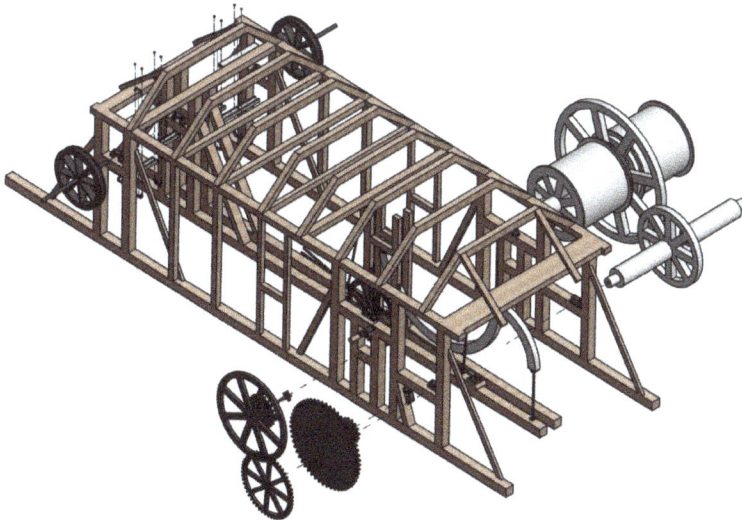

Figure 5. Exploded view of the 3D CAD model.

Both figures show the final result of the modeling of the invention after the assembly of all its elements. The isometric views of the inclined plane (Figure 4) are taken from the point of intersection of the three main ortographic views at a sufficient distance to see the ensemble. On the other hand, the exploded view of the model (Figure 5) has been obtained as follows: The first step has been to eliminate the components that are not part of the mechanism: stone factory, tug boats, ropes, tracks, and plane. After this, on this simplified view of the mechanism, the main components (shafts, gears, and pulleys) are separated indicating the direction of extraction. Thus, this view allows a better understanding of the position in which they work and also the visualization of elements that are hidden in the isometric view.

In the process of geometric modeling, the elements with the greatest difficulty were fundamentally the chains and the ropes. The problem with the chains has only an approximate solution, given that links are modeled whose behavior is approximate to that of a real link.

On the other hand, the ropes are not static elements and the software used does not characterize well the physical properties of these elements, neither when they behave under tension nor when they are suspended. For this reason, the ropes are defined as rigid elements of a material similar to hemp fiber, but with a static behavior that does not correspond to the actual behavior of the element.

One of the purposes of the article is the detailed explanation of how the mechanism works in order to understand the complexity of the geometric modeling process.

To explain its operation (both in the ascending and descending movement), we will use the plan of the ensemble with an indicative list of the elements and their materials (Figure 6), along with a longitudinal section of the Hay inclined plane (Figure 7) and a profile view of the same to visualize the position and difference of height of the different elements of the mechanism (Figure 8).

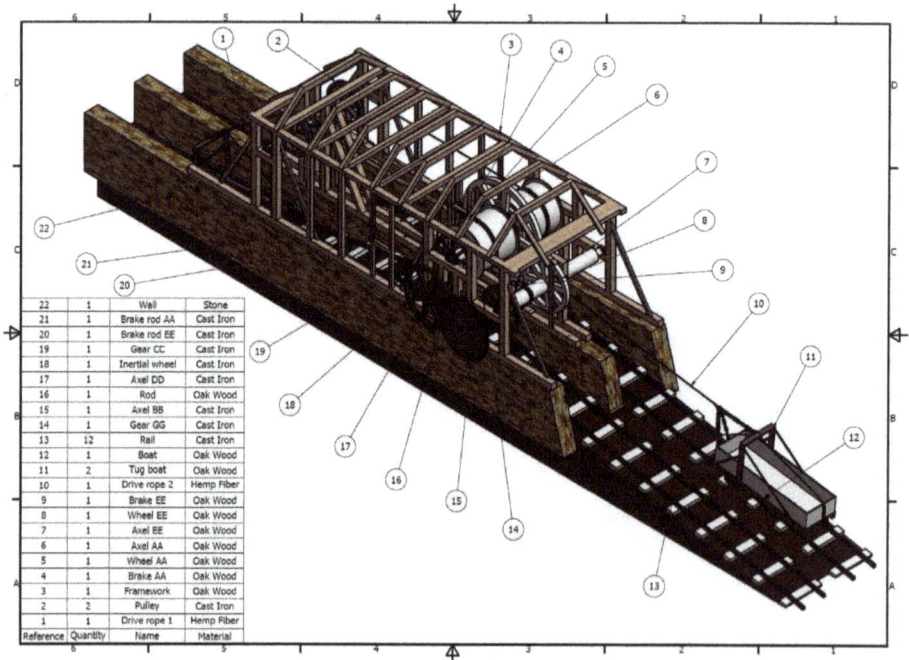

Reference	Quantity	Name	Material
22	1	Wall	Stone
21	1	Brake rod AA	Cast Iron
20	1	Brake rod EE	Cast Iron
19	1	Gear CC	Cast Iron
18	1	Inertial wheel	Cast Iron
17	1	Axel DD	Cast Iron
16	1	Rod	Oak Wood
15	1	Axel BB	Cast Iron
14	1	Gear GG	Cast Iron
13	12	Rail	Cast Iron
12	1	Boat	Oak Wood
11	2	Tug boat	Oak Wood
10	1	Drive rope 2	Hemp Fiber
9	1	Brake EE	Oak Wood
8	1	Wheel EE	Oak Wood
7	1	Axel EE	Oak Wood
6	1	Axel AA	Oak Wood
5	1	Wheel AA	Oak Wood
4	1	Brake AA	Oak Wood
3	1	Framework	Oak Wood
2	2	Pulley	Cast Iron
1	1	Drive rope 1	Hemp Fiber
Reference	Quantity	Name	Material

Figure 6. Plan of the ensemble of the Hay inclined plane with an indicative list of the elements.

Figure 7. Longitudinal section of the Hay inclined plane.

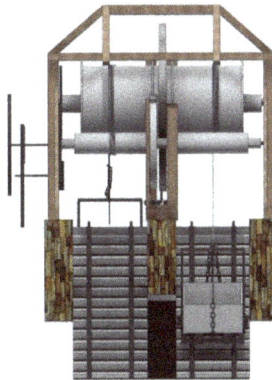

Figure 8. Profile view of the Hay inclined plane.

3.1. Upward Movement

At the bottom of the Hay inclined plane was the loading port on the River Severn (Coalport). Goods were loaded and unloaded at this point of the river and moved from large ships to smaller boats (12) of up to 5 t. These boats navigated the canal to the embouchure of the inclined plane. At this point, an empty tug boat (11) was waiting, which was partially submerged, facilitating the boat's placement on the trailer. Thus, an operator tied the boat to the trailer with two chains and began the ascension process. The tug boat was a structure of wood and iron supported on six iron wheels. The wheels were designed to facilitate transport by rails (13). Four were aligned on the tracks and two were outside the structure, presenting a function that will be explained later. The tug boat, through a chain, was tied to a drive rope (10) responsible for pulling from the lower area to the upper area of the inclined plane.

The traction rope went from the tug boat to the AA axle (6) through a pulley (2) that facilitated traction. As the AA axle rotated clockwise, the rope coiled and the trailer with the boat climbed the inclined plane. Obviously, this operation required a system of gears and an engine to move it, which is presented below.

An inertia wheel (18) (called 'wheel of fire' by Agustín de Betancourt) was the element responsible for moving the entire mechanism, and was attached to a large steam engine that can be seen in the old photographs of the inclined plane [19], and from which remains the chimney and the foundations of the building that housed it (in ruins). This inertia wheel was solidly connected by means of a DD shaft (17) to two gears, a close one of 20 teeth and a distal one of 5 teeth, which was in turn supported on a bearing joined to a moving wooden rod (16), whose function was to rapidly uncouple the two gears attached to the inertia wheel from the rest of the mechanism. This action caused the immediate stopping of the mechanism in case of accident or breakdown without the need to stop the steam engine.

The axle of inertia DD was coupled to an intermediate shaft BB (15) with two gears, one larger than 56 teeth, coupled to the gear of 20 teeth, and one smaller than 16 teeth, coupled to the gear shaft EE (7).

Finally, axes AA and EE were moved by the same wheel AA (5). The AA axle was driven through the coupling of the 5-tooth gear and the 51-tooth DC gear (19) and rotated clockwise (in the opposite direction of the inertia wheel) at a rate of 0.098 turns for each revolution it gave the flywheel of inertia. The EE axle (7), driven through the coupling between the 16-tooth gear of the intermediate shaft and of the GG gear of 51 teeth (14), rotated counter-clockwise (in the direction of the inertia wheel) at the rate of 0.14 turns for each turn that the flywheel made (Figure 9).

Figure 9. Direction of rotation when the transmission system is activated.

As mentioned previously, when the movement was ascending, the traction rope (10) was wound in the form of a coil on the axle AA (6). This axle had the shape of a drum and its diameter measured 2.37 m. Knowing the previously explained relation of teeth, it is easy to calculate that for each turn that the inertia wheel gave the coil collected approximately 0.73 m of rope. The tug boat climbed at that rate to the upper point of the inclined plane. At this point, the rope was disengaged from the trailer and was placed on the upper part, dropping the boat by its own weight onto the short inclined plane that communicated with the upper channel. This inclined plane measured just 10 m and was less steep than the one that ascended from the River Severn ending in a flooded area (upper channel), so there was no need for any braking system to stop it. However, if the merchandise was delicate and the other channel was free the intermediate shaft gear BB (15) could be decoupled, leaving the AA axle (6) without movement. From the intermediate shaft BB, another drive rope (1) could be attached to the chain at the rear of the trailer, allowing the operator to lower the loaded boat to the upper channel by moving the brake bar EE (20) that activated the brake EE (9) on the flywheel (8) of this axle. When descending towards the upper channel, the two unaligned wheels of the trailer, which have been alluded to above, were located on a section of exterior road with less slope, which made the trailer adopt an almost horizontal position facilitating the release of the vessel in the upper channel.

The system was designed to be able to use the two climbing (or descending) routes simultaneously (Figure 10), although the usual way was to use one for ascending and the other for descending.

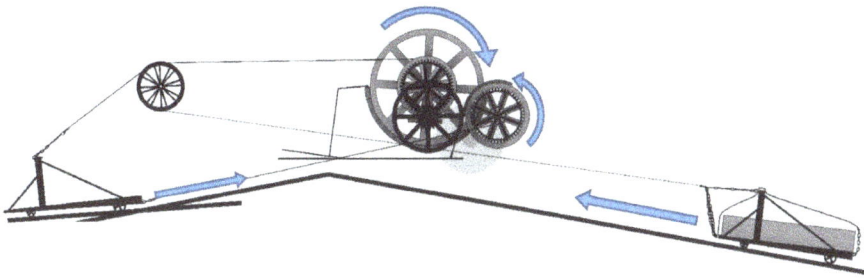

Figure 10. Sequence of movements in the ascent of the boats.

3.2. Downward Movement

The empty tug boat awaited the arrival of the boat in the upper channel in an almost horizontal position. When this arrived, it was hooked to the trailer and the intermediate shaft gear BB was coupled to give movement to the EE axle. This axle made the trailer ascend with the boat along the short section of the inclined plane to the upper area at a rate of 0.30 m per turn of the inertia wheel. Upon reaching this upper area, it disengaged from the rope (1) and the traction rope (10) disengaged from the upper part and hooked onto the trailer towbar. On the downward hemisphere of the axle AA (6), the rope was wound in the opposite direction to that of the rising hemisphere, and thus, when the drum rotated with axle AA, the rope was wound on one side and unwound on the opposite side. In this way, the load was dropped little by little down the inclined plane to the lower channel.

In this case, the movement of the AA axle could not be disconnected from the movement of the inertia wheel for safety reasons, since this side of the plane had a steep slope and had to be lowered mechanically with great care. In case of failure of the steam engine or due to excessive inertia wheel speed, the AA axle could also be braked by a maneuver of the AA brake rod (21) that operated the AA brake (4) on the wheel (5) of the AA axle (Figure 11). Once the lower channel was reached, the boat was released and continued its itinerary to the loading port on the River Severn.

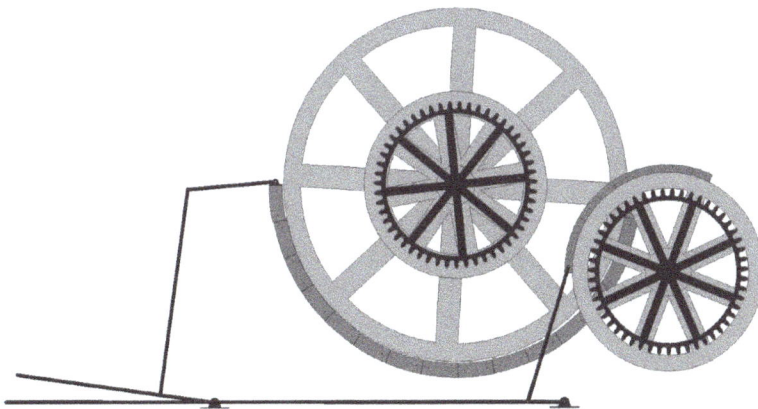

Figure 11. Braking system of the drive shafts.

The inclined plane is an alternative solution to the use of locks to raise and lower vessels from a higher channel to another lower one when the difference in level is very great. However, the mechanism and the maneuvers necessary to carry out the operation of raising or lowering are complex given that many elements intervene.

Among the remarkable features that should be noted of this ingenious device, it is especially relevant its capacity to ascend and descend boats at the same time without interrupting the movement. The mechanism was designed with a longitudinal symmetry that meant that the same ascending movement that wound the rope (10) on the axle AA (6) was the one that facilitated the downward movement of the trailer (11), unrolling the rope of the left hemisphere located on the same AA axle (6). For the successive ascents and descents, it was necessary to alternate the channel of ascent of one hemisphere to the other: When finishing the ascent of a boat by one of the channels, this automatically became the channel of descent, and vice versa. The ingenious mechanism had a longitudinal symmetry that doubled the capacity of transporting goods along the inclined plane.

4. Conclusions

This article shows the complex process followed to obtain the geometric modeling and virtual reconstruction of the Hay inclined plane located in Coalbrookdale (Shropshire, England) as it was originally built, thanks to the parametric software Autodesk Inventor Professional. For this, the investigation has been based on the manuscript notes of Agustín de Betancourt that he took in his visit to the place between 1793 and 1796, shortly after the machine was put into operation by William Reynolds.

The methodology used has been based on knowledge of descriptive geometry and the use of empirical techniques of direct measurement of the dimensions on the printing of the illustrations available from the different sources of information. Also, due to the lack of detailed information it has been necessary to establish a series of geometrical, dimensional, and movement restrictions (degrees of freedom) in order to obtain a coherent and functional 3D CAD model from the point of view of mechanical engineering, which will be the starting point for a future static analysis with CAE techniques.

From this 3D CAD model, it has been possible to obtain the detailed geometric documentation of each of the elements that make up the historical invention, as well as various perspectives such as an isometric view, an exploded perspective, an overall plan with an indication of all the elements of the set and its materials, a longitudinal section, and a profile view, as well as various graphic details, which help to better understand some aspects of its operation (detail of the shaft of transmission and braking system of the drive axles). All this has allowed us to explain easily and in detail the maneuvers of ascent and descent of the boats on the inclined plane.

Thus, this historical invention was a considerable industrial advance in the economy of the area, because thanks to the help of a simple steam engine, the mechanism designed by William Reynolds was used to raise a boat of 5 tons by a difference of 207 feet in only 6 minutes. This supposed much less work than that necessary to unload the cargo in the port of the River Severn and transport it by wagons and horsepower to Coalbrookdale.

Furthermore, a rigorous analysis of his ingenuity has shown that William Reynolds designed a mechanism seeking a longitudinal symmetry whereby, from a single continuous movement, the mechanism would allow two vessels to ascend and descend simultaneously. This engineering solution helped double the working capacity of the device and even reduce the energy needs of the system.

The influence that the design of the inclined plane had on two of William Reynolds' contemporaries, Agustín de Betancourt and Robert Fulton, is noteworthy since both brought this inclined plane model to France and the rest of Europe, adapting it to other types of boats and improving on some of its deficiencies. In particular, Robert Fulton took this more successful inclined plane to the United States. The study of their patents could be the object of further research.

Finally, this methodology would be equally usable for the 3D modeling of a multitude of devices belonging to the industrial heritage of the time. There are a large number of inventions on which is stored the memory of the techniques of engineering that run the risk of being forgotten. Much of the knowledge that the human being has about the mechanisms of the first industrial revolution has to be put back into value. For example, the Canarian Foundation Orotava of the History of Science houses an immense legacy of the Spanish engineer Agustín de Betancourt where inventions not studied are detailed: Those related to the mining methods of the Almadén mines, looms, or printing presses, among others. Similarly, the material present in the Archive of the Ironbridge Gorge Museums related to a large number of English engineers who developed in Shropshire a great activity also linked to the first industrial revolution can be studied.

Author Contributions: Formal analysis, E.D.l.M.-D.l.F.; funding acquisition, J.I.R.-S.; investigation, J.I.R.-S.; methodology, E.D.l.M.-D.l.F.; project administration, J.I.R.-S.; supervision, J.I.R.-S.; validation, E.D.l.M.-D.l.F.; visualization, J.I.R.-S.; writing—original, draft, J.I.R.-S. and E.D.l.M.-D.l.F.; writing—review and editing, J.I.R.-S. and E.D.l.M.-D.l.F.

Funding: This research was funded by the Spanish Ministry of Economic Affairs and Competitiveness, under the Spanish Plan of Scientific and Technical Research and Innovation (2013–2016), and European Fund Regional Development (EFRD) under grant number [HAR2015-63503-P].

Acknowledgments: We are very grateful to the Fundación Canaria Orotava de Historia de la Ciencia for permission to use the material of Project Betancourt available at their website. Also, we sincerely appreciate the work of the reviewers of this article.

Conflicts of Interest: The authors declare no conflicts of interest.

References

1. Muñoz Bravo, J. *Biografía Cronológica de Don Agustín de Betancourt y Molina en el 250 Aniversario de su Nacimiento*; Acciona Infraestructuras: Murcia, Spain, 2008. (In Spanish)
2. Martín Medina, A. *Agustín de Betancourt y Molina*; Dykinson: Madrid, Spain, 2006. (In Spanish)
3. Rojas-Sola, J.I.; Galán-Moral, B.; De la Morena-de la Fuente, E. Agustín de Betancourt's double-acting steam engine: Geometric modeling and virtual reconstruction. *Symmetry* **2018**, *10*, 351. [CrossRef]
4. Rojas-Sola, J.I.; De la Morena-de la Fuente, E. Digital 3D reconstruction of Agustin de Betancourt's historical heritage: The dredging machine of the port of Krondstadt. *Virtual Archaeol. Rev.* **2018**, *9*, 44–56. [CrossRef]
5. Rojas-Sola, J.I.; De la Morena-de la Fuente, E. Agustin de Betancourt's wind machine for draining marshy ground: Approach to its geometric modeling with Autodesk Inventor Professional. *Technologies* **2017**, *5*, 2. [CrossRef]
6. Principles of Seville. Available online: http://smartheritage.com/wp-content/uploads/2016/06/PRINCIPIOS-DE-SEVILLA.pdf (accessed on 9 April 2019).
7. London Charter. Available online: http://www.londoncharter.org (accessed on 9 April 2019).
8. Brown, P. *The Shropshire Union Canal: From the Mersey to the Midlands and Mid-Wales*; Railway & Canal History Society: London, UK, 2018.
9. Morriss, R.K. *Canals of Shropshire*; Shropshire Books: Shrewsbury, UK, 1991.
10. Trinder, B. *The Industrial Archaeology of Shropshire*; Phillimore & Co Ltd.: Chichester, UK, 1996.
11. Clarke, N. *Waterways of East Shropshire Through Time*; Amberley Publishing: Columbia, MD, USA, 2015.
12. Shropshire Canals. Available online: http://shropshirehistory.com/canals/canals/shropshire.htm (accessed on 9 April 2019).
13. Hadfield, C. *Thomas Telford's Temptation*; M. & M. Baldwin: Cleobury Mortimer, UK, 1993.
14. Betancourt, A. Mémoire sur un nouveau système de navigation intérieure. Available online: http://fundacionorotava.es/pynakes/lise/betan_memoi_fr_01_1807/ (accessed on 9 April 2019).
15. Betancourt, A. Dessin de la machine pour faire monter et descendre les bateaux d'un canal inferieur a un superieur et reciproquement sur deux plans inclines. Available online: http://fundacionorotava.org/pynakes/lise/betan_dessi_fr_01_179X/0/?zoom=large (accessed on 9 April 2019).
16. Fulton, R. *A Treatise on the Improvement of Canal Navigation*; I. and J. Taylor: London, UK, 1796.
17. Betancourt, A. Moyen de franchir les chutes dans les canaux, etc. par M. Recicourt. Available online: http://fundacionorotava.es/pynakes/lise/recic_moyen_fr_01_1795/ (accessed on 9 April 2019).
18. Ironbridge Gorge Museums. Available online: https://www.ironbridge.org.uk/ (accessed on 9 April 2019).
19. Hay Inclined Plane (from the Bottom Canal) in 1879. Available online: https://captainahabswaterytales.blogspot.com/search?q=hay+plane (accessed on 9 April 2019).

20. Fundación Canaria Orotava de Historia de la Ciencia. Available online: http://fundacionorotava.org (accessed on 9 April 2019).
21. Fundación Agustín de Betancourt. Available online: http://www.fundacionabetancourt.org (accessed on 9 April 2019).
22. Proyecto Digital Betancourt. Available online: http://fundacionorotava.es/betancourt/ (accessed on 9 April 2019).
23. Moreno Fernández, J. *Prehistoria del ferrocarril*; Fundación de los ferrocarriles españoles: Madrid, Spain, 2018. (In Spanish)
24. Wu, Y.; He, F.; Han, S. Collaborative CAD synchronization based on a symmetric and consistent modeling procedure. *Symmetry* **2017**, *9*, 59. [CrossRef]
25. Shih, R.H. *Parametric Modeling with Autodesk Inventor 2016*; SDC Publications: Mission, KS, USA, 2015.

symmetry

MDPI

Article

Anomaly Detection Based on Mining Six Local Data Features and BP Neural Network

Yu Zhang [1], Yuanpeng Zhu [2,*], Xuqiao Li [2], Xiaole Wang [2] and Xutong Guo [2]

[1] School of Mechanical & Automotive Engineering, South China University of Technology, Guangzhou 510641, China; 201630065258@mail.scut.edu.cn

[2] School of Mathematics, South China University of Technology, Guangzhou 510641, China; maxqli@mail.scut.edu.cn (X.L.); w820095324@163.com (X.W.); g648412727@sian.com (X.G.)

* Correspondence: ypzhu@scut.edu.cn

Received: 8 March 2019; Accepted: 15 April 2019; Published: 19 April 2019

Abstract: Key performance indicators (KPIs) are time series with the format of (timestamp, value). The accuracy of KPIs anomaly detection is far beyond our initial expectations sometimes. The reasons include the unbalanced distribution between the normal data and the anomalies as well as the existence of many different types of the KPIs data curves. In this paper, we propose a new anomaly detection model based on mining six local data features as the input of back-propagation (BP) neural network. By means of vectorization description on a normalized dataset innovatively, the local geometric characteristics of one time series curve could be well described in a precise mathematical way. Differing from some traditional statistics data characteristics describing the entire variation situation of one sequence, the six mined local data features give a subtle insight of local dynamics by describing the local monotonicity, the local convexity/concavity, the local inflection property and peaks distribution of one KPI time series. In order to demonstrate the validity of the proposed model, we applied our method on 14 classical KPIs time series datasets. Numerical results show that the new given scheme achieves an average F_1-score over 90%. Comparison results show that the proposed model detects the anomaly more precisely.

Keywords: anomaly detection; local data features; BP neural network; local monotonicity; convexity/concavity; local inflection; peaks distribution

---

1. Introduction

Key performance indicators (KPIs) are time series with the format of (timestamp, value), which can be collected from network traces, syslogs, web access logs, SNMP, and other data sources [1]. Table 1 shows the description of 14 classical KPIs and Figure 1 shows these 14 classical KPIs, which can be downloaded at http://iops.ai/dataset_list/. For example, KPI1 is a typical periodic data series [2], which is very common in our daily life. KPI5 is a classical stable data series [3], which may indicate the enterprise production index of one company. KPI11 is an unstable data series [4], in which the distribution of anomalies is very irregular. KPI10 and KPI14 belong to continuous fluctuation data series [5], of which the variation degree is dramatic so that anomalies could be detected very arduously. Furthermore, in KPI2, KPI3, KPI6, KPI8, and KPI12, the distribution between the normal data and the anomalies is extraordinarily unbalanced, which also results in the low accuracy of KPIs anomaly detection.

Table 1. Description of 14 classical KPIs.

	KPI1	KPI2	KPI3	KPI4	KPI5	KPI6	KPI7
Description	Periodic series	Periodic and fluctuation	Unstable series	Unstable series	Stable series	Unstable series	Unstable series
	KPI8	KPI9	KPI10	KPI11	KPI12	KPI13	KPI14
Description	Stable series	Unstable series	Continuous fluctuation series	Unstable series	Periodic and fluctuation series	Stable series	Continuous fluctuation series

Anomaly detection is purposed to find "the variation", as the so-called anomaly, from the norm KPI dataset. In recent years, anomaly detection plays an increasingly important role in some big data analysis areas. For example, in the field of finance, anomaly detection technology is used to detect fraud [6] and network intrusion in network security [7].

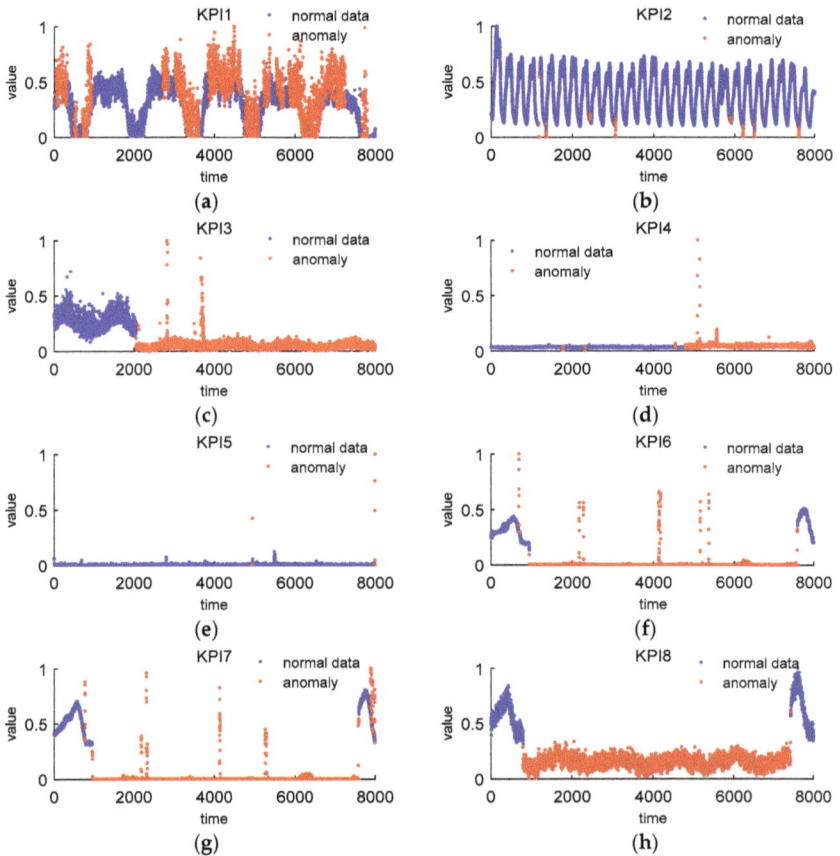

Figure 1. *Cont.*

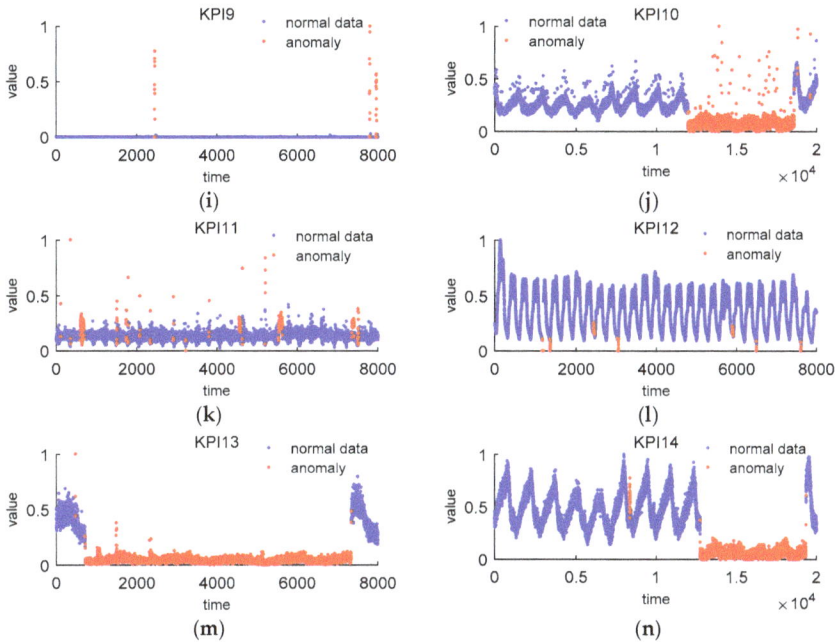

Figure 1. Fourteen classical key performance indicators (KPIs). (**a**): Periodic time series; (**b**): Periodic and continuous fluctuation time series; (**c**): Unstable time series; (**d**): Unstable time series; (**e**): Stable time series; (**f**): Unstable time series; (**g**): Unstable time series; (**h**): Stable time series; (**i**): Unstable time series; (**j**): Continuous fluctuation time series; (**k**): Unstable time series; (**l**): Periodic and continuous fluctuation time series; (**m**): Stable time series; (**n**): Continuous fluctuation time series.

Up to now, many anomaly detection approaches have been proposed. In [8], Hu et al. proposed an anomaly detection method known as Robust SVM (RSVM). By neglecting noisy data and using averaging technique, the RSVM makes the decision surface smoother and controls regularization automatically. In [9], Kabir et al. proposed a Least Square SVM (LS-SVM) method. Compared with the standard SVM, this method behaves more sensitive to anomalous and noise in training set. By using an optimum allocation scheme and selecting samples depending on variability, the algorithm is optimized to produce an effective result. Since Bayesian Network can be used for an event classification scheme, it can also be used for anomaly detection. In [10], Kruegel et al. identified two reasons for a large amount of false alarms. The first reason is the simplistic aggregation of model outputs, which leads to high false positives. The second is that anomaly detection system may misjudge some unusual but legitimate behaviors. To solve these problems, an anomaly detection approach based on Bayesian Network was proposed in [10]. Neutral network is also applicable for detecting anomaly. In [11], Hawkins et al. presented a Replicator Neural Network (RNN). By providing an outlyingness factor for anomaly, the method reproduces the input data pattern at output layer after training and achieves high accuracy without class labels. For the statistics-based approaches, Shyu et al. proposed an effective method based on robust principal component analysis in [12]. The method was developed from two principal components. One of the principal components explains about half of the total variation, while the other minor component's eigenvalues are less than 0.2. This technique has benefits of reducing dimension of data without losing important information and having low computational complexity.

One of the essential keys to develop anomaly detection models to detect the KPIs anomalies efficiently is time-series feature mining technique, which may affect the superior limit of the models. In previous studies, sliding window-based strategy was widely used for time series analysis, see for example [13–16] and the references therein. However, the prediction performance of this method relies on the description of similarity metrics between two sub-sequences. Moreover, in this method, similarity metrics are just represented by the calculation of the distance. In order to avoid the problem, Hu et al. proposed a meta-feature-based approach in [17], in which six statistics data characteristics including kurtosis, coefficient of variation, oscillation, regularity, square waves, and trend are mined. Nevertheless, these six statistics data characteristics are the features only representing the entire variation of the sequence described, and the relationship between several adjacent points are not revealed subtly (in other words, local variation situation between a few adjacent points could not be well described). We take the following coefficient of variation as an example, which describes the degree of dispersion of one time series

$$C = \frac{\sigma}{\mu},\tag{1}$$

where C denotes the coefficient of variation of one time series, σ denotes the standard deviation of this series, and μ is the mean value of this series. From Equation (1), we know that the coefficient of variation reflects the variation situation from an overall perspective of one time sequence, and thus the local variation situation could not be well reflected.

In the field of anomaly detection, generally, many anomalous events may have not happened successively or the probability of occurrence in succession is very small, which means one anomalous event usually appears suddenly and rarely. Therefore, due to the low frequency of abnormal events [18], we are not able to confirm an anomaly just using some characters describing the entire variation situation of one sequence, and we could not locate or predict the coming time of the next unknown anomaly precisely. In this situation, the subtle insight of local dynamics of the described sequence is particularly needed.

The major innovations of this work could be summarized as follows: we mine six local data features on behalf of the real-time dynamics of described time series. By means of vectorization description between every four adjacent points, the local geometric characteristics of one time series curve could be well described in a precise mathematical way. For example, local monotonicity, local convexity/concavity, and local inflection properties could be well revealed. Then input these six local data features into supervised back-propagation (BP) neural network, a new anomaly detection scheme is proposed. Numerical examples on the above 14 typical KPIs show that, taking advantage of the six local features as the inputs of the BP neural network, the new given scheme achieves an average F_1-score over 90%. Compared with the traditional statistics data characteristics used in [19], our method has a higher score, which means that our six local data features can be well described in the local dynamics of one KPI time series. Compared with SVM method [20] and SVM + PCA [21] method, our method based on BP neural network also has a higher average F_1-score.

The rest of this paper is organized as follows. Section 2 gives the basic concept of BP neural network. Besides, analysis of six local geometric characteristics is discussed in detail. Several numerical examples are given in Section 3 to argue the validity of our model. Discussion is given in the Section 4, and conclusion is summarized in Section 5.

2. Materials and Methods

Figure 2a shows the framework of our anomaly detection method. Figure 2b is the semantic drawing of six local data features spaces. By means of vectorization description on a normalized training/verifying dataset innovatively, the local geometric characteristics of one time series curve could be well described in a precise mathematical way. Thus six local data features have been mined to describe the local monotonicity, convexity/concavity, and the local inflection properties of one KPI series curve. Then input these six features into BP neural network, after multiple training processes, a new anomaly detection model is established.

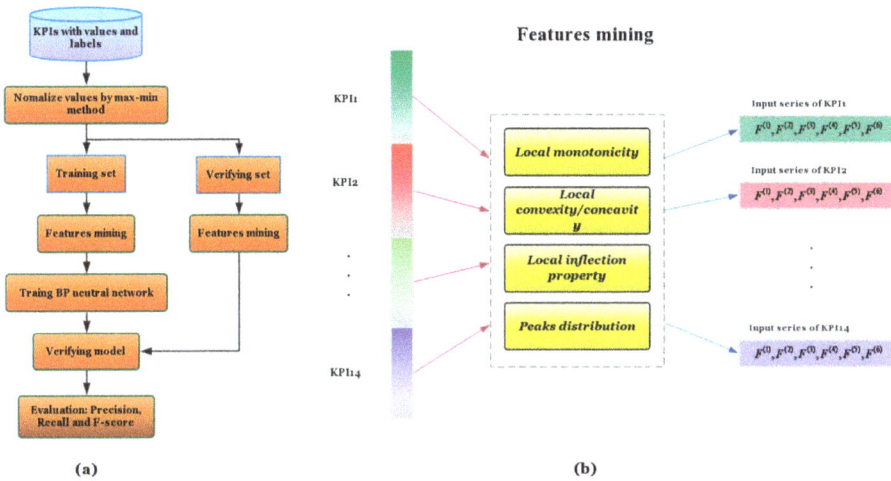

Figure 2. (**a**): The flowchart of the proposed approach for KPIs time series; (**b**): the semantic drawing of six local data feature space.

2.1. BP Neural Network Method

In this subsection, we shall give a few necessary backgrounds on back-propagation (BP) neural network. We will merely mention a few mathematical statements necessary for a good understanding for the present paper, and more details can be found in [22–26].

BP neural network is a kind of artificial neural networks on the basis of error back-propagation algorithm. Usually, BP neural network consists of one input layer, one or more hidden layer, and one output layer.

Let m, k, respectively, denote the neural number of input layer and the neural number of output layer, and L denotes the number of hidden layers. Additionally, $Label = (l_1, l_2, \cdots, l_k)$ denotes the target vector, $value = (v_1, v_2, \cdots, v_m)$ denotes the input vector of BP neural network, $a^L = \left(a_1^L, a_2^L, \cdots a_k^L\right)$ denotes the output vector of BP. BP uses $f_l(x)$ as the neuron activation function in the lth layer, $l = 1, 2, \ldots, L$. The 1st layer of the neural network is input layer, from the 2nd layer to the $(L-1)$th layer are hidden layers, and layer Lth is the output layer. Let w_{ij}^l denotes the weight from node i of layer $(l-1)$th to node j of layer lth, and b_j^l denotes the bias of node j in layer lth.

In BP neural network, the neurons just in adjacent layers are fully connected; nevertheless, there is no connection in the same neurons' layer. After each training process, the output value (the vector of predicted labels) is compared with the target value (the vector of correct labels), and then we can amend weights and thresholds of the input layer and the hidden layer with error feedback. With a hidden layer, BP neural network can express any continuous function accurately.

Let a_j^l denotes the output of node j in layer lth, and let z_j^l denotes the assemble of inputs in node j of layer lth, and it can be expressed as follows [23]

$$z_j^l = \sum_k w_{kj}^l a_k^{l-1} + b_j^l. \tag{2}$$

Therefore, the output a_j^l of node j in layer lth is expressed as follows

$$a_j^l = f_l(z_j^l) = f_l(\sum_k w_{kj}^l a_k^{l-1} + b_j^l), \tag{3}$$

where $f_l(x)$ is the activation function of layer lth.

There are three transfer functions in the BP neural network such that tan-sigmod, log-sigmoid, and purelin. Tan-sigmod or purelin transfer function maps any input value into an output value between −1 and 1. Log-sigmoid transfer function maps any input value into an output value between 0 and 1. The transfer functions in neural network can mix freely without unifying, so that we can reduce the network's parameters and hidden layer's nodes during the establishment of BP.

Since *Label* is the target vector, and a^L is output vector, the error function $E(w, b)$ can be expressed as follows [23]

$$E(w, b) = \|Label - a^L\|^2 = \sum_{i=1}^{k} (l_i - y_i)^2,$$
(4)

where k denotes the number of output layer nodes.

In this paper, we use the following mean square error (MSE) as the error output function of BP neural network [23]

$$MSE = \frac{1}{2p} \sum_{n=1}^{p} \|Label(x_n) - a^L(x_n)\|^2,$$
(5)

where x_n denotes the input of each train sample, and P denotes the number of train samples. It can decrease the global error of training dataset and the local error when each data point inputs.

In order to reduce the *MSE* gradually so that the predicted output value can be closer and closer to expectations booked in advance, BP neural network needs to adjust its weights and bias values constantly [24].

The classification accuracy of BP neural network is heavily dependent on the selected topology and on the selection of the training algorithm [25]. In this paper, we use Widrow-Hoff LMS method [26] to adjust the weight w_{ij}^l and bias b_j^l, that is

$$w_{ij}^l = w_{ij}^l - \eta \left(\frac{\partial MSE}{\partial w_{ij}^l} \right),$$
(6)

$$b_j^l = b_j^l - \eta \left(\frac{\partial MSE}{\partial b_j^l} \right),$$
(7)

where η is used to control its amendment speed, which can be variable or constant, generally speaking $0 < \eta < 1$.

According to the basic principle of BP neural network, we can obtain the update formula of weight and bias in each layer.

We write δ_j^L for the value of $\partial MSE / \partial z_j^L$, which can be expressed as follows

$$\delta_j^L = \frac{\partial MSE}{\partial z_j^L} = \frac{\partial MSE}{\partial a_j^L} \frac{\partial a_j^L}{\partial z_j^L} = \frac{\partial MSE}{\partial a_j^L} f_L'(z_j^L),$$
(8)

where f' of the formula above is the first-order partial derivatives of the activation function of layer *l*th $f_l(x)$.

And we write δ_j^l for the value of $\partial MSE / \partial z_j^l$, which can be expressed as follows

$$\delta_j^l = \frac{\partial MSE}{\partial z_j^l} = \sum_k \frac{\partial MSE}{\partial z_k^{l+1}} \frac{\partial z_k^{l+1}}{\partial z_j^l} = \sum_k \frac{\partial z_k^{l+1}}{\partial z_j^l} \delta_k^{l+1},$$
(9)

since

$$z_k^{l+1} = \sum_i w_{ik}^{l+1} a_i^l + b_k^{l+1} = \sum_i w_{ik}^{l+1} f_l(z_i^l) + b_k^{l+1},$$
(10)

we have $\partial z_k^{l+1}/\partial z_j^l = w_{jk}^{l+1} f'_l(z_j^l)$, then δ_j^l can be defined by recurrence as follows:

$$\delta_j^l = \sum_k w_{jk}^{l+1} f'_l(z_j^l) \delta_k^{l+1}. \tag{11}$$

Similarly, we can prove that [23]

$$\frac{\partial MSE}{\partial b_j^l} = \frac{\partial MSE}{\partial z_j^l} \frac{\partial z_j^l}{\partial b_j^l} = \delta_j^l, \tag{12}$$

$$\frac{\partial MSE}{\partial w_{kj}^l} = \frac{\partial MSE}{\partial z_j^l} \frac{\partial z_j^l}{\partial w_{kj}^l} = a_k^{l-1} \delta_j^l = f_{l-1}(z_k^{l-1}) \delta_j^l. \tag{13}$$

Consequently, the basic idea of BP neural network is summarized as follows. Firstly, input training data into neural network. Then during the processing of continuous learning and training, BP neural network will modify the weights and threshold values step by step, and when it reaches the precision error setup in advance, it will stop the learning. Finally, the output value is acquired.

2.2. Features Mining Method

By means of vectorization description on a normalized KPIs dataset innovatively, the local geometric characteristics of one time series curve could be well described in a precise mathematical way. We shall mine six local data features to describe the local monotonicity, convexity/concavity, the local inflection properties of one series curve.

2.2.1. Normalization by Max–Min Method

For a KPIs data with value set $V = \{V_1, V_2, V_3, \cdots V_n, \cdots V_{n+m}\}$, we firstly use a max–min method to normalize each of the values as follows:

$$v_i = \frac{V_i - V_{\min}}{V_{\max} - V_{\min}}, \tag{14}$$

where $V_{\max} = \max_i V_i$, $V_{\min} = \min_i V_i$, $i = 1, 2, \ldots, n + m$. The purpose of normalization is to avoid large differences between different values in a KPI time series.

2.2.2. The Definition of Six Local Data Features

For a resulting normalized value dataset $v = \{v_1, v_2, v_3, \cdots v_n, \cdots v_{n+m}\}$, we divide it into a train part $V_{train} = \{v_1, v_2, v_3, \cdots v_n\}$ and a verifying or test part $V_{test} = \{v_{n+1}, v_{n+2}, v_{n+3}, \cdots v_{n+m}\}$. We shall use the train part to establish the model while use the verifying part to test the performance of the model.

Local monotonicity, convexity/concavity, local inflection properties, and peaks distribution are four essential features of a given data set, which describe the local increasing/decreasing rates of the data set. With this in mind, we mine the following six features of the resulting normalized value dataset $v = \{v_1, v_2, v_3, \cdots v_n, \cdots v_{n+m}\}$

$$
\begin{cases}
F_i^{(1)} = v_i, i = 1, 2, \ldots, n+m, \\
F_i^{(2)} = v_{i+1} - v_i, i = 1, \ldots, n+m-1, \\
F_i^{(3)} = v_{i+2} - 2v_{i+1} + v_i, i = 1, 2, \ldots, n+m-2, \\
F_i^{(4)} = (v_{i+2} - v_{i+1})(v_{i+1} - v_i), i = 1, 2, \ldots, n+m-2, \\
F_i^{(5)} = v_{i+3} - 3v_{i+2} + 3v_{i+1} - v_i, i = 1, 2, \ldots, n+m-3, \\
F_i^{(6)} = (v_{i+3} - 2v_{i+2} + v_{i+1})(v_{i+2} - 2v_{i+1} + v_i), i = 1, 2, \ldots, n+m-3.
\end{cases}
\tag{15}
$$

We give some geometric explanations on the six mined features. The feature $F_i^{(1)}$ can describe peaks distribution of the normalized value data. As shown in Figures 3 and 4, the feature $F_i^{(2)}$ and $F_i^{(3)}$ are in fact the first and second difference of the normalized value data, respectively, which can describe the local monotonicity and convexity/concavity of the normalized value data. For example, with $F_i^{(2)} > 0, F_{i+1}^{(2)} > 0$ and $F_i^{(3)} > 0$, the normalized value data is both monotonically increasing and convex locally (in other words, the normalized value data has a faster and faster increasing rate locally).

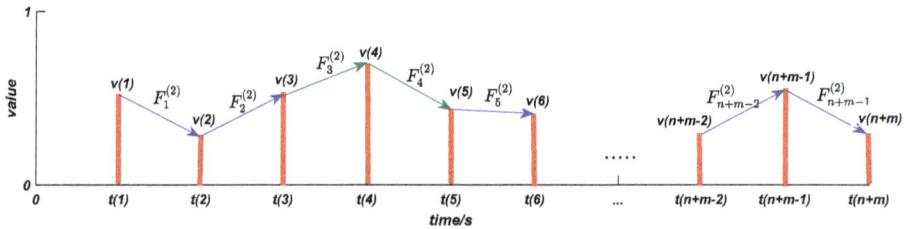

Figure 3. Schematic illustration of the feature $F_i^{(2)}$.

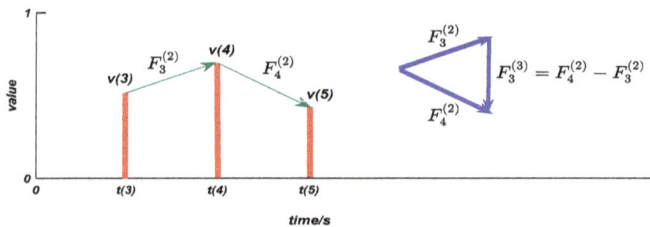

Figure 4. Schematic illustration of the feature $F_i^{(2)}$ and $F_i^{(3)}$.

The feature $F_i^{(4)}$ can describe the local inflection property of the normalized value data. For example, with $F_i^{(4)} < 0$, that is $F_i^{(2)} > 0, F_{i+1}^{(2)} < 0$ or $F_i^{(2)} < 0, F_{i+1}^{(2)} > 0$, it implies that the normalized value data has a local switch between "increasing" and "decreasing" values. The feature $F_i^{(5)}$ is the third difference of the normalized value data, and the feature $F_i^{(6)}$ can describe the local switch of the sign of $F_i^{(4)}$.

Figure 5 shows the numerical results of the six features mined of 14 KPIs. From this figure, we can see that the first, second, and third difference $F_i^{(2)}, F_i^{(3)}$ and $F_i^{(5)}$ distinguish anomalies and normal data significantly. The point whose values of $F_i^{(2)}, F_i^{(3)}$ and $F_i^{(5)}$ differ from that of the other points extraordinarily may be considered as an anomaly. The features $F_i^{(4)}$ and $F_i^{(6)}$ reveal the anomalies in a subtle way, which can prevent the misjudgments given by $F_i^{(2)}, F_i^{(3)}$, and $F_i^{(5)}$.

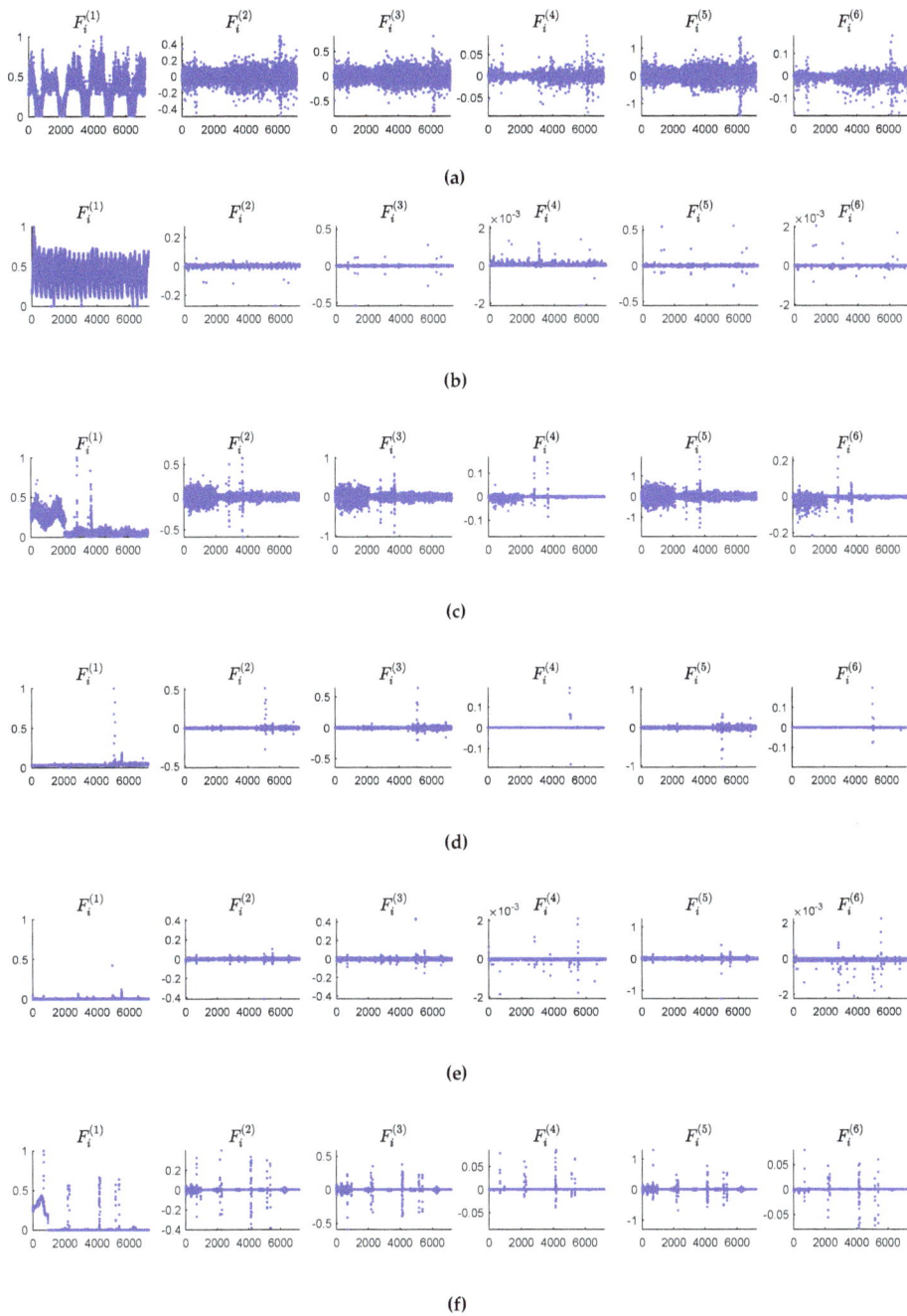

(a)

(b)

(c)

(d)

(e)

(f)

Figure 5. *Cont.*

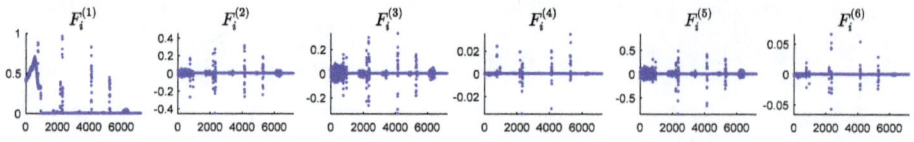

(g)

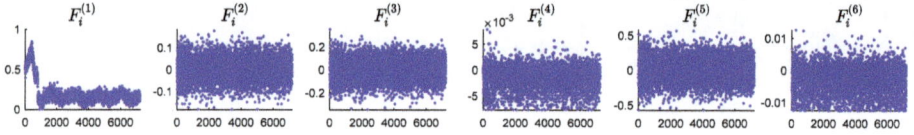

(h)

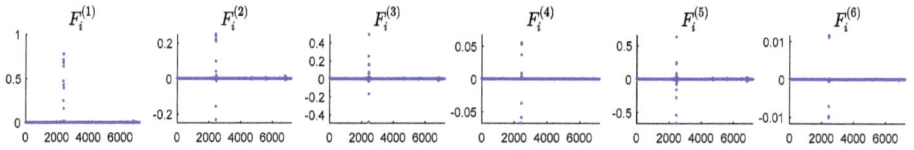

(i)

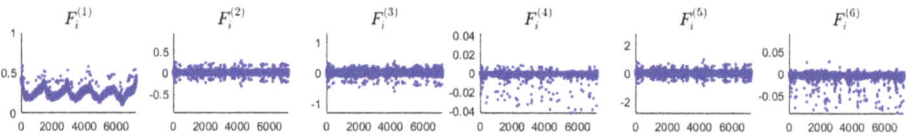

(j)

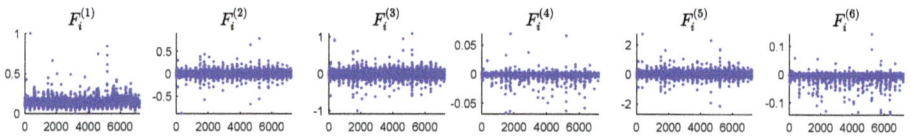

(k)

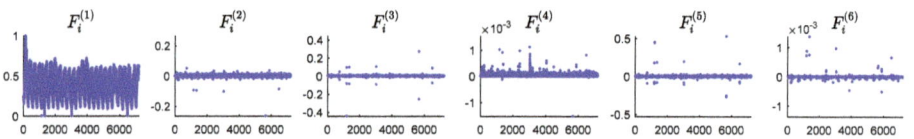

(l)

Figure 5. *Cont.*

(m)

(n)

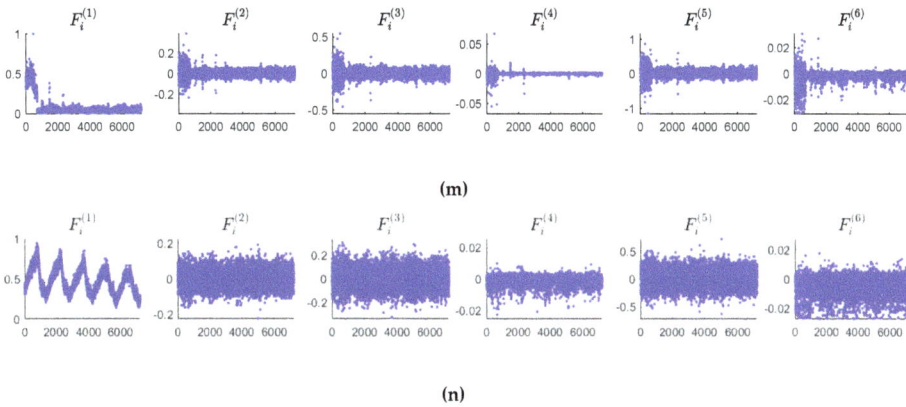

Figure 5. Six features mined of the KPIs. (**a**) Six features mined of the KPI1; (**b**) Six features mined of the KPI2; (**c**) Six features mined of the KPI3; (**d**) Six features mined of the KPI4; (**e**) Six features mined of the KPI5; (**f**) Six features mined of the KPI6; (**g**) Six features mined of the KPI7; (**h**) Six features mined of the KPI8; (**i**) Six features mined of the KPI9; (**j**) Six features mined of the KPI10; (**k**) Six features mined of the KPI11; (**l**) Six features mined of the KPI112; (**m**) Six features mined of the KPI13; (**n**) Six features mined of the KPI14.

2.3. Algorithm Description

Input:

In training model, we input

$$F^{(1)} = \left\{ F_4^{(1)}, F_5^{(1)}, \cdots F_n^{(1)} \right\},$$

$$F^{(2)} = \left\{ F_3^{(2)}, F_4^{(2)}, \cdots F_{n-1}^{(2)} \right\},$$

$$F^{(3)} = \left\{ F_2^{(3)}, F_3^{(3)}, \cdots F_{n-2}^{(3)} \right\},$$

$$F^{(4)} = \left\{ F_2^{(4)}, F_3^{(4)}, \cdots F_{n-2}^{(4)} \right\},$$

$$F^{(5)} = \left\{ F_1^{(5)}, F_2^{(5)}, \cdots F_{n-3}^{(5)} \right\},$$

$$F^{(6)} = \left\{ F_1^{(6)}, F_2^{(6)}, \cdots F_{n-3}^{(6)} \right\}.$$

In verifying model, we input

$$F^{(1)} = \left\{ F_{n+1}^{(1)}, F_{n+2}^{(1)}, \cdots F_{n+m}^{(1)} \right\},$$

$$F^{(2)} = \left\{ F_n^{(2)}, F_{n+1}^{(2)}, \cdots F_{n+m-1}^{(2)} \right\},$$

$$F^{(3)} = \left\{ F_{n-1}^{(3)}, F_n^{(3)}, \cdots F_{n+m-2}^{(3)} \right\},$$

$$F^{(4)} = \left\{ F_{n-1}^{(4)}, F_n^{(4)}, \cdots F_{n+m-2}^{(4)} \right\},$$

$$F^{(5)} = \left\{ F_{n-2}^{(5)}, F_{n-1}^{(5)}, \cdots F_{n+m-3}^{(5)} \right\},$$

$$F^{(6)} = \left\{ F_{n-2}^{(6)}, F_{n-1}^{(6)}, \cdots F_{n+m-3}^{(6)} \right\}.$$

Output:

The output is the predicted label vector;

Step 1: normalize the values of KPIs series data;
Step 2: separate the KPI into training dataset and verifying dataset;
Step 3: calculate the value of six local data features according to Equations (14) and (15);
Step 4: input features vector and target vector into BP algorithm;
Step 5: BP neural network outputs the detecting results.

2.4. Evaluation Method of Model Performance

In this experiment, confusion matrices (TP, TN, FP, and FN) have been applied to define the evaluation criterion. The meaning corresponding to confusion matrices are categorized in Table 2, where true positive (TP) means the number of anomalies precisely diagnosed as anomalies, whereas true negative (TN) means the number of normal data correctly diagnosed as normal. In the same way, false positive (FP) means the number of normal data diagnosed as anomalous by mistake, and false negative (FN) means the number of anomalies inaccurately diagnosed as normal.

Table 2. The meaning of confusion matrices.

		Actual Value	
		Anomaly	Normal
Predication Value	Anomaly	TP	FP
	Normal	FN	TN

In order to give the evaluations of the performance of the proposed model, evaluation criteria such as Recall, Precision, and F_1-score are considered [18]

$$Recall = \frac{TP}{TP + FN}, \tag{16}$$

$$Precision = \frac{TP}{TP + FP}, \tag{17}$$

$$F_1\_score = 2 \times \frac{Precision \times Recall}{Precision + Recall}. \tag{18}$$

Recall, which is computed by Equation (16), denotes the number of anomalies detected by the anomaly detection technology. Precision, which is computed by Equation (17), denotes the numbers of the values being accurately categorized as anomalies. It is the most intuitive performance evaluation criterion. F_1-score, which is computed by Equation (18), consists of a harmonic mean of precision and recall while accuracy is the ratio of correct predictions of a classification model [27,28]. In the next numerical experiments, we shall adopt the F_1-score to evaluate the performance of the model.

3. Results

In next experiments, we shall use the computer with 8 GB memory as well as core i5 inside. The model is established by MATLAB 2016a.

3.1. Explore Different Topology Structures of BP Network

Inputting six mined local data features into BP neural network, a novel anomaly detection model is proposed. In order to find out the best-performing topology structure of BP network, we have done five experiments to explore the optimal combination of different layers and neural nodes. Figure 6 shows the F_1-scores of different topology structures of BP network for each of 14 KPIs. Table 3 shows the average score of different topology structures of BP network. From these, we can see that the topology structure of $6 \rightarrow 10 \rightarrow 10 \rightarrow 10 \rightarrow 1$ has the highest average F_1-score among the five topology structures. The topology structure of $6 \to 10 \to 10 \to 10 \to 1$ means 6 input nodes, 10 nodes of each hidden layer, and 1 output node. We use the log-sigmoid function as the transfer function in the BP neural network. It should be noted that when the predicted label is no smaller than 0.5, it will be set as 1, otherwise 0. In other words, a data point with the predicted label above 0.5 is regarded as an anomaly while under 0.5 is regarded as a normal data. In the next compared experiments, we shall use the best structure of $6 \to 10 \to 10 \to 10 \to 1$ to establish the BP model.

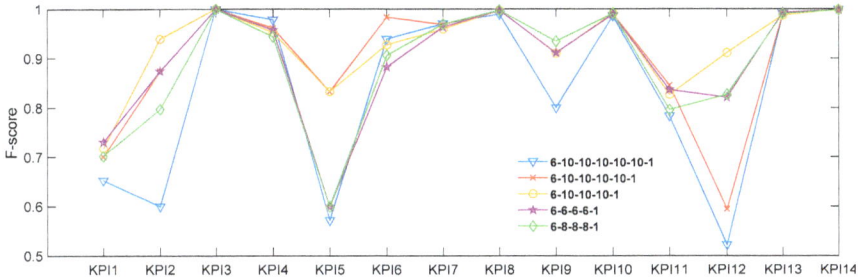

Figure 6. F_1-scores of different topology structures of BP network for each of 14 KPIs.

Table 3. Comparative results of different topology structures of back-propagation (BP) network.

	$6 \to 6 \to 6$ $\to 6 \to 1$	$6 \to 8 \to 8$ $\to 8 \to 1$	$6 \to 10 \to 10$ $\to 10 \to 1$	$6 \to 10 \to 10 \to$ $10 \to 10 \to 1$	$6 \to 10 \to 10 \to 10$ $\to 10 \to 10 \to 1$
$Precision|_{average}$ (%)	96.50	96.59	96.80	97.68	94.76
$Recall|_{average}$ (%)	85.58	84.41	89.33	85.64	88.64
$F_1 - score|_{average}$ (%)	89.66	88.93	92.92	90.33	91.60

3.2. Results Presentation

We show the numerical results of the structure of $6 \to 10 \to 10 \to 10 \to 1$ on each of 14 KPIs. Table 4 shows the values of three evaluation criteria of the verifying dataset of each of 14 KPIs. From the results, we can see that the detection effects on these 14 KPIs are good, especially for KPI 3. All the anomalies had been detected and there is no misjudgments happened in KPI 3. According to Equation (19), the new given scheme achieves an average F_1-score over 90%, which verifies the remarkable anomaly detection effects.

$$F_1 - score|_{average} = 2 \times \frac{Precision|_{average} \times Recall|_{average}}{Precision|_{average} + Recall|_{average}} = 92.92\%. \tag{19}$$

Table 4. Values of evaluation criteria using our method.

	KPI1	KPI2	KPI3	KPI4	KPI5	KPI6	KPI7	KPI8	KPI9	KPI10	KPI11	KPI12	KPI13	KPI14
Precision (%)	86.25	99.50	100	95.25	99.75	88.25	94.87	99.88	99.38	99.55	97.38	97.25	98.00	99.90
Recall (%)	61.41	88.89	100	95.25	71.43	97.69	97.00	99.55	84.00	98.62	71.83	85.71	99.32	99.85
F_1-score (%)	71.74	93.90	100	95.25	83.25	92.73	95.92	99.71	91.04	99.08	82.67	91.12	98.65	99.88

Figure 7 shows the numerical results of the structure of $6 \to 10 \to 10 \to 10 \to 1$ on each of 14 KPIs. In the figure, the red points are original anomalies of one KPI. The circles represent the predicted anomalies. When the circle coincides in position with one red point, it means that this abnormal data point has been detected by our method. From Figure 7, we know that on the left of the dotted line, the detection results of the train models achieve a higher accuracy, while there are a few misjudgments taking place in this process. On the right of the dotted line, the detection results about verifying data are shown. For KPI1, which is a periodic time series, our method is not capable to achieve satisfactory performance. There are some anomalies that have not been detected and some normal data are misjudged as anomalies. For KPI2–KPI10, numerical results show a remarkable detection effect. For KPI11, although there are some anomalies that have not been detected, misjudgments are rare, which means that once a point is diagnosed as an anomaly, this point may well be an original anomaly. For KPI12–KPI14, numerical results also show a remarkable detection effect.

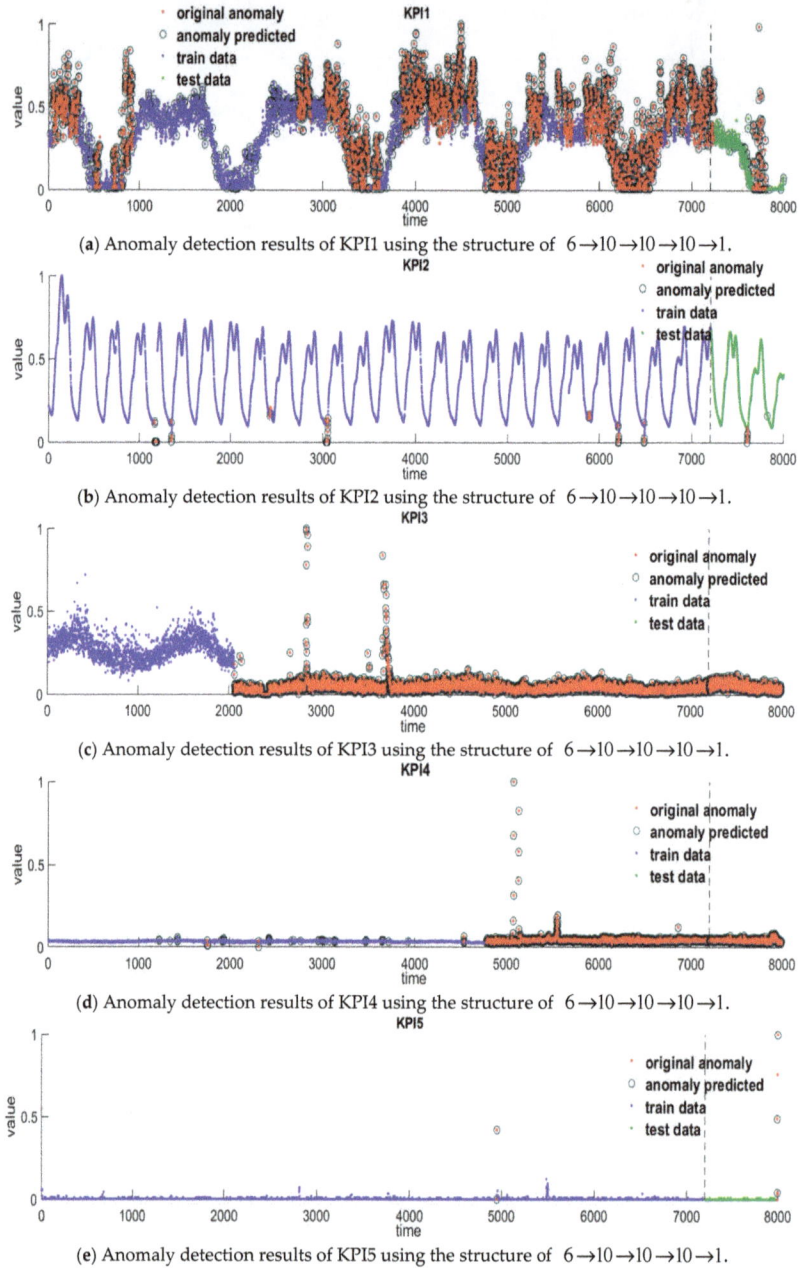

(**a**) Anomaly detection results of KPI1 using the structure of $6 \rightarrow 10 \rightarrow 10 \rightarrow 10 \rightarrow 1$.

(**b**) Anomaly detection results of KPI2 using the structure of $6 \rightarrow 10 \rightarrow 10 \rightarrow 10 \rightarrow 1$.

(**c**) Anomaly detection results of KPI3 using the structure of $6 \rightarrow 10 \rightarrow 10 \rightarrow 10 \rightarrow 1$.

(**d**) Anomaly detection results of KPI4 using the structure of $6 \rightarrow 10 \rightarrow 10 \rightarrow 10 \rightarrow 1$.

(**e**) Anomaly detection results of KPI5 using the structure of $6 \rightarrow 10 \rightarrow 10 \rightarrow 10 \rightarrow 1$.

Figure 7. *Cont.*

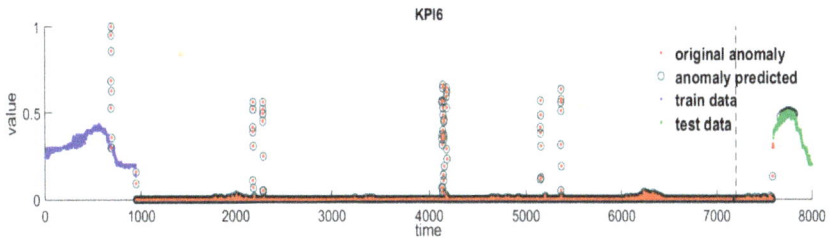

(**f**) Anomaly detection results of KPI6 using the structure of 6→10→10→10→1.

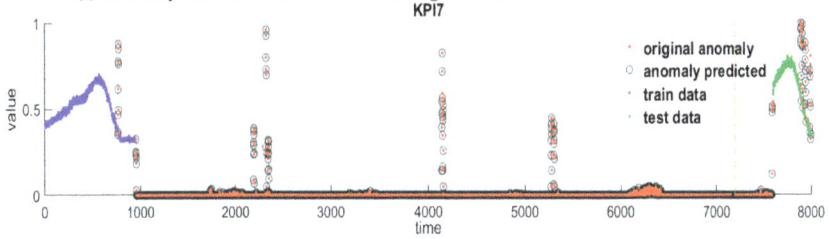

(**g**) Anomaly detection results of KPI7 using the structure of 6→10→10→10→1.

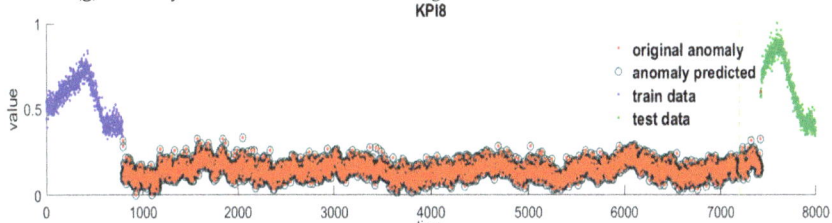

(**h**) Anomaly detection results of KPI8 using the structure of 6→10→10→10→1.

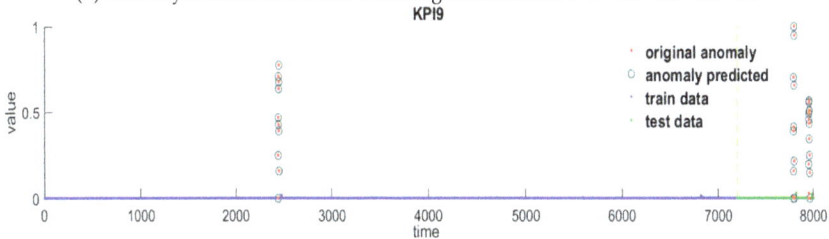

(**i**) Anomaly detection results of KPI9 using the structure of 6→10→10→10→1.

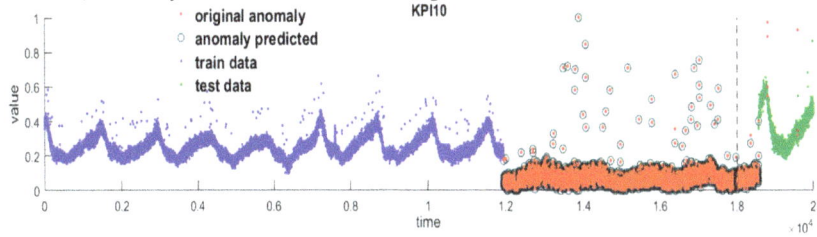

(**j**) Anomaly detection results of KPI10 using the structure of 6→10→10→10→1.

Figure 7. *Cont.*

(k) Anomaly detection results of KPI11 using the structure of $6 \rightarrow 10 \rightarrow 10 \rightarrow 10 \rightarrow 1$.

(l) Anomaly detection results of KPI12 using the structure of $6 \rightarrow 10 \rightarrow 10 \rightarrow 10 \rightarrow 1$.

(m) Anomaly detection results of KPI13 using the structure of $6 \rightarrow 10 \rightarrow 10 \rightarrow 10 \rightarrow 1$.

(n) Anomaly detection results of KPI14 using the structure of $6 \rightarrow 10 \rightarrow 10 \rightarrow 10 \rightarrow 1$.

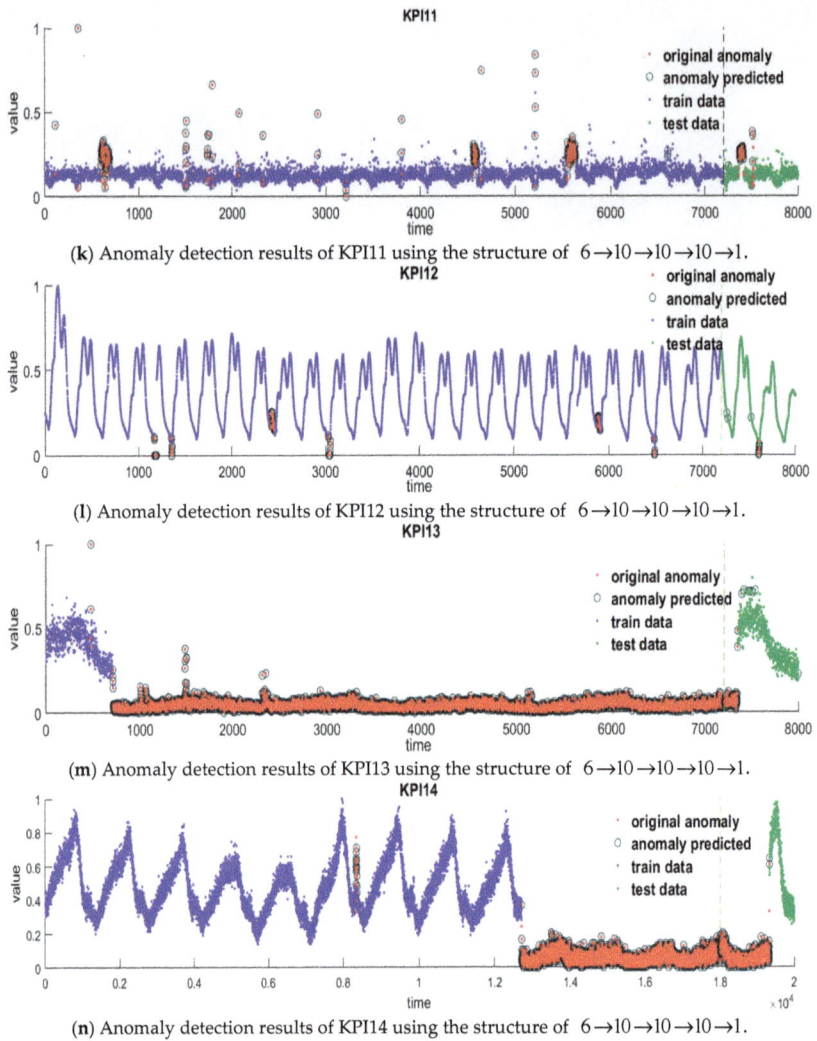

Figure 7. Anomaly detection results using the structure of $6 \rightarrow 10 \rightarrow 10 \rightarrow 10 \rightarrow 1$. (a): Anomaly detection results of KPI1 using the structure of $6 \rightarrow 10 \rightarrow 10 \rightarrow 10 \rightarrow 1$; (b): Anomaly detection results of KPI2 using the structure of $6 \rightarrow 10 \rightarrow 10 \rightarrow 10 \rightarrow 1$; (c): Anomaly detection results of KPI3 using the structure of $6 \rightarrow 10 \rightarrow 10 \rightarrow 10 \rightarrow 1$; (d): Anomaly detection results of KPI4 using the structure of $6 \rightarrow 10 \rightarrow 10 \rightarrow 10 \rightarrow 1$; (e): Anomaly detection results of KPI5 using the structure of $6 \rightarrow 10 \rightarrow 10 \rightarrow 10 \rightarrow 1$; (f): Anomaly detection results of KPI6 using the structure of $6 \rightarrow 10 \rightarrow 10 \rightarrow 10 \rightarrow 1$; (g): Anomaly detection results of KPI7 using the structure of $6 \rightarrow 10 \rightarrow 10 \rightarrow 10 \rightarrow 1$; (h): Anomaly detection results of KPI8 using the structure of $6 \rightarrow 10 \rightarrow 10 \rightarrow 10 \rightarrow 1$; (i): Anomaly detection results of KPI9 using the structure of $6 \rightarrow 10 \rightarrow 10 \rightarrow 10 \rightarrow 1$; (j): Anomaly detection results of KPI10 using the structure of $6 \rightarrow 10 \rightarrow 10 \rightarrow 10 \rightarrow 1$; (k): Anomaly detection results of KPI11 using the structure of $6 \rightarrow 10 \rightarrow 10 \rightarrow 10 \rightarrow 1$; (l): Anomaly detection results of KPI12 using the structure of $6 \rightarrow 10 \rightarrow 10 \rightarrow 10 \rightarrow 1$; (m): Anomaly detection results of KPI13 using the structure of $6 \rightarrow 10 \rightarrow 10 \rightarrow 10 \rightarrow 1$; (n): Anomaly detection results of KPI14 using the structure of $6 \rightarrow 10 \rightarrow 10 \rightarrow 10 \rightarrow 1$.

4. Discussion

In this section, firstly, we use the traditional statistics data features given in [19] as the input of BP network, and apply this model on the same KPIs. Secondly, we also explore SVM [20] and SVM + PCA [21] methods and the results are presented as well. Finally, we analyze the performance of these models.

4.1. Traditional Statistics Data Features and BP Network

We performed an experiment using the traditional statistics data features given in [19] and BP network with topology structure of $6 \to 10 \to 10 \to 10 \to 1$. These traditional statistics data features included average value, maximum value, minimum value, standard deviation, and variance of one time series. The results are presented in Table 5. According to Equations (16)–(18), we have

$$Precision|_{average} = 84.29\%, \; Recall|_{average} = 86.14\%, \; F_1 - score|_{average} = 85.20\%.$$

Table 5. Values of evaluation criteria using the method in [19].

	KPI1	KPI2	KPI3	KPI4	KPI5	KPI6	KPI7	KPI8	KPI9	KPI10	KPI11	KPI12	KPI13	KPI14
Precision (%)	66.12	54.55	100	100	50.00	90.57	97.90	100	90.0	98.29	95.92	37.50	99.33	99.93
Recall (%)	64.36	66.67	100	95.75	66.67	97.46	96.11	97.80	72.0	98.80	66.20	85.71	98.67	99.78
F_1-score (%)	65.23	60.00	100	97.83	57.14	93.89	97.00	98.89	80.0	98.55	78.33	52.17	99.00	99.85

4.2. Explore Different Machine Learning Models

In this subsection, we shall use SVM [20] and SVM + PCA [21] methods to further verify the validity of the six new mined features given in Equation (15).

- SVM method

Table 6 shows the anomaly detection results using SVM method with the six new mined features given in Equation (15) as the input. From the results, it is observed that SVM-based method is not able to find any anomaly in KPI2, but it has a high score on the other KPIs. The average score on the other 13 KPIs are calculated as follows:

$$Precision|_{average} = 96.98\%, \; Recall|_{average} = 85.93\%, \; F_1 - score|_{average} = 91.12\%.$$

Table 6. Values of evaluation criteria using SVM method.

	KPI1	KPI2	KPI3	KPI4	KPI5	KPI6	KPI7	KPI8	KPI9	KPI10	KPI11	KPI12	KPI13	KPI14
Precision (%)	69.94	0	100	100	100	97.96	96.80	100	100	100	100	96.00	100	100
Recall (%)	61.96	0	100	95.50	14.29	98.46	97.69	99.55	84.00	98.27	70.42	100	98.63	98.73
F_1-score (%)	65.71	NaN	100	97.70	25.00	98.21	97.24	99.78	91.30	99.13	82.64	97.96	99.31	99.36

- SVM + PCA Method

Table 7 shows the anomaly detection results using SVM + PCA method with the six new mined features given in Equation (15) as the input. The detection results for the combined SVM and PCA methods have some improvements. However, as for KPI5, this method shows a poor performance. The average score has been calculated as follows:

$$Precision|_{average} = 93.29\%, \; Recall|_{average} = 79.54\%, \; F - score|_{average} = 85.87\%.$$

Table 7. Values of evaluation criteria using SVM + PCA method.

	KPI1	KPI2	KPI3	KPI4	KPI5	KPI6	KPI7	KPI8	KPI9	KPI10	KPI11	KPI12	KPI13	KPI14
Precision (%)	46.91	100	100	100	100	74.03	85.60	99.55	100	100	100	100	100	100
Recall (%)	61.96	66.67	99.88	92.50	14.28	97.95	96.07	99.55	80.00	98.10	67.61	42.86	98.63	97.53
F_1-score (%)	53.40	80.00	99.94	96.10	25.00	84.33	90.53	99.55	88.89	99.04	80.67	60.00	98.63	98.75

4.3. Performance Analysis of Different Models

Table 8 shows the comparative results on the same 14 KPIs using different methods. Our method, SVM method, and SVM + PCA method all use the six new mined features given in Equation (15) as the input. And our method is established by using BP network with the structure of $6 \rightarrow 10 \rightarrow 10 \rightarrow 10 \rightarrow 1$. Besides, the method in [19] is also established by using BP network with the same structure, which the traditional statistics data characteristics are inputted into. As can be seen from Table 8, compared with the traditional statistics data characteristics used in [19], our method has a higher score, which means that our six local data features can well describe the local dynamics of the KPIs. Compared with SVM and SVM + PCA methods, our method also has a higher score, which means that BP network has a better anomaly detection effect. In the whole, our method is capable for anomaly detection on some complexity KPIs.

Table 8. Comparative results of different methods.

	Our Method	Method in Literature [19]	SVM Method	SVM + PCA Method
$Precision|_{average}(\%)$	96.80	84.29	96.98	93.29
$Recall|_{average}(\%)$	89.33	86.14	85.93	79.54
$F_1 - score|_{average}(\%)$	92.92	85.20	91.12	85.87

5. Conclusions

We have proposed six local data features to mine the local monotonicity, the local convexity/concavity, the local inflection properties, and peaks distribution of KPI time series data. With these six local data features as the input of BP network, we have established a new anomaly detection model.

Compared with the traditional statistics data characteristics method given in [19], our scheme shows a higher accuracy and universality which demonstrates the remarkable detection effects. Our experiments also show that BP neural network has a better universality and accuracy degree than SVM and SVM + PCA methods. In the future, some other neural network algorithms will be explored to further this study. In addition, the classification accuracy of BP neural network is heavily dependent on the selected topology and on the selection of the training algorithm, and the performance of our proposed methodology could be further improved by selecting more sophisticated training algorithms in the future work.

Since our method is based on mining six local data features, as for periodic data series like KPI1, these local data features are not adequate enough to characterize the periodic data series. In the future study, we shall mine some features describing the periodic time series.

Author Contributions: Conceptualization, Y.Z. (Yu Zhang), Y.Z. (Yuanpeng Zhu), and X.L.; methodology, Y.Z. (Yu Zhang), Y.Z. (Yuanpeng Zhu), and X.L.; software, Y.Z. (Yu Zhang) and X.L.; validation, Y.Z. (Yu Zhang) and X.L.; formal analysis, Y.Z. (Yu Zhang), Y.Z. (Yuanpeng Zhu), and X.L.; investigation, X.L. and X.W.; resources, Y.Z. (Yuanpeng Zhu); data curation, Y.Z. (Yu Zhang) and X.L.; writing—original draft preparation, Y.Z. (Yu Zhang), Y.P.Z. (Yuanpeng Zhu), and X.L.; writing—review and editing, Y.Z. (Yuanpeng Zhu) and X.L.; visualization, X.L., X.W., and X.G.; supervision, Y.Z. (Yuanpeng Zhu); project administration, Y.Z. (Yuanpeng Zhu); funding acquisition, Y.Z. (Yuanpeng Zhu) and Y.Z. (Yuanpeng Zhu).

Funding: The research is supported by the National Natural Science Foundation of China (No. 61802129), the Postdoctoral Science Foundation of China (No. 2015M571931), the Fundamental Research Funds for the Central

Universities (No. 2017MS121), the Natural Science Foundation Guangdong Province, China (No. 2018A030310381), and the National Training Program of Innovation and Entrepreneurship for Undergraduates (201810561174).

Acknowledgments: This work was supported by South China University of Technology.

Conflicts of Interest: The authors declare no conflict of interest.

References

1. Pérez-Álvarez, J.M.; Maté, A.; Gómez-López, M.T.; Trujillob, J. Tactical Business-Process-Decision Support based on KPIs Monitoring and Validation. *Comput. Ind.* **2018**, *102*, 23–39. [CrossRef]
2. Yang, J.; Wan, W.; Yu, P.S. Mining Asynchronous Periodic Patterns in Time Series Data. *IEEE Trans. Knowl. Data Eng.* **2003**, *15*, 613–628. [CrossRef]
3. Kruczek, P.; Wyłomańska, A.; Teuerle, M.; Gajda, J. The modified Yule-Walker method for α-stable time series models. *Phys. A Stat. Mech. Appl.* **2017**, *469*, 588–603. [CrossRef]
4. Grillenzoni, C. Forecasting unstable and nonstationary time series. *Int. J. Forecast.* **1998**, *14*, 469–482. [CrossRef]
5. Pierini, J.; Telesca, L. Fluctuation analysis of monthly rainfall time series. *Fluct. Noise Lett.* **2010**, *20*, 219–228. [CrossRef]
6. Ahmed, M.; Mahmood, A.N.; Islam, M.R. A survey of anomaly detection techniques in financial domain. *Future Gener. Comput. Syst.* **2016**, *55*, 278–288. [CrossRef]
7. Hong, J.H.; Liu, C.C.; Govindarasu, M. Integrated Anomaly Detection for Cyber Security of the Substations. *IEEE Trans. Smart Grid* **2014**, *5*, 1643–1653. [CrossRef]
8. Hu, W.J.; Liao, Y.; Vemuri, V.R. Robust support vector machines for anomaly detection in computer security. In Proceedings of the International Conference Machine Learning & Applications-ICMLA, Los Angeles, CA, USA, 23–24 July 2003.
9. Kabir, E.; Hu, J.; Wang, H.; Zhuo, G. A novel statistical technique for intrusion detection systems. *Future Gener. Comput. Syst.* **2018**, *79*, 303–318. [CrossRef]
10. Kruegel, C.; Mutz, D.; Robertson, W.; Valeur, F. Bayesian event classification for intrusion detection. In Proceedings of the 19th Annual Computer Security Applications Conference, Las Vegas, NV, USA, 8–12 December 2003.
11. Hawkins, S.; He, H.; Williams, G.; Baxter, R. Outlier detection using replicator neural networks. In *Data Warehousing and Knowledge Discovery, Proceedings of the International Conference on Data Warehousing and Knowledge Discovery, Aix-en-Provence, France, 4–6 September 2002*; Lecture Notes in Computer Science; Kambayashi, Y., Winiwarter, W., Arikawa, M., Eds.; Springer: Berlin/Heidelberg, Germany, 2002; Volume 2454, pp. 170–180.
12. Shyu, M.L.; Chen, S.C.; Kanoksri, S.; Chang, L.W. A novel anomaly detection scheme based on principal component classifier. In *IEEE Foundations and New Directions of Data Mining Workshop*; Miami Univ Coral Gables Fl Dept of Electrical and Computer Engineering: Coral Gables, FL, USA, 2003; pp. 171–179.
13. Zhang, T.; Yue, D.; Gu, Y.; Wang, Y.; Yu, G. Adaptive correlation analysis in stream time series with sliding windows. *Comput. Math. Appl.* **2008**, *57*, 937–948. [CrossRef]
14. Ding, Z.; Fei, M. An anomaly detection approach based on isolation forest algorithm for streaming data using sliding window. *IFAC Proc.* **2013**, *46*, 12–17. [CrossRef]
15. Ren, H.; Ye, Z.; Li, Z. Anomaly detection based on a dynamic Markov model. *Inf. Sci.* **2017**, *411*, 52–65. [CrossRef]
16. Chou, J.S.; Ngo, N.T. Time series analytics using sliding window metaheuristic optimization-based machine learning system for identifying building energy consumption patterns. *Appl. Energy* **2016**, *177*, 751–770. [CrossRef]
17. Hu, M.; Ji, Z.W.; Yan, K.; Guo, Y.; Feng, X.W.; Gong, J.H.; Zhao, X. Detecting Anomalies in Time Series Data via a Meta-Feature Based Approach. *IEEE Access* **2018**, *6*, 27760–27776. [CrossRef]
18. Liu, D.; Zhao, Y.; Xu, H.; Sun, Y.; Pei, D.; Luo, J.; Jing, X.; Feng, M. Opprentice: Towards practical and automatic anomaly detection through machine learning. In Proceedings of the Internet Measurement Conference AMC, Tokyo, Japan, 28–30 October 2015.

19. Kumar, P.H.; Patil, S.B.; Sandya, H.B. Feature extraction, classification and forecasting of time series signal using fuzzy and garch techniques. In Proceedings of the National Conference on Challenges in Research & Technology in the Coming Decades National Conference on Challenges in Research & Technology in the Coming Decades (CRT 2013) IET, Ujire, India, 27–28 September 2013.

20. Amraee, S.; Vafaei, A.; Jamshidi, K.; Adibi, P. Abnormal event detection in crowded scenes using one-class SVM. *Signal Image Video Proc.* **2018**, *12*, 1115–1123. [CrossRef]

21. Li, Z.C.; Zhitang, L.; Bin, L. Anomaly detection system based on principal component analysis and support vector machine. *Wuhan Univ. J. Nat. Sci.* **2006**, *11*, 1769–1772.

22. Dong, X.F.; Lian, Y.; Liu, Y.J. Small and multi-peak nonlinear time series forecasting using a hybrid back propagation neural network. *Inf. Sci.* **2018**, *424*, 39–54. [CrossRef]

23. Maren, A.J.; Harston, C.T.; Pap, R.M. *Handbook of Neural Computing Applications*; Academic Press: San Diego, CA, USA, 1990.

24. Hagan, M.T.; Beale, M.H.; Demuth, H.B. *Neural Network Design*; PWS Pub: Boston, MA, USA, 1996.

25. Livieris, I. Improving the Classification Efficiency of an ANN Utilizing a New Training Methodology. *Informatics* **2018**, *6*, 1. [CrossRef]

26. Livieris, I.; Pintelas, P.E. *A Survey on Algorithms for Training Artificial Neural Networks*; Technical Report TR08-01; Department of Math, University of Patras: Patras, Greece, 2008.

27. Livieris, I.; Kiriakidou, N.; Kanavos, A.; Tampakas, V.; Pintelas, P. On Ensemble SSL Algorithms for Credit Scoring Problem. *Informatics* **2018**, *5*, 40. [CrossRef]

28. Powers, D. Evaluation: From Precision, Recall and F-Measure to ROC, Informedness, Markedness & Correlation. *J. Mach. Learn. Technol.* **2011**, *2*, 37–63.

symmetry

MDPI

Article

Adaptive Edge Preserving Weighted Mean Filter for Removing Random-Valued Impulse Noise

Nasar Iqbal [1], Sadiq Ali [1], Imran Khan [1] and Byung Moo Lee [2],*

[1] Department of Electrical Engineering, University of Engineering and Technology, P.O. Box. 814, Peshawar 25120, Pakistan; nasar@uetpeshawar.edu.pk (N.I.); sadiqali@uetpeshawar.edu.pk (S.A.); imran_khan@uetpeshawar.edu.pk (I.K.)
[2] School of Intelligent Mechatronics Engineering, Sejong University, Seoul 05006, Korea
* Correspondence: blee@sejong.ac.kr

Received: 23 January 2019; Accepted: 14 March 2019; Published: 18 March 2019

Abstract: This paper proposes an adaptive noise detector and a new weighted mean filter to remove random-valued impulse noise from the images. Unlike other noise detectors, the proposed detector computes a new and adaptive threshold for each pixel. The detection accuracy is further improved by employing edge identification stage to ensure that the edge pixels are not incorrectly detected as noisy pixels. Thus, preserving the edges avoids faulty detection of noise. In the filtering stage, a new weighted mean filter is designed to filter only those pixels which are identified as noisy in the first stage. Different from other filters, the proposed filter divides the pixels into clusters of noisy and clean pixels and thus takes into only clean pixels to find the replacement of the noisy pixel. Simulation results show that the proposed method outperforms state-of-the-art noise detection methods in suppressing random valued impulse noise.

Keywords: adaptive threshold; clustering; edge preserving; noise detector; random value impulse noise; weighted mean filter

---

1. Introduction

For image analysis, de-noising is an important pre-processing step. Digital images are oftentimes corrupted by impulse noise during acquisition, transmission and impaired camera sensors [1], which degrades image features such as edges, sharpness, depth, etc. These degradations can severely hamper the performance of some post-processing steps such as image segmentation, edge detection, feature extraction, target detection, recognition and classification. Therefore, the removal of impulse noise from these images is one of the most fundamental problems in the field of digital image processing. It refers to the removal of noise from corrupted images and the preservation of useful information such as edges and discontinuities. Unlike Gaussian noise, Impulse noise does not corrupt every pixel in the image. It corrupts only certain number of pixels based on noise density. Impulse noise is of two types: Salt & Pepper Noise (SPN) and Random Valued Impulse Noise (RVIN) [2]. SPN corrupts image with two fixed extreme Values while RVIN takes any arbitrary value in range $[n_{min}, n_{max}]$. For 8-bit images, $n_{min} = 0$ and $n_{max} = 255$. Therefore, detection of RVIN is much more complex and difficult as compare to SPN [3]. For simplicity, let $x_{i,j}$ and $c_{i,j}$ be the intensities of pixels of noisy and original images, respectively. Then, the noise model for RVIN [4] can be described as:

$$x_{i,j} = \begin{cases} c_{i,j}, & \text{for } p = 1 - p_0, \\ n_{i,j}, & \text{for } p = p_0, \end{cases} \qquad (1)$$

where $n_{i,j}$ shows gray level of uniformly distributed noise in range $[n_{min}, n_{max}]$ while p_0 is the probability of noise.

To suppress impulse noise, different filters have been used. However, most of these filters have a disadvantage of losing some important information such as edges [5]. These are simple low pass filters where, for every pixel, some operation is performed on all the neighboring pixels in the predefined window and the result replaces the central pixel. The simplest filter to remove impulse noise is median filter (MF) [6]. This filter replaces every pixel in the image by the median value of the given window, irrespective of whether the pixel is noisy or non-noisy. Though MF is computationally simple and less expensive, but as the clean pixels are also replaced by median value of the selected window, this degrades the quality of image, and also produces edge jitters and streaking. Similarly, some patches are also created at higher noise density [7].

To make a trade-off between edges preservation and complexity, multiple variants of MF [8] have been proposed. Some proposed variants are Tri-State Median filter (TSMF) [9], Recursive Weighted Median Filter (RWMF) [10], the Multi-state Median Filter [11] the Central Weighted Median Filter (CWMF) [12], the Rank-Order Mean Filter and the Stack Filter [13]. However, these filters still degrade the quality of images as they replace all pixels in the image without considering whether the test pixel is noisy or not [14]. To overcome this drawback, several filtering schemes have been developed which are integrated with noise detectors. In these techniques, image de-noising is performed in two stages. In the first stage, the noisy pixels are detected by approximation from neighborhood pixels. In the second stage, only those pixels which are detected as noisy are cleaned while remaining pixels are left unchanged. The performance of detection stage heavily depends on the estimation of proper threshold value. A prior threshold is used by most of the techniques, where the predefined threshold is used throughout the image. Some notable examples of the filtering techniques which use a fixed threshold for noise detection are: adaptive center-weighted median (ACWM) filter [15], the optimal direction median filter [2], the two-pass switching rank-ordered arithmetic mean (TSRAM) [16] filter, Progressive Switching Median Filter (PSMF) [17], Luo-Iterative Median Filter (Luo) [18], Directional Weighted Median (DWM) filter [19], the contrast Enhancement-based Filter (CEF) [20] and robust outlyingness ratio non-local mean (ROR-NLM) [21], etc. The ACWM supresses the RVIN by using the difference between the output of CWM and the current pixel. In CEF, exponential function is used to enlarge the differences, which are then summed to detect the noisy pixels. The DWM only considers pixels along four certain directions to detect noise. Once the noise is detected, the noisy pixel is replaced by weighted median value in the optimal direction. In ROR-NLM [21], the weights of non-local means filter (NLM) [22] were used as noise detectors on initial denoised image. These filters usually perform well at lower noise density. However, at higher noise density, these filters have limited performance in terms of image details and edges preservation. This is because the statistics of the image are not uniform for the whole image, so the threshold value suitable for specific region may not be adapted well to other region in same image. In addition, the threshold value adjusted for a particular image may not work well for other image.

In order to reduce the shortcomings of the fixed threshold scheme, an Adaptive Switching Median Filter (ASWM) [23] was proposed. This filter automatically defines a new threshold for every pixel based on the local standard deviation, and does not require a priori knowledge for the threshold selection. Though this filter provides better detection results, when the noise is not too much high. However, due to random noise, some intensity values of noisy pixels may be very different from other neighbor pixels inside a current sliding window—in which case, the standard deviation of current window is very large, which can result a very large threshold. Thus, the higher standard deviation can cause inaccurate calculation of the local threshold. Secondly, the detected noisy pixel is replaced with simple median value of the whole current window. As the current sliding window also contains noisy pixels, the median value of such noisy window is thus not always accurate. This leads to incorrect replacement of a noisy pixel. Thus, the image details are not preserved. In order to overcome the shortcomings of this technique, we propose an adaptive two staged technique, where, for each pixel processing, a more accurate threshold is calculated. Hence, the main contributions of our proposed technique are:

1. We propose an adaptive noise detector and a new weighted mean filter to remove random valued impulse noise from the images.
2. In contrast to other state-of-the-art noise detectors, the proposed detector computes a novel adaptive threshold for each pixel. The accuracy of the detection scheme is further enhanced by employing additional steps to avoid faulty detection of edge pixels, which are noisy pixels. This step indeed ensures the preservation of the edges avoiding faulty detection of noise.
3. In the filtering stage, a novel weighted mean filter is designed to filter only those pixels which are identified as noisy in the detection stage. In contrast to the existing filtering techniques, the proposed filter first divides the pixels into clusters of clean and noisy pixels. Then, to design the filter, only clean pixels are utilized to find the replacement of the noisy pixel.

Extensive simulation results are presented that show that the proposed RVIN suppression mechanism is superior compared to the state-of-the-art schemes.

2. Proposed Noise Detection Scheme

In this section, we present the proposed efficient noise detector for the detection of noisy pixels. Before the detection process, window size is adjusted for the whole image. The window size is selected adaptively as it is set automatically based on the noise density. To estimate the noise density, we adopt the idea of the universal threshold [24]. In this method, wavelet transform is used to estimate the noise density σ of the noisy signal using Robust Median Estimator as:

$$\sigma = \frac{\text{median}(HF)}{0.6745},$$ (2)

where HF is the high frequency sub-band. In this technique, the noisy image or signal is divided into sub-bands of low and high frequencies. The edges, sharp details and sudden changes are found in the high frequencies [25]. Thus, mostly, the high frequency components contain noise. Therefore, only a high frequency sub-band (HF) is used to estimate the noise of the signal, as illustrated in Equation (2). For example, if σ comes to be 50, the image has 50% noise and so on. The size of the window depends on σ, such that, for low σ, a smaller window is selected, while a larger window is selected for high σ.

Once the window is adjusted, then the detection process starts. Figure 1 illustrates the block diagram of the proposed noise detector. Each block given in Figure 1 is then elaborated with example in the following subsections.

2.1. Calculation of Separating Threshold

Once the window size is defined, the first step of the proposed noise detection process is the selection of separating threshold γ_s. The algorithm to find separating threshold is illustrated with the help of example in Figure 2 and elaborated in the following steps.

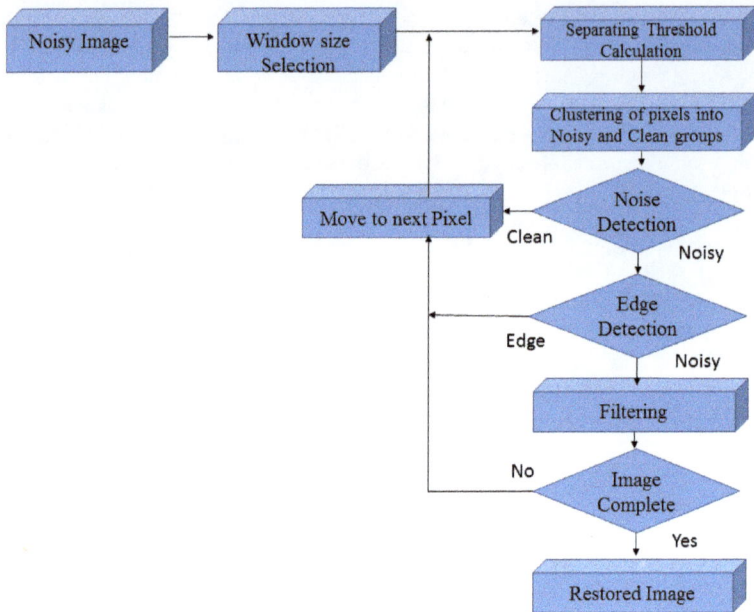

Figure 1. Proposed noise detection scheme.

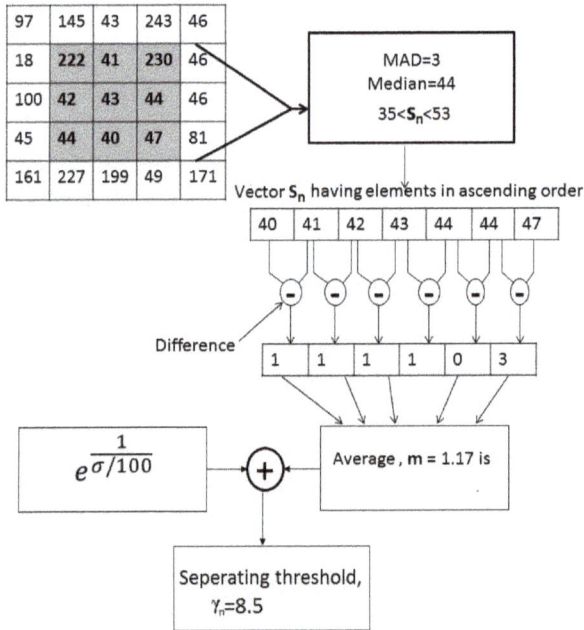

Figure 2. Separating threshold calculation.

1. Let $x_{i,j}$ be the pixel under observation at location (i, j) to know whether it is noise or clean, then it is considered as a central pixel. Window of the size found in Equation (2) is set around the central pixel.

2. The corresponding pixel values of of the current window are sorted in ascending order in vector **s**. The outliers are removed from **s** by adopting the "three sigma rule" with median absolute deviation (MAD) [21]. The MAD for **s** is found as:

$$\text{MAD} = \text{median} \mid \mathbf{s} - \text{median}(\mathbf{s}). \mid \tag{3}$$

Vector **s** is searched, and, if the difference between any value in **s** and the median value is greater than $3 \times \text{MAD}$, it is considered as an outlier and is not taken into account for further calculation. After removal of outliers, let $\mathbf{s_n}$ be the resultant vector, where n denotes the size of $\mathbf{s_n}$.

3. Compute the absolute difference d_k between every two adjacent pixels in $\mathbf{s_n}$ as:

$$d_k = \mid s_k - s_{k+1} \mid, \tag{4}$$

where $k = 1, 2, \ldots, n - 1$.

4. The separating threshold γ_s is found as:

$$\gamma_s = m + e^{\frac{1}{\sigma/100}}, \tag{5}$$

where

$$m = \frac{1}{n-1} \sum_{k=1}^{n-1} d(k). \tag{6}$$

We can observe that the term $e^{\frac{1}{\sigma/100}}$ depends upon the noise density σ. It can be observed that, for lower noise density, this term is large and, for higher noise density, this term is small. This term adjusts the threshold to ensure that clean pixels are not considered noisy pixels at very low noise density. Similarly, it also adjusts the threshold to make sure noisy pixels are not grouped into clean pixels, as this threshold will be used in grouping discussed in the following subsection.

2.2. Separation of Pixels into Clusters

Using the separating threshold γ_s, in this section, we divide the pixels in the sorted vector $\mathbf{s_n}$ in groups as follows:

1. Initialize from the first pixel in sorted vector $\mathbf{s_n}$. If the difference between current pixel s_i and next pixel s_{i+1} is less than or equal to γ_s, keep the pixel s_{i+1} in the current group together with s_i.

2. However, if the difference is greater than γ_s, finish the current group and start a new group, such that the first element of the new group will be s_{i+1}. Apply the same procedure until the last element of vector $\mathbf{s_n}$.

3. The clean and noisy groups are decided based on the size of the group. The first two largest clusters are considered as clean clusters while all other clusters are considered as noisy and are discarded. The clean clusters are denoted as $\mathbf{g_1}$ and $\mathbf{g_2}$. Cluster $\mathbf{g_1}$ is used in the detection phase while $\mathbf{g_1}$ and $\mathbf{g_2}$ are used in filtering.

2.3. Noise Detection

Using the results of the group stage, in this section, the noise detection stage is presented. The noise detector is:

$$x_{i,j} = \begin{cases} \text{Clean,} & \text{if } \mid x_{i,j} - \mu \mid \leq \gamma, \\ \text{Noisy,} & \text{Otherwise,} \end{cases} \tag{7}$$

where threshold γ is

$$\gamma = 3 \times \text{MAD}(g_1), \tag{8}$$

and g_1 is the largest group discussed in the grouping stage.

Similarly, $x_{i,j}$ is the central pixel and μ is the mean value of g_1. If the $x_{i,j}$ is detected as noisy, filtering is required to replace the $x_{i,j}$ with the filtered output. On the other hand, if $x_{i,j}$ is a clean pixel, it is left unchanged.

2.4. Edge Pixel Identification

In this section, we present a mechanism to avoid the faulty detection of edge pixels as noisy pixels. This step is performed only when the pixel is identified as a noisy pixel. This will make the detection scheme more robust by verification of the noisy pixels to avoid situations where the edge pixel is wrongly considered as a noisy pixel. This process is performed as follows:

1. Keeping $x_{i,j}$ as a reference pixel, pixels present along four directions are taken into account. These four directions θ_A, θ_B, θ_C and θ_D are illustrated in Figure 3.
2. θ'_A contains the pixels from the direction θ_A excluding $x_{i,j}$. Similarly, θ'_B, θ'_C and θ'_D contain the pixels from the directions θ_B, θ_C and θ_D, respectively, excluding $x_{i,j}$.
3. Find the absolute difference between central pixel $x_{i,j}$ with all pixels along direction θ'_A. The difference vector is defined as $d^A = \mid x_{i,j} - \theta'_A \mid$.
4. d^A is searched to find the values less than or equal to m. Indexes of pixels corresponding to these values are identified in θ_A. These pixels are stacked in vector **v**. The vector **v** is considered to be an edge if total number of pixels in **v** is at least 50% of direction θ_A. Step 4 is only performed if this condition holds true. Otherwise, discard the **v** and go to step 5.
5. Take the standard deviation of central pixel $x_{i,j}$ with **v**. The pixel is identified as an edge if standard deviation of **e** and $x_{i,j}$ is less than or equal to threshold γ:

$$x_{i,j} = \begin{cases} \text{Edge,} & \text{if std}(\mathbf{v}, x_{i,j}) \leq \gamma, \\ \text{Go to next Direction,} & \text{Otherwise.} \end{cases} \tag{9}$$

6. If $x_{i,j}$ is not identified an edge in direction θ_A, then the three other directions θ_B, θ_C and θ_D are checked for an edge identification by applying the five steps above on each direction.

If $x_{i,j}$ is not identified an edge in all four directions, it is considered a noisy pixel and filtering is required to de-noise it.

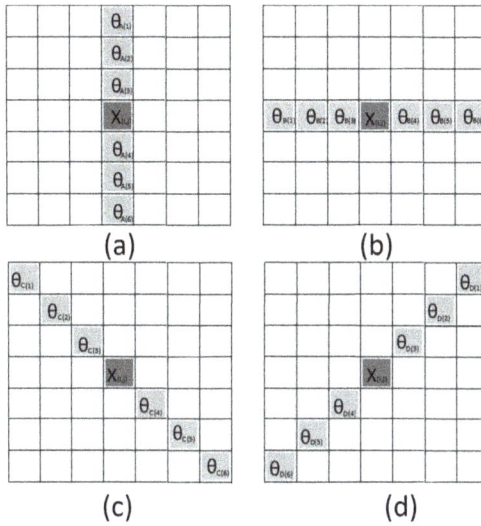

Figure 3. Edges in different directions (a) θ_A; (b) θ_B; (c) θ_C; (d) θ_D.

3. Illustration

This section illustrates the noise detection of the proposed method with example. Consider the image patch given in Figure 4 that is taken from a 40% noisy "lena" image. The size of the window depends on the noisy density σ such that:

$$\text{Size of window} = \begin{cases} 7 \times 7, & \text{if } \sigma \geq 60, \\ 5 \times 5, & \text{otherwise.} \end{cases} \qquad (10)$$

To check whether the specified central pixel is noisy, the following steps are performed.

1. Using Equation (2), the noise density of the whole image is calculated. As the noisy density is 40%, therefore, according to Equation (10), the window size is selected 5×5.
2. Sort all the pixels inside the selected patch in ascending order as x = [17 64 64 65 65 65 65 66 66 67 68 68 69 69 83 84 112 112 133 143 152 199 227 236 252].
3. The extreme outliers are removed by using MAD given in Equation (3). The MAD value of vector **x** is 5, while the median value of **x** is 69. Therefore, according to the "Three Sigma Rule", any value below 54 (69 − 15) and above 84 (69 + 15) is discarded and the vector **x** becomes [64 64 65 65 65 65 66 66 67 68 68 69 69 83 84].
4. Next, the deference between every two consecutive pixels is found to be **d** = [0 0 0 0 1 0 1 1 0 1 0 14 2].
5. The average value m of the vector **d** is equal to 1.4. Putting this value of m in Equation (5), γ_s becomes 13.59.
6. Using the separating threshold, all pixels of **x** are divided into groups as [64 64 65 65 65 65 66 66 67 68 68 69 69], [83 84].
7. The largest group is [64 64 65 65 65 65 66 66 67 68 68 69 69], whose size is much greater compared to other group. Thus, this group is considered the clean group, while the other group is considered the noisy group.
8. Using Equation (8), threshold γ becomes 3. Similarly, mean value becomes μ = 66.
9. By employing Equation(9), any value greater than 69 and less than 63 is considered to be noisy.

In this example, the value of the central pixel is 67; therefore, it is a clean pixel and does not need filtering. In case the pixel comes out to the noisy pixel, then we need filtering to replace the noisy pixel with estimated clean pixel. In the next section, we present the filtering scheme.

Figure 4. A patch from Lena image: (**a**) original image; (**b**) 40% noisy image.

4. Proposed Filtering Scheme

After the detection of the noise, the appropriate filter is required to replace the noisy pixel. For this, we design a novel Weighted Mean Filter (WMF), which is more robust compared to the existing median and mean filters. In a smooth region, the absolute difference among all neighbor pixels is very small and the whole region is flat [26], so, in most cases, only one large clean group is formed. On the other hand, the region that contains the edge is not as flat, in which case more than one cluster is required to replace the noisy pixel. Therefore, in the proposed filtering scheme, the second largest group g_2 is also considered. We remark that the g_2 is not required in detection stage. Instead, there is a separate section for edge detection. The two largest groups are assigned weights based on standard deviation and size of the group. Supposing that $x_{i,j}$ is the noisy pixel at location (i, j), then the filtered pixel $\hat{x}_{i,j}$ is given as:

$$\hat{x}_{i,j} = \text{mean}\{w_1 \diamond g_1, w_2 \diamond g_2\},\tag{11}$$

where $\diamond$ is the replication operator [27] that shows the repetition of any term. For example, if w_1 is 2, and w_2 is 1, then g_1 is repeated two times while g_2 is repeated 1 time, such as $\hat{x}_{i,j} = \text{mean}\{g_1, g_1, g_2\}$, as the size of g_1 is greater than g_2. Therefore, to decide the weights, we also consider the standard deviation of both groups. The weights w_1 and w_2 are determined by a ratio R as:

$$R = \frac{s_2}{s_1} \times \frac{sd_1}{sd_2},\tag{12}$$

where s_1, s_2 represents the size, while sd_1 and sd_2 represent the standard deviation of g_1 and g_2, respectively. This relation defines the weights of g_2 with respect to g_1. As the higher standard deviation means less correlation, it is inversely proportional to the R, while the larger size shows larger priority; therefore, it is in direct relation to R. Based on the value of R, the following weights are assigned in Equation (11).

1. If the size of g_1 is very large compared to the size of g_2 or $R \leq 0.5$, then $w_1 = 1$, $w_2 = 0$.
2. Similarly, when $0.5 \leq R \leq 1$, the assigned weights are: $w_1 = 1$, $w_2 = 1$.
3. Finally, for $1 \leq R \leq 2$, the assigned weights are: $w_1 = 1$, $w_2 = 2$.

The filtering is further improved by only considering the closest neighbor pixels of the central pixel from the selected groups, such that, if there are at least three pixels from the selected groups which are in a 3 × 3 window around the central pixel, then take the mean value of only these closest pixels.

5. Summary of the Proposed Denoising Scheme

Although the proposed scheme gives good results in single iterations, with second iterations, it gives the best performance and there is no need to apply further iterations as in other iterative methods. The larger threshold value is used in the first iteration while a smaller threshold value is used in the second iteration [2]. Therefore, if the threshold in the first iteration is $3 \times MAD$, in the second iteration, the threshold is set to $1.5 \times MAD$. We summarize our proposed technique in Algorithm 1 as:

Algorithm 1: Proposed algorithm for random valued impulse noise removal.

    Input :The noisy image X, with size of $L_1 \times L_2$

1 Set iteration $i=1$.
2 Outer loop: for $l_1 = 1, ...L_1$, Inner loop: for $l_2 = 1, ...L_2$
3 Set a sliding window centered at $x_{i,j}$, the size of window is determined by σ.
4 Divide the window into groups of clean and noisy pixels using separating threshold computed in Equation (5).
5 Find mean μ and threshold γ from clean cluster, while discard the noisy clusters.
6 Using Equation (7) check $x_{i,j}$ for noise. If $| x_{i,j} - \mu | \le \gamma$, go to step (2).
7 If the pixel is identified as noisy, check $x_{i,j}$ for an edge using Equation (9). Such that if $std(\mathbf{e}, x_{i,j}) \le \gamma$, go to step (2).
8 Filter $x_{i,j}$ as in Equation (11).
9 $i=i+1$, if $i > i_{max}$, stop iteration, else X $\longleftarrow$ Y, go to next iteration.
    Output: De-noised image Y

---

6. Experimental Results

To assess the capability the proposed noise removal procedure, in this section, comprehensive numerical results are compared with other state-of-the-art techniques.

6.1. Comparison of Noise Detection

We compare our proposed noise detection scheme with different state-of-the-art techniques such as (ACWM) [15], the (DWM) [19], (TSM) [9] (PSMF [17], (Luo) [18], (ROR-NLM) [21], SBF [28] and (ASWM) [23].

In Table 1, the noise detection results of noisy "lena" image with noise densities 40%, 50% and 60% are given. The table lists three results of the noise detection: the number of undetected noisy pixels (False Negative), the number of clean pixels that are falsely detected as noisy (False Positive) and the Total number (False Negative + False Positive).

From Table 1, it is clear that the proposed detection scheme has generated less 'Total' number than all of the above techniques especially at higher noise density. At lower noise density, only ROR-NLM produces a lower total number, but they produce much higher false negative terms, indicating that they miss most of the noisy pixels, and most of noise patches are left unchanged. Thus, the proposed detection technique shows superiority over other detection schemes. It is because a noise detector is considered to be efficient and robust, if it reduces both false positive and false negative numbers. When the noise density increases, the superiority of the proposed method is clear as it generates far less 'Total' terms than all other methods. This means that our technique is more efficient and robust, even at high noise density.

Table 1. Detection comparison of different filters at different noise densities.

Methods	40%			50%			60%		
	FN	FP	Total	FN	FP	Total	FN	FP	Total
ACWM	14,590	2367	16,857	21,897	3706	25,603	30,198	6526	36,724
Luo's	14,679	1831	16,510	21,665	3068	24,733	33,987	2780	36,767
TSM	18,921	5201	24,122	23,921	6218	30,139	28,123	8903	37,026
PSM	18,672	4982	23,654	23,762	6123	29,885	27,987	8393	36,380
ASWM	7489	11,564	19,053	11,779	12,786	24,565	19,982	16,482	36,464
DWM	11,786	8931	20,717	15,321	8756	24,077	15,728	14,816	30,544
SD-OOD	13,500	10,675	24,175	11,987	15,827	27,814	17,821	18,261	36,082
ROR-NLM	12,890	3328	16,218	15,297	3487	18,784	21,827	7808	29,635
Propoposed	10,908	7973	18,881	11,668	9613	21,241	13,571	9760	23,331

6.2. Comparison of Image Restoration

The de-noising performance of the proposed method is tested on different standard images such as Lena, pepper, Bridge, and Boat with different noise densities. All these images are 8-bit gray scale and have the same size of 512 × 512 pixels. It should be noted that the size of the window is defined only once based on the noise density using Equation (2), which is used on all pixels of the current image. In our case, for noise density less than 60%, window size of 5 × 5 is used, while window size of 7 × 7 is used otherwise. Peak Signal to Noise Ratio (PSNR) is used as quantitative measurement to compare the proposed method with different well known techniques. The PSNR is defined as:

$$\text{PSNR (dB)} = 10\log_{10}\left(\frac{255^2}{\text{MSE}}\right), \tag{13}$$

where MSE is mean squared error and is defined as:

$$\text{MSE} = \frac{1}{MN}\sum_{i=1}^{m}\sum_{j=1}^{n}\left(O_{i,j} - R_{i,j}\right)^2, \tag{14}$$

where O is the original clean image while R is the output restored image. Tables 2 and 3 illustrate comparison of the proposed method with other state-of-the-art methods.

Table 2. Peak Signal-to-Noise Ratio (dB) values of different filters for Lena and Pepper image corrupted by random valued impulse noise of different noise densities.

Methods	Lena			Pepper		
(Noise Density)	40%	50%	60%	40%	50%	60%
TSM	24.91	22.22	18.98	24.02	22.56	17.78
ACWM	28.12	24.64	20.32	28.12	26.09	21.52
LOU'S	29.62	25.82	22.69	28.41	26.33	23.89
PSM	29.12	25.31	24.23	28.23	25.41	23.83
ASWM	30.91	28.42	26.25	29.02	27.12	25.33
DWM	30.92	28.91	26.51	29.13	27.79	25.48
SBF	30.12	27.12	23.13	28.89	27.02	24.67
ROR-NLM	31.42	29.21	25.61	29.61	27.89	25.45
PROPOSED	31.77	30.01	28.03	29.75	28.11	26.62

Tables 2 and 3 list the PSNR values of different test images that are contaminated by RVIN of noise densities ranges from 40% to 60%. From both tables, it is clear that our method has performed better as compared to all other listed methods in terms of PSNR. Figures 5 and 6 show de-noised results of Lena and Pepper images, respectively. From these figures, it is obvious that TSM, ACWM, PSM and Luo have created artifacts and damaged image details. While DWM, ASWM and ROR-NLM have performed better at lower noise densities, but their performance is degraded at higher noise densities. This is due to the fact that these filters replace the noisy pixel by considering all the window pixels. On the other hand, the proposed method only considers clean pixels to replace the noisy pixel. As the noise density increases, the number of corrupted pixels in the window are also increased, which increases the error when the central noisy pixel is replaced by mean or median value of noisy neighbor pixels. Finally, in Figure 7, we present performance results of different filters in restoring Lena image corrupted by different noise densities. These results confirm the results given in Tables 2 and 3 and clearly show that the proposed method outperforms other state of the art techniques.

Table 3. Peak Signal -to- Noise Ratio (dB) values of different filters for Bridge and Boat image corrupted by random valued impulse noise of different noise densities.

Methods	Bridge			Boat		
(Noise Density)	40%	50%	60%	40%	50%	60%
TSM	21.55	19.72	17.26	22.89	20.72	17.55
ACWM	23.72	22.19	19.12	25.32	23.56	21.45
LOU'S	23.84	22.78	19.17	25.45	23.78	21.61
PSM	23.65	21.91	19.39	24.98	22.98	20.87
ASWM	24.02	22.69	21.12	26.76	25.26	23.23
DWM	24.07	22.72	21.19	26.83	25.31	23.57
SBF	22.12	21.31	20.15	25.33	24.88	22.67
ROR-NLM	24.27	22.91	21.21	27.23	25.43	24.21
PROPOSED	24.35	23.08	21.75	27.85	26.61	24.87

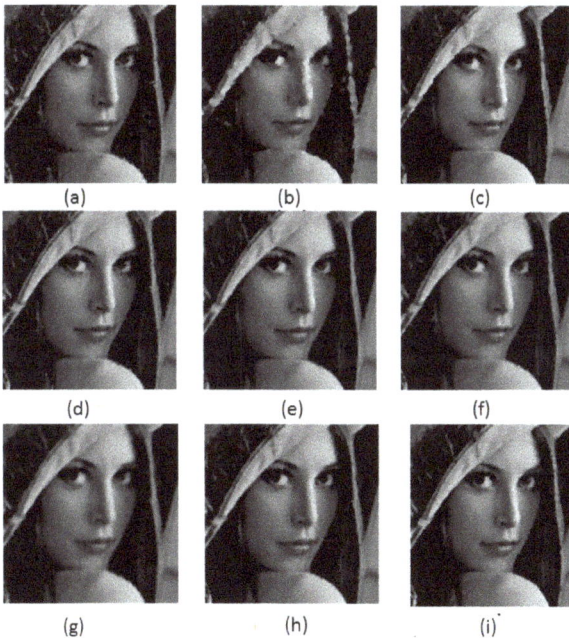

Figure 5. Results of different filters in restoring 40% corrupted Lena image: (**a**) adaptive switching median filter (ASWM); (**b**) tri-state median filter (TSM); (**c**) progressive switching median filter (PSM); (**d**) luo-iterative median filter (Luo); (**e**) directional weighted median filter (DWM); (**f**) Adaptive Switching Median Filter (ASWM); (**g**) Switching Bilateral Filter (SBF); (**h**) Robust Outlyingness Ratio Non-Local Mean (ROR-NLM); (**i**) proposed method.

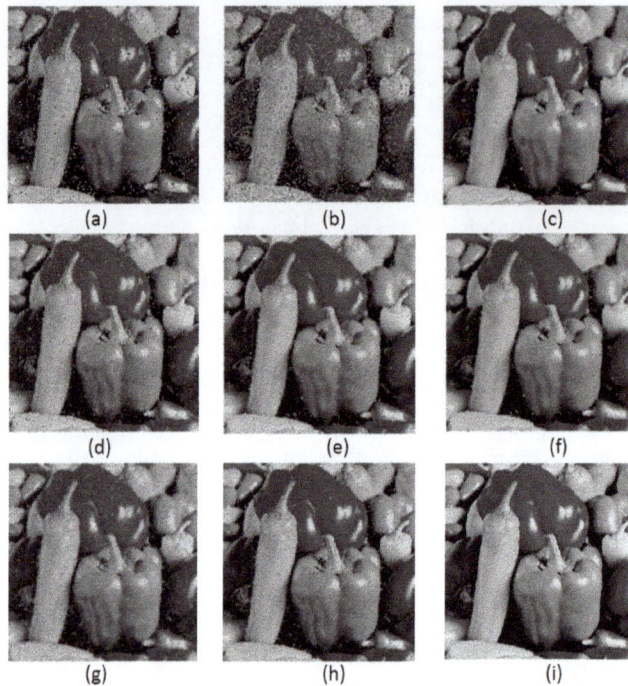

Figure 6. Results of different filters in restoring 60% corrupted pepper image: (**a**) adaptive switching median filter (ASWM); (**b**) tri-state median filter (TSM); (**c**) progressive switching median filter (PSM); (**d**) luo-iterative median filter (Luo); (**e**) directional weighted median filter (DWM); (**f**) adaptive switching median filter (ASWM); (**g**) switching bilateral filter (SBF); (**h**) robust outlyingness ratio non-local mean (ROR-NLM); (**i**) proposed method.

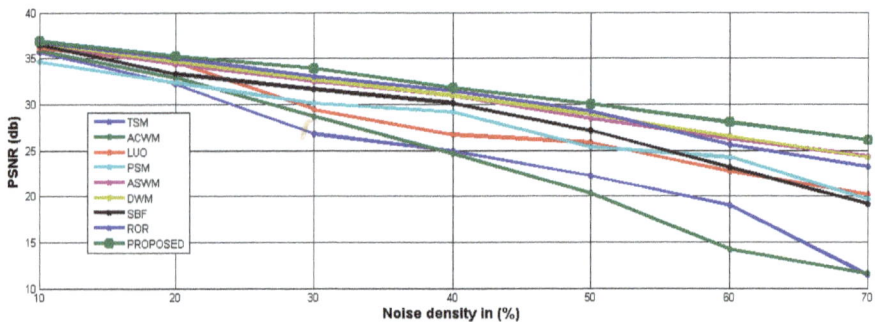

Figure 7. Results of different filters in restoring Lena image corrupted by different noise levels.

7. Conclusions

In this paper, a new algorithm is presented to remove random valued impulse noise. The proposed approach is based on calculation of new and adaptive thresholds for every pixel. The pixels are divided into noisy and clean clusters. Then, only a clean cluster is used to calculate the threshold. A new edge identification step is proposed to ensure that a clean pixel is not considered as a noisy pixel falsely. If the pixel is identified as noisy, a novel weighted mean filter is designed to filter the noisy pixel. Instead of taking the mean of all pixels in the selected window, the proposed filter considers only clean pixels to replace the noisy pixel. Simulation results show that the proposed method outperforms state-of-the-art de-noising techniques, both visually and quantitatively (PSNR), even when the noise density is high.

Author Contributions: Conceptualization, S.A., I.K. and B.M.L.; Data curation, N.I. and I.K.; Formal analysis, S.A., I.K. and N.I.; Funding acquisition, I.K. and B.M.L.; Investigation, S.A., I.K. and N.I.; Methodology, S.A., I.K., B.M.L. and N.I.; Project administration, N.I. and I.K.; Software, S.A., I.K. and N.I.; Writing—original draft, S.A., I.K. and B.M.L.; Writing—review & editing, S.A., I.K., B.M.L. and N.I.

Funding: This work was supported by the Basic Science Research Program through the National Research Foundation of Korea (NRF) funded by the Ministry of Education (Grant No.: NRF-2017R1D1A1B03028350).

Conflicts of Interest: The authors declare no conflict of interest.

References

1. Gao, G.; Liu, Y.; Labate, D. A two-stage shearlet-based approach for the removal of random-valued impulse noise in images. *J. Vis. Commun. Image Represent.* **2015**, *32*, 83–94. [CrossRef]
2. Awad, A.S. Standard deviation for obtaining the optimal direction in the removal of impulse noise. *IEEE Signal Process. Lett.* **2011**, 18, 407–410. [CrossRef]
3. Ilke, T. A new method to remove random-valued impulse noise in images. *AEU-Int. J. Electron. Commun.* **2013**, *67*, 771–779
4. Chan, R.H.; Hu, C.; Nikolova, M. An iterative procedure for removing random-valued impulse noise. *IEEE Signal Process. Lett.* **2004**, *11*, 921–924. [CrossRef]
5. Jayasree, S.; Bodduna, K.; Pattnaik, P.K.; Siddavatam, R. An expeditious cum efficient algorithm for salt-and-pepper noise removal and edge-detail preservation using cardinal spline interpolation. *J. Vis. Commun. Image Represent.* **2014**, *25*, 1349–1365. [CrossRef]
6. Gonzalez, R.C.; Woods, R.E. *Digital Image Processing*; Prentice Hall: Upper Saddle River, NJ, USA, 2002.
7. Astola, J.; Kuosmaneen, P. *Fundamentals of Nonlinear Digital Filtering*; CRC: BocaRaton, FL, USA, 1997.
8. Singh, N.; Thilagavathy, T.; Lakshmipriya, R.T.; Umamaheswari, O. Some studies on detection and filtering algorithms for the removal of random valued impulse noise. *IET Image Process.* **2017**, *11*, 953–963 [CrossRef]
9. Chen, T.; Ma, K.; Chen, L. Tri-state median filter for image denoising. *IEEE Trans. Image Process.* **1999**, *8*, 1834–1838. [CrossRef] [PubMed]
10. Arce, G.R.; Paredes, J.L. Recursive weighted median filters admitting negative weights and their optimization. *IEEE Trans. Signal Process.* **2000**, *48*, 768–779. [CrossRef]
11. Ari, N.; Heinonen, P.; Neuvo, Y. A new class of detail-preserving filters for image processing. *IEEE Trans. Pattern Anal. Mach. Intell.* **1987**, *1*, 74–90.
12. Ko, S.-J.; Lee, Y.H. Center weighted median filters and their applications to image enhancement. *IEEE Trans. Circuits Syst.* **1991**, *38*, 984–993. [CrossRef]
13. Coyle, E.J.; Lin, J.; Gabbouj, M. Optimal stack filtering and the estimation and structural approaches to image processing. *IEEE Trans. Acoust. Speech Signal Process.* **1989**, *37*, 2037–2066. [CrossRef]
14. Vikas, G.; Chaurasia, V.; Shandilya, M. Random-valued impulse noise removal using adaptive dual threshold median filter. *J. Vis. Commun. Image Represent.* **2015**, *26*, 296–304.
15. Chen, T.; Hong, R.W. Adaptive impulse detection using center-weighted median filters. *IEEE Signal Process. Lett.* **2001**, *8*, 1–3. [CrossRef]
16. Lin, T.-C. Switching-based filter based on Dempster's combination rule for image processing. *Inf. Sci.* **2010**, *180*, 4892–4908. [CrossRef]
17. Zhou, W.; Zhang, D. Progressive switching median filter for the removal of impulse noise from highly corrupted images. *IEEE Trans. Circuits Syst. II Analog Digit. Signal Process.* **1999**, *46*, 78–80. [CrossRef]
18. Luo, W. A new efficient impulse detection algorithm for the removal of impulse noise. *IEICE Trans. Fund. Electron. Commun. Comput. Sci.* **2005**, *88*, 2579–2586. [CrossRef]
19. Dong, Y.; Xu, S. A new directional weighted median filter for removal of random-valued impulse noise. *IEEE Signal Process. Lett.* **2007**, *14*, 193–196. [CrossRef]
20. Ghanekar, U.; Singh, A.K.; Pandey, R. A contrast enhancement-based filter forremoval of random valued impulse noise. *IEEE Signal Process. Lett.* **2010**, *1*, 47–50.

21. Bo, X.; Yin, Z. A universal denoising framework with a new impulse detector and nonlocal means. *IEEE Trans. Image Process.* **2012**, *21*, 1663–1675.

22. Antoni, B.; Coll, B.; Morel, J.-M. A non-local algorithm for image denoising. In Proceedings of the 2005 IEEE Computer Society Conference on Computer Vision and Pattern Recognition (CVPR'05), San Diego, CA, USA, 20–25 June 2005.

23. Akkoul, S.; Ledee, R.; Leconge, R.; Harba, R. A new adaptive switching median filter. *IEEE Signal Process. Lett.* **2010**, *17*, 587–590. [CrossRef]

24. Johnstone, I.M.; Silverman, B.W. Wavelet threshold estimators for data with correlated noise. *J. R. Stat. Soc. Ser. B (Stat. Methodol.)* **1997**, *59*, 319–351. [CrossRef]

25. Cihan, T.; Akinlar, C. Edge drawing: a combined real-time edge and segment detector. *J. Vis. Commun. Image Represent.* **2012**, *23*, 862–872.

26. Chen, J.; Zhan, Y.; Cao, H.; Wu, X. Adaptive probability filter for removing salt and pepper noises. *IET Image Process.* **2018**, *12*, 863–871. [CrossRef]

27. Bovik, A.C. *Handbook of Image and Video Processing*; Academic Press: Cambridge, MA, USA, 2010.

28. Lin, C.H.; Jia, S.T.; Ching, T.C. Switching bilateral filter with a texture/noise detector for universal noise removal. *IEEE Signal Process. Soc.* **2010**, *19*, 2307–2320.

symmetry

MDPI

Article

Parametric Fault Diagnosis of Analog Circuits Based on a Semi-Supervised Algorithm

Ling Wang [1], Dongfang Zhou [2], Hui Tian [1], Hao Zhang [1] and Wei Zhang [1,*]

[1] College of Mechanical and Electrical Engineering, Henan Agricultural University, Zhengzhou 450002, China; wangling0351@126.com (L.W.); th407@163.com (H.T.); hao.zhang2016@hotmail.com (H.Z.)

[2] Department of Communication, National Digital Switching System Engineering and Technology R&D Center (NDSC), Zhengzhou 450002, China; 13598028188@139.com

* Correspondence: zwz04@163.com; Tel.: +86-371-6355-8040

Received: 10 January 2019; Accepted: 12 February 2019; Published: 14 February 2019

Abstract: The parametric fault diagnosis of analog circuits is very crucial for condition-based maintenance (CBM) in prognosis and health management. In order to improve the diagnostic rate of parametric faults in engineering applications, a semi-supervised machine learning algorithm was used to classify the parametric fault. A lifting wavelet transform was used to extract fault features, a local preserving mapping algorithm was adopted to optimize the Fisher linear discriminant analysis, and a semi-supervised cooperative training algorithm was utilized for fault classification. In the proposed method, the fault values were randomly selected as training samples in a range of parametric fault intervals, for both optimizing the generalization of the model and improving the fault diagnosis rate. Furthermore, after semi-supervised dimensionality reduction and semi-supervised classification were applied, the diagnosis rate was slightly higher than the existing training model by fixing the value of the analyzed component.

Keywords: fault diagnosis; lifting wavelet; local preserving projection; Fisher linear discriminant analysis; semi-supervised random forest

1. Introduction

Analog circuits are extensively used in consumer electronics, industrial systems, and aerospace applications. However, services provided by the analog circuits are severely threatened by parametric faults. Therefore, parametric fault diagnosis and fault location in the analog circuits are now some of the hottest fields. Feature extraction, dimensionality reduction, and selection of classification algorithms are the main research contents of the parametric fault diagnosis in the analog circuits.

Fault feature extraction is the precondition and foundation for the design of subsequent classifiers. Due to tolerance and nonlinearity of electronic components, the original signal overlaps in both the traditional time domain and frequency domain. The fault feature extraction based on signal processing is one of the hot topics, where Hilbert–Huang Transform (HHT) [1], wavelet [2–4], and wavelet packet transform [5] can obtain the time-frequency features for fault diagnosis in analog circuits. Rényi's entropy [6], conditional entropy [4,7] and cross-wavelet singular entropy [8] are used for fault feature extraction, since the entropy can be used to measure the uncertainty and variation of information. In order to reflect the faulty information from different perspectives, the statistical properties of the fractional transform signals are proposed as the fault features [9], for example, distance, mean, standard deviation, skewness, kurtosis, entropy, median, third central moment, and centroid. The modified binary bat algorithm (MBBA) with chaos and Doppler Effect is used to utilize the optimized feature subset [10].

Due to the high dimensionality of fault features and the complexity of classifier, it is necessary to reduce the dimension before inputting the fault feature to the classifier. The current dimensionality reduction methods can be classified into two groups: linear dimensionality reduction and nonlinear dimensionality reduction. For the linear dimensionality reduction, principal component analysis (PCA) [11] and linear discriminant analysis (LDA) [12] are commonly used, where PCA is mainly for maximizing the mutual information between the original high-dimensional data and the post-projection low-dimensional data. The LDA, also called Fisher linear discriminant analysis (FDA), is utilized to obtain the optimal projection vector by maximizing the trace ratio of between-class scatter and within-class scatter.

The key to parametric fault diagnosis is the selection and optimization of the classifier. The classification algorithm, back propagation (BP) neural network (NN) [13], neuromorphic analyzers [14], extreme learning machine (ELM) [15–17], decision tree support vector machine (DTSVM) [18], quantum clustering-based multi-valued quantum fuzzification decision tree (QC-MQFDT) [19], and Gaussian Bernoulli deep belief network (GB-DBN) [20] were used in fault diagnosis of the analog circuits. The average fault diagnosis rate is more than 90% in the fixed training samples and test samples.

All the above mentioned methods were handled for electronic component parameter deviation of ±50%. The fault diagnosis rate of the current methodology achieved by the above given references is very low, where the models were trained by a single fixed parameter. In order to improve the fault diagnosis rate in analog circuits, and improve the generalization ability of the training models, this paper presents a new method that randomly selects component parameters in the range of parametric variation as unlabeled samples, where the representative samples are labeled by experts. Semi-supervised learning (SSL) received significant attention over the past decade from computer vision and machine learning research communities [21,22]. During the dimensionality reduction, both labeled and unlabeled samples are considered for semi-supervised dimensionality reduction. A semi-supervised dimensionality reduction algorithm based on local preserving mapping (LPP) that optimizes FDA is proposed to extract circuit features. Then, the semi-supervised cooperative training algorithm is used to diagnose the fault.

The remainder of this paper is organized as follows. First, we start with outlining the feature extraction method of lifting wavelet in Section 2. Then, the semi-supervised LPP optimization FDA algorithm is introduced in Section 3. This in followed by the semi-supervised random forest algorithm, which is elaborated on in Section 4. Afterward, the framework for analog circuit fault diagnosis, the detailed experimental process, and results analysis are given in Section 5. Finally, the paper is concluded.

2. Lifting Wavelet Transform

The lifting wavelet transform, also known as the second-generation wavelet transform, is used to improve the Laurent polynomial convolution algorithm associated with the Euclidean algorithm [23,24]. The lifting wavelet transform uses simple scalar multiplication to replace the convolution operation of the original wavelet transform. It can simplify the computation, realize the integer wavelet transform, and solve the boundary problem. The lifting scheme divides the transformation process into three phases: split/merge, prediction, and update [25–27].According to the parity, the split stage divides the input signal s_i into two groups, including s_{i-1} and d_{i-1}, where the split function is $F(s_i) = (s_{i-1}, d_{i-1})$;the prediction stage uses the forecast sequence $P(s_{i-1})$ of odd sequence s_{i-1} to predict d_{i-1}, following the value of the residual signal $P(s_{i-1}) - d_{i-1}$ placed with d_{i-1}, i.e., $d_{i-1} = d_{i-1} - P(s_{i-1})$.Then, the decomposition signal carrying decompose information is used to repeat the decomposition and the prediction process. The original signal s_i can be represented as $\{s_n, d_n, \ldots, s_1, d_1\}$. In order to maintain the global characteristics of the signal s_i in the updating stage,$Q(s_{i-1}) = Q(s_i)$, an operator U and d_{i-1} are introduced to update s_{i-1}, i.e., $s_{i-1} = s_{i-1} + U(d_{i-1})$.The reconfiguration process is just the opposite, which is shown in Figure 1.

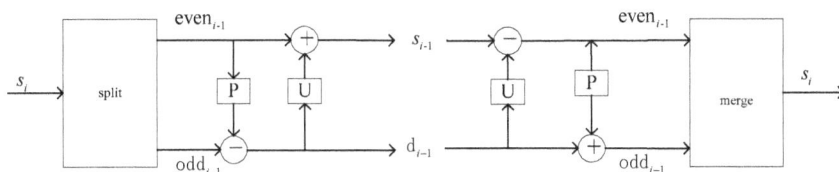

Figure 1. The decomposition and reconstruction of the lifting wavelet transform.

In the lifting algorithm, the split operator F, the prediction operator P, and the update operator U are expressed as follows:

$$F(s_i) = (even_{i-1}, odd_{i-1}),$$
$$d_{i-1} = odd_{i-1} - P(even_{i-1}), \qquad (1)$$
$$s_{i-1} = even_{i-1} + U(odd_{i-1}).$$

There steps of the reconstruction process of lifting transformation, i.e., restore update, restore prediction, and merge, are given as follows:

$$even_{i-1} = s_{i-1} - U(odd_{i-1}),$$
$$odd_{i-1} = d_{i-1} + P(even_{i-1}), \qquad (2)$$
$$s_i = merge(even_{i-1}, odd_{i-1}).$$

In this model, different prediction operators P and update operators U can be used to construct the required wavelet functions, for example, the Harr wavelet, db2 wavelet, and so on.

3. Local Fisher Discriminant Analysis (LFDA)

$x_i \in R^d$ ($i = 1, 2, \ldots, n$) represents the d-dimensional sample, where $y_i \in \{1, 2, \ldots, c\}$ is related to the label set. According to the definition of between-class scatter matrix and within-class scatter matrix, we can obtain

$$S_b = \sum_{i=1}^{c} n_i (\mu_i - \mu)(\mu_i - \mu)^T, \qquad (3)$$

$$S_w = \sum_{i=1}^{c} \sum_{x_k \in classi} (\mu_i - x_k)(\mu_i - x_k)^T, \qquad (4)$$

where S_b is a between-class scatter matrix, S_w is a within-class scatter matrix, i is the number of samples, x_i denotes the sample of I, $\mu_i = \frac{1}{n} \sum_{x \in classi} x$ is the mean of the samples in I, $\mu = \frac{1}{m} \sum_{i=1}^{m} x_i$ is the mean of the overall sample, and $(\mu_i - \mu)(\mu_i - \mu)^T$ is the covariance matrix describing the relationship between the sample x_i and the overall samples. The function in the diagonal line of the matrix represents the variance of the class related to all the samples. Similarly, the non-diagonal elements represent the covariance of the sample population means, i.e., the degree of the correlation or the redundancy between the sample and the overall samples. The lower the coupling degree is between classes, the higher the degree of polymerization will be within the class or the smaller the value will be of the within-class scatter matrix; thus, the larger the value will be of the between-class scatter matrix.

The Fisher discriminant expression is shown as

$$J_{fisher}(\varphi) = \frac{\varphi^T S_b \varphi}{\varphi^T S_w \varphi}, \qquad (5)$$

where φ is an n-dimensional vector; FDA is used to select the vector φ when $J_{fisher}(\varphi)$ reaches the maximum value as the projection direction. The meaning is that the projected samples have the maximum between-class scatter and the minimum within-class scatter.

$$W_{opt} = \text{argmax}\frac{|w^T S_b w|}{|w^T S_w w|} = [w_1, w_2, \ldots, w_n]. \tag{6}$$

The formulas findout a set of projection matrices consisting of optimal discriminant vectors W_{opt}, which are also the eigenvectors corresponding to the maximum eigenvalues of $S_b \varphi = \lambda S_w \varphi$. The number of projection axes is $d \leq c - 1$.

FDA is a traditional linear supervised dimensionality reduction method [28–30]; however, the dimensionality reduction effect is not suitable for multi-peak sample data. For dimensionality reduction of a multi-peak sample, the first thing needed is to preserve the local structure of the data. LPP can achieve a good dimensionality reduction effect by preserving the local structure of the data [31], but it can be only used in unsupervised situations. The label information of the sample cannot be taken into account, because the between-class scatter matrix is not full rank. Therefore, FDA can only map the data to a low-dimensional space, where the dimension is less than the number of classes.

LPP is a more classical manifold algorithm. The main idea is to study the local domain structure of the samples in high-dimensional space, and to preserve this manifold structure after dimensionality reduction [32]. That is, LPP is to minimize the weighted square sum of the distance between adjacent samples in low-dimensional space. The solution is to calculate the generalized eigenvalues. The distance of the projected samples is the same as that before projection.

$$\min\frac{1}{2}\sum_{i,j}(y_i - y_j)^2 S_{ij}. \tag{7}$$

Let A be the total sample; then, $A_{i,j}$ represents the correlation matrix between the two sets of samples x_i and x_j, when the sample sum is in the k-nearest neighbors.

$$A_{i,j} = \exp(-\frac{\|x_i - x_j\|^2}{\sigma_i \sigma_j}), \tag{8}$$

where σ_i represents the domain scale of the sample x_i determined by $\sigma_i = \|x_i - x_i^{(k)}\|$, and $x_i^{(k)}$ represents the k-nearest neighbors of the samples x_i. If $A_{i,j} \in [0, 1]$, the closer x_i and x_j are, the bigger $A_{i,j}$ will be.

When there are many scattered aggregation points in the same class of the sample space due to the integrity of samples, a mapping error might occur in the FDA algorithm. Since LPP is an unsupervised dimensionality reduction method which does not consider the class information, there will be an overlap while dealing with the samples with similar positions but different classes. In order to overcome the shortcomings of these two methods, a local Fisher linear discriminant analysis (LFDA) is proposed to calculate the local between-class and within-class scatter [33–35]. Considering the ability to preserve the local information, the LPP is applied to FDA to ensure the dimensionality reduction effect of the multi-peak data, and also to improve the efficiency of feature extraction.

The formulas of FDA are expressed as

$$S^{(W)} = \frac{1}{2}\sum_{i=1}^{n}\sum_{j=1}^{n} w_{i,j}^{(w)}(x_i - x_j)(x_i - x_j)^T, \tag{9}$$

$$S^{(b)} = \frac{1}{2}\sum_{i=1}^{n}\sum_{j=1}^{n} w_{i,j}^{(b)}(x_i - x_j)(x_i - x_j)^T,$$

where

$$w_{i,j}^{(w)} \equiv \begin{cases} \frac{1}{n_l} & if \quad y_i = y_j = l \\ 0 & if \quad y_i \neq y_j \end{cases}$$

$$w_{i,j}^{(b)} = \begin{cases} \frac{1}{n} - \frac{1}{n_l} & if \quad y_i = y_j = l \\ \frac{1}{n} & if \quad y_i \neq y_j \end{cases}. \tag{10}$$

The expressions of LFDA are defined as

$$\bar{S}^{(W)} = \frac{1}{2}\sum_{i=1}^{n}\sum_{j=1}^{n}\bar{w}_{i,j}^{(w)}(x_i - x_j)(x_i - x_j)^T, \tag{11}$$

$$\bar{S}^{(b)} = \frac{1}{2}\sum_{i=1}^{n}\sum_{j=1}^{n}\bar{w}_{i,j}^{(b)}(x_i - x_j)(x_i - x_j)^T,$$

where

$$\bar{w}_{i,j}^{(w)} \equiv \begin{cases} \frac{A_{i,j}}{n_l} & if \quad y_i = y_j = l \\ 0 & if \quad y_i \neq y_j \end{cases},$$

$$\bar{w}_{i,j}^{(b)} = \begin{cases} A_{i,j}(\frac{1}{n} - \frac{1}{n_l}) & if \quad y_i = y_j = l \\ \frac{1}{n} & if \quad y_i \neq y_j \end{cases}. \tag{12}$$

LFDA can be converted into a projection vector as

$$T_{LFDA} \equiv \underset{T \in R^{d \times r}}{\text{argmax}}[\text{tr}(T^T \bar{S}^{(w)})^{-1} T^T \bar{S}^{(b)} T]. \tag{13}$$

4. Semi-Supervised Random Forest Algorithm

The semi-supervised cooperative training algorithm [36,37] assumes that there are two independent groups of data with the same distribution. These two groups of labeled data are trained to obtain two classifiers, which are used to label each sample to finalize the semi-supervised learning.

Random forest is an integrated classifier composing of multiple decision trees [38–41], which is a strong classifier formed by the combination of several weak classifiers in the form of voting. In practice, it is difficult to divide the data sets into two disjoint subsets. The training data subset is sampled from the training data by the bootstrap method, and the attribute subset is randomly selected to keep the randomness of the trees and the nodes in the decision tree. Semi-supervised random forest introduces the cooperative training idea of the semi-supervised learning to the random forest algorithm, training the random forest classifier, such as H_1 and H_2, from the labeled data, shown as Figure 2. The two classifiers are used to predict the unlabeled samples, where the consistency of prediction label is taken as the confidence degree of the samples. The unlabeled samples with a confidence degree greater than the default threshold are added into the training samples of the other side; then, the classifier is retrained and iterated over and over until all samples are labeled.

The training sample set X consists of the labeled sample set $X_L = \{x_1, x_2, \ldots x_l\}$ and the unlabeled set $X_U = \{x_{l+1}, x_{l+2}, \ldots x_u\}$. The threshold of the classification model is defined as θ. The unlabeled samples whose confidence degrees greater than θ are going to be added to the new training set. The number of decision trees in a random forest is set as odd, and the characteristic attribute subset of decision tree is given as $\log_2 M + 1$, where M denotes the number of attributes of the dataset. The unlabeled samples whose prediction consistency is greater than the threshold of decision trees in random forests are added into the other labeled samples and iterated repeatedly until all samples are labeled. Finally, the semi-supervised random forest classification models $H_1(x)$ and $H_2(x)$ are established.

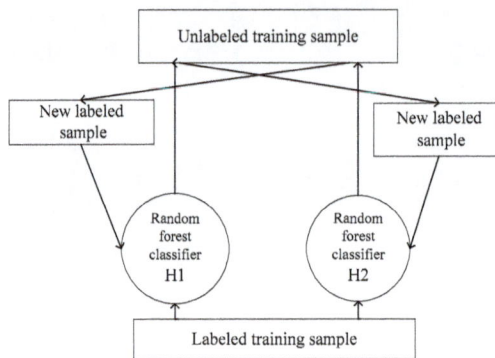

Figure 2. Semi-supervised random forest classifier.

5. Experimental Results and Discussion

The electronic components in analog circuits mainly include resistors, capacitors, inductors, and integrated circuits. When the electronic components change within a certain range, the topology of the circuit will not be changed, but the output features of the circuit will be changed. These include the voltage and current of time domain response, and the corresponding amplitude and phase of the frequency domain. In fact, the parametric fault diagnosis of the analog circuits is the fault location and separation of the electronic components with the changed parameters. Parametric fault refers to the degradation parameters of the electronic components to a certain extent. The degradation parameters are continuous values. There are a large number of component parameters in theory; therefore, a large number of unlabeled component values are generated randomly in the parametric fault zone of electronic components, where some samples are labeled by experts. After extracting the features of the tested circuits using lifting wavelet, the unlabeled samples are used as k-nearest neighbors of the labeled samples to reduce the dimension of features. Then, the semi-supervised random forest algorithm is used to train the fault classifier, and the test samples are put into the classifier after the feature extraction to locate the fault components. The fault diagnosis process is illustrated in Figure 3.

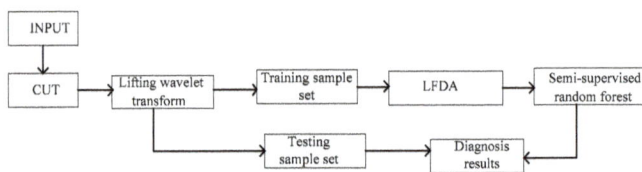

Figure 3. Parametricfault diagnosis process of analog circuit based on semi-supervised learning.

5.1. Sallen–Key Band-Pass Filter Circuit

The Sallen–Key band-pass filter circuit was the same as with References [3,8,9,14,17], and was composed of five resistive, two capacitive, and one operational amplifier. Its nominal value with a central frequency of 25 kHz is shown in Figure 4. The tolerances of the resistors and capacitors were set to 5% and 10%, respectively A component sensitivity analysis was performed on the circuit under test (CUT) to identify the critical single faults. The sensitivity ranking of the discrete components to center frequency was R3, C2, R2, and C1 [14]. Therefore, the parametric faults of four components were mainly considered as in References [3,8,9,14,17]. The input signal was a monopulse signal with the amplitude of 5V and the pulse width of 10 μs. Eight single-fault modes were considered as shown in Table 1.

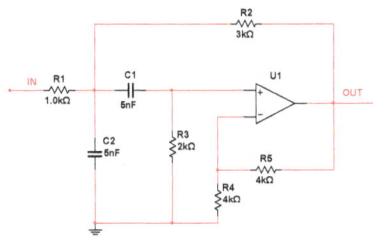

Figure 4. Sallen–Key band-pass filter.

Table 1. Single fault in Sallen–Key band-pass filter.

Fault ID	Fault Mode	Nominal	Faulty Value and Variation Percentage
F0	normal	—	—
F1	C1↑	5 nF	5 nF (1 + 50%) 5 nF (1 + 100%)
F2	C1↓	5 nF	5 nF (1 − 80%) 5 nF (1 − 50%)
F3	C2↑	5 nF	5 nF (1 + 50%) 5 nF (1 + 100%)
F4	C2↓	5 nF	5 nF (1 − 80%) 5 nF (1 − 50%)
F5	R2↑	3 kΩ	3 kΩ (1 + 50%) 3 kΩ (1 + 100%)
F6	R2↓	3 kΩ	3 kΩ (1 − 80%) 3 kΩ (1 − 50%)
F7	R3↑	2 kΩ	2 kΩ (1 + 50%) 2 kΩ (1 + 100%)
F8	R3↓	2 kΩ	2 kΩ (1 − 80%) 2 kΩ (1 − 50%)

As seen from the parameter sweep curve in Figure 5, the response curve of the component parameter was different in the parametric fault range. From the sensitivity analysis, the higher the sensitivity of the component is, the greater the change of the output will be. It can be seen from Figure 5a that the changes of C1 parameter had little effect on the output; however, the parameters in Figure 5b,c had a great influence on the output. Therefore, it was necessary to randomly select the parameters in the parametric fault range. From the time domain response curve of output, it can be seen that some response characteristic curves were very similar; thus, the degree of distinction was low. The method of lifting wavelet transform was used to extract the fault feature. By selecting three cases with similar time domain response curves, the feature extraction of three-layer Harr lifting wavelet was carried out. As seen in Figure 6, these three kinds of time domain original signals were very similar; thus, it was very difficult to classify them by using the time domain feature extraction method. However, through the three-layer lifting wavelet transform, we can see that the approximate and detail coefficients of these three layers were surely different. Therefore, three layers of detail coefficients were selected for feature extraction.

In the parametric range of Table 1 shown, eight types of single faults and 100 component parameters were randomly selected. The transient response curve in the time domain had 2000 dimensions, meaning that the data amount was $9 \times 100 \times 2000$. After lifting wavelet transform, three-layer detail coefficients were selected as the feature, which had 250 dimensions. Experts randomly labeled 40% of the faulty data, selecting k-nearest neighbors of unlabeled data. In this experiment, LFDA was used to reduce the dimensions to eight, where the data amount also reduced to $9 \times 100 \times 8$. Then, the data were put into the semi-supervised random forest for classification, where the number of attributes of the dataset was classified as eight. By setting the decision tree number of the random forest as odd and the feature attribute subset of the decision tree as 4,two semi-supervised random forest classification models on the labeled sample set $H_1(x)$ and $H_2(x)$ were established, which were used to predict the same unlabeled sample. The fault diagnosis rate and its comparison with existing methods are shown in Table 2.

Table 2. The single-fault diagnosis results of Sallen–Key band-pass filter.

Fault ID	Fault Type	Nominal	Method 1 [9]		Method 2 [3]		Method 3 [17]		Proposed Method	
			Fault Value	Accuracy	Fault Value	Accuracy	Fault Value	Accuracy	Fault Value	Accuracy
F0	normal	—	—	97.2%	—	99%	—	100%	—	100%
F1	C1↑	5 nF	7.5 nF	99%	10 nF	100%	7.5 nF	95%	7.5 nF 10 nF	100%
F2	C1↓	5 nF	2.5 nF	100%	2.5 nF	100%	2.5 nF	100%	1 nF 2.5 nF	100%
F3	C2↑	5 nF	7.5 nF	96%	10 nF	100%	7.5 nF	90%	7.5 nF 10 nF	100%
F4	C2↓	5 nF	2.5 nF	97%	2.5 nF	100%	2.5 nF	100%	1 nF 2.5 nF	100%
F5	R2↑	3 kΩ	4.5 kΩ	98%	6 kΩ	99.3%	4.5 kΩ	100%	4.5 kΩ 6 kΩ	98%
F6	R2↓	3 kΩ	1.5 kΩ	100%	1.5 kΩ	99.3%	1.5 kΩ	100%	0.6 kΩ 1.5 kΩ	95%
F7	R3↑	2 kΩ	3 kΩ	100%	4 kΩ	100%	3 kΩ	95%	3 kΩ 4 kΩ	100%
F8	R3↓	2 kΩ	1 kΩ	98.6%	1 kΩ	100%	1 kΩ	100%	0.4 kΩ 1 kΩ	100%

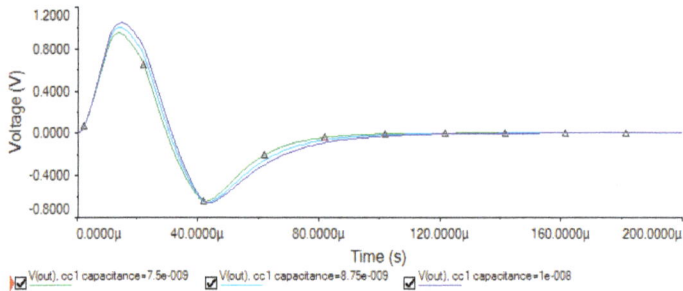

(a)

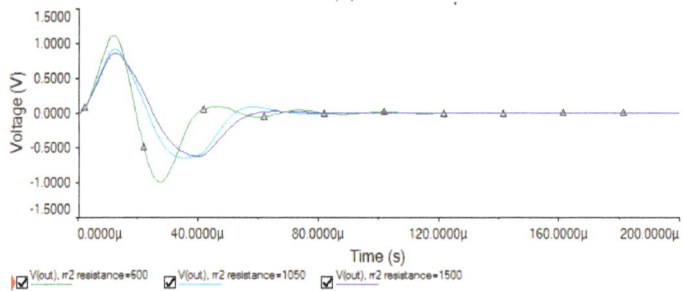

(b)

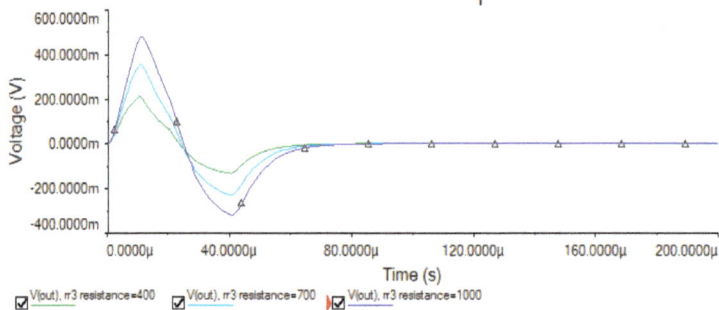

(c)

Figure 5. The representative response curves of parametric fault in different fault modes;(**a**) F1,three representative response curve of C1 within the range of 5 nF (1 + 50%) 5 nF (1 + 100%); (**b**) F6,three representative response curve of R2 within the range of 3 kΩ (1 − 80%) 3 kΩ (1 − 50%); (**c**) F8,three representative response curve of R3 within the range of 3 kΩ (1 − 80%) 3 kΩ (1 − 50%).

According to the engineering statistics, eight types of double faults of electronic components were considered as shown in Table 3. In the range of the two fault elements, 100 groups of fault components were randomly generated, and each type of fault value was further analyzed using the Monte-Carlo method, where the instance value was 5% of resistance and 10% of capacitance, and the distribution was Gaussian. After extracting the time domain response signals of the 100 × 100 set, the transient response curve is depicted in Figure 7. The fault diagnosis rate is shown in Table 4, where the average was 98.6%.

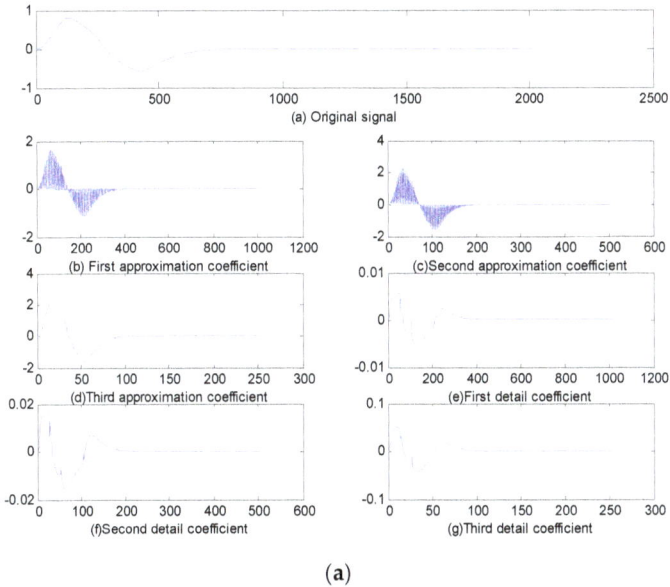

(a)

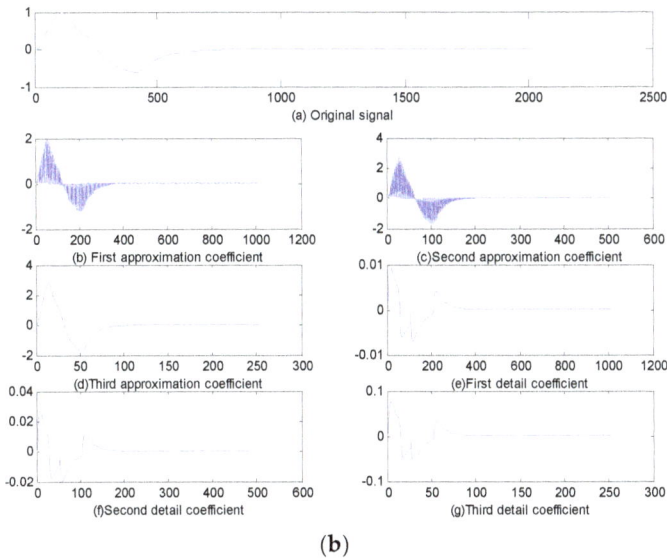

(b)

Figure 6. *Cont.*

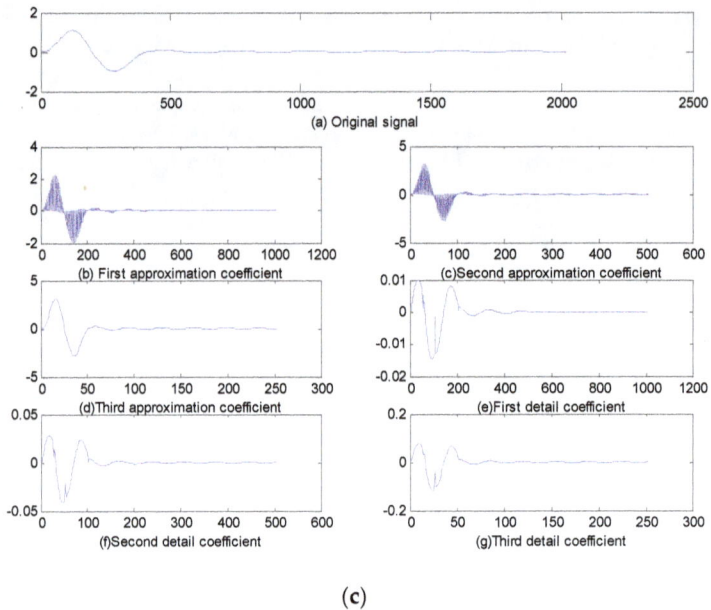

(c)

Figure 6. Three types of fault signals with lower fault differentiation and three-layer Haar lifting wavelet decomposition; (**a**) lifting wavelet feature extraction of F0, (**b**) lifting wavelet feature extraction of F4, and (**c**) lifting wavelet feature extraction of F6.

Table 3. Double faults of Sallen–Key band-pass filter.

Fault ID	Fault Mode	Nominal	Faulty Value and Variation Percentage
F0	—	—	—
F1	C1↑C2↑	5 nF 5 nF	5 nF (1 + 50%) 5 nF (1 + 100%)
			5 nF (1 + 50%) 5 nF (1 + 100%)
F2	C1↓C2↓	5 nF 5 nF	5 nF (1 − 80%) 5 nF (1 − 50%)
			5 nF (1 − 80%) 5 nF (1 − 50%)
F3	R2↑R3↑	3 kΩ 2 kΩ	3 kΩ (1 + 50%) 3 kΩ (1 + 100%)
			2 kΩ (1 + 50%) 2 kΩ (1 + 100%)
F4	R2↓R3↓	3 kΩ 2 kΩ	3 kΩ (1 − 80%) 3 kΩ (1 − 50%)
			2 kΩ (1 − 80%) 2 kΩ (1 − 50%)
F5	R2↑C1↑	3 kΩ 5 nF	3 kΩ (1 + 50%) 3 kΩ (1 + 100%)
			5 nF (1 + 50%) 5 nF (1 + 100%)
F6	R2↑C2↓	3 kΩ 5 nF	3 kΩ (1 + 50%) 3 kΩ (1 + 100%)
			5 nF (1 − 80%) 5 nF (1 − 50%)
F7	R3↓C1↑	2 kΩ 5 nF	2 kΩ (1 − 80%) 2 kΩ (1 − 50%)
			5 nF (1 + 50%) 5 nF (1 + 100%)
F8	R3↓C2↓	2 kΩ 5nf	2 kΩ (1 − 80%) 2 kΩ (1 − 50%)
			5 nF (1 − 80%) 5 nF (1 − 50%)

Table 4. The double fault diagnosis result of Sallen–Key band-pass filter.

Fault ID	F0	F1	F2	F3	F4	F5	F6	F7	F8
Accuracy	100%	100%	100%	100%	96%	100%	100%	98%	94%

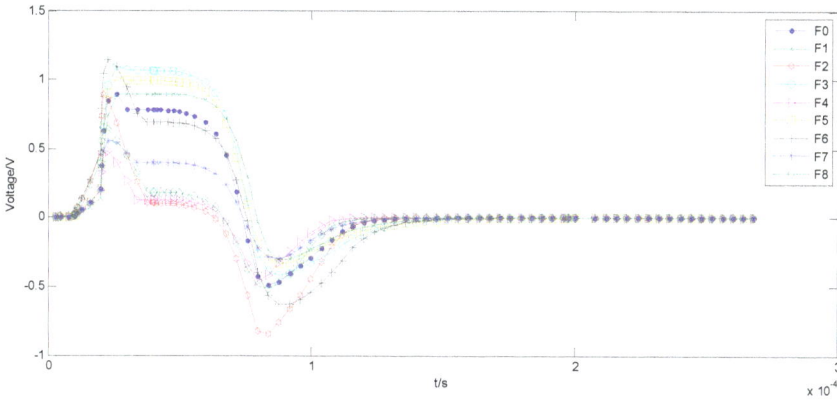

Figure 7. Typical transient response curves of the double faults.

5.2. Three-Opamp Active Band-Stop Filter Circuit

The three operational amplifiers active band-stop filter circuit was the same as References [18,42], and was composed of 12 resistors, four capacitors, and three operational amplifiers. Its nominal value is shown in Figure 8, with tolerance ranges of 5% and 10% for resistors and capacitors, respectively. The input was a monopulse signal with a 5-V peak and a 10-μs pulse width. In Reference [18], only the single faults were diagnosed. Thus, the single, double, and mixed fault models were taken from Reference [42], and eight common single failures and three combinations of them were selected, as shown in Table 5.

Table 5. The fault modes of three-opamp active band-stop filter circuit.

Fault ID	Fault Mode	Nominal	Faulty Value and Variation Percentage
F0	normal	—	—
F1	C4open	10 nF	100 MΩ
F2	R1↑	15 kΩ	15 kΩ (1 + 20%) 15 kΩ (1 + 50%)
F3	R2↑	15 kΩ	15 kΩ (1 + 20%) 15 kΩ (1 + 50%)
F4	C2↓	10 nF	10 nF (1 − 50%) 10 nF (1 − 20%)
F5	C3↑	10 nF	10 nF (1 + 50%) 10 nF (1 + 100%)
F6	R8↓	5.65 kΩ	5.65 kΩ (1 − 80%) 5.65 kΩ (1 − 50%)
F7	R9↑	10 kΩ	10 kΩ (1 + 50%) 10 kΩ (1 + 100%)
F8	R10↓	10 kΩ	10 kΩ (1 − 80%) 10 kΩ (1 − 50%)
F9	R11↑	10 kΩ	10 kΩ (1 + 50%) 10 kΩ (1 + 100%)
	R5↓	31 kΩ	31 kΩ (1 − 50%) 31 kΩ (1 − 20%)
F10	and R6↑	31 kΩ	31 kΩ (1 + 20%) 31 kΩ (1 + 50%)
	and C2↓	10 nF	10 nF (1 − 50%) 10 nF (1 − 20%)
	R8↓	5.65 kΩ	5.65 kΩ (1 − 50%) 5.65 kΩ (1 − 20%)
F11	and R9↑	10 kΩ	10 kΩ (1 + 20%) 10 kΩ (1 + 50%)
	and C3↑	10 nF	10 nF (1 + 20%) 10 nF (1 + 50%)
F12	R10↓	10 kΩ	10 kΩ (1 − 50%) 10 kΩ (1 − 20%)
	and R11↑	10 kΩ	10 kΩ (1 + 20%) 10 kΩ (1 + 50%)

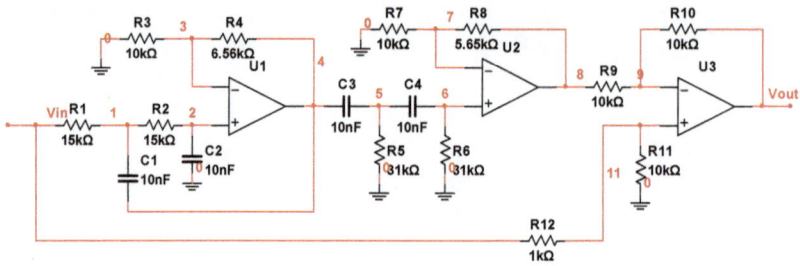

Figure 8. Three-opamp active band-stop filter circuit.

For a certain fault range, 100 fault values were randomly selected, components of which were analyzed using the Monte-Carlo method to generate 100 combinations; the instance value was 5% of other resistances and 10% of other capacitances, and the distribution was Gaussian. By following this, 100×100 sets of transient response signals were extracted in the time domain, where the transient response curve had 2000 dimensions, meaning that the data amount was $13 \times 100 \times 100 \times 2000$. Through the three-layer lifting wavelet transform, three layers of detail coefficients were selected for feature extraction, each of which had 250 dimensions. Thus, the amount of data could be reduced to $13 \times 100 \times 100 \times 250$. Experts randomly labeled 30% of the faulty data, selecting k-nearest neighbors of the unlabeled data. LFDA was used to reduce the dimensions to eight, where the data amount also reduced to $13 \times 100 \times 100 \times 8$. Then, the data were put into the semi-supervised random forest for classification, where the number of attributes of the dataset was classified as 13. By setting the decision tree number of the random forest and the feature attribute subset as 4, two semi-supervised random forest classification models on the labeled sample sets $H_1(x)$ and $H_2(x)$ were established, which were used to predict the same unlabeled sample. As illustrated in Table 6, the average fault diagnosis rate of the proposed method was 98.2%, which is higher than that of 93.08% achieved by Reference [42].

Table 6. The fault diagnosis results of three-opamp active band-stop filter circuit.

Fault ID	Fault Type	Nominal	Method1 [42] Fault Value	Proposed Method Fault Value
F0	normal	—	—	—
F1	C4open	10 nF	C4open	100MΩ
F2	R1↑	15 kΩ	15 kΩ (1 + 20%)	15 kΩ (1 + 20%) 15 kΩ (1 + 50%)
F3	R2↑	15 kΩ	15 kΩ (1 + 20%)	15 kΩ (1 + 20%) 15 kΩ (1 + 50%)
F4	C2↓	10 nF	10 nF (1 − 20%)	10 nF (1 − 50%) 10 nF (1 − 20%)
F5	C3↑	10 nF	10 nF (1 + 50%)	10 nF (1 + 50%) 10 nF (1 + 100%)
F6	R8↓	5.65 kΩ	5.65 kΩ (1-50%)	5.65 kΩ (1 − 80%) 5.65 kΩ (1 − 50%)
F7	R9↑	10 kΩ	10 kΩ (1 + 50%)	10 kΩ (1 + 50%) 10 kΩ (1 + 100%)
F8	R10↓	10 kΩ	10 kΩ (1-50%)	10 kΩ (1 − 80%) 10 kΩ (1 − 50%)
F9	R11↑	10 kΩ	10 kΩ (1 + 50%)	10 kΩ (1 + 50%) 10 kΩ (1 + 100%)
F10	R5↓	31 kΩ	31 kΩ (1 − 20%)	31 kΩ (1 − 50%) 31 kΩ (1 − 20%)
	and R6↑	31 kΩ	31 kΩ (1 + 20%)	31 kΩ (1 + 20%) 31 kΩ (1 + 50%)
	and C2↓	10 nF	10 nF (1 − 20%)	10 nF (1 − 50%) 10 nF (1 − 20%)
F11	R8↓	5.65 kΩ	5.65 kΩ (1 − 80%)	5.65 kΩ (1 − 50%) 5.65 kΩ (1 − 20%)
	and R9↑	10 kΩ	10 kΩ (1 + 20%)	10 kΩ (1 + 20%) 10 kΩ (1 + 50%)
	and C3↑	10 nF	10 nF (1 + 20%)	10 nF (1 + 20%) 10 nF (1 + 50%)
F12	R10↓	10 kΩ	10 kΩ (1 − 20%)	10 kΩ (1 − 50%) 10 kΩ (1 − 20%)
	and R11↑	10 kΩ	10 kΩ (1 + 20%)	10 kΩ (1 + 20%) 10 kΩ (1 + 50%)
Average fault diagnosis			93.08%	98.2%

6. Conclusions

In this paper, a semi-supervised random forest algorithm for parametric fault diagnosis in analog circuits was proposed. The difficulty in diagnosing analog circuit parametric fault is the successive changes of the parameter values. The existing fault diagnosis models trained by fixing fault component values cannot adapt to the engineering applications. The change in fault parameters produces a

Symmetry **2019**, *11*, 228

large number of fault samples. However, the labeled fault samples are also limited. Therefore, a semi-supervised learning algorithm was used to aid unlabeled samples through the labeled ones. In order to improve the accuracy of the semi-supervised classification algorithm, LDFA was utilized after feature extraction with lifting wavelet for feature dimensionality reduction, in consideration of labeled and unlabeled samples. Then, two circuits were used to validate the proposed method, which diagnosed single, multiple, and mixed faults, under the premise of improving generalization ability, where by the fault diagnosis rate is slightly higher than existing methods. Future work will cover implementation of the complex analog circuits and development of the test assemblies.

Author Contributions: L.W. and D.Z. conceived and designed this work; H.Z. and W.Z. collected and analyzed the data; L.W. drafted the manuscript; H.T. revised the manuscript. All authors read and approved the final manuscript.

Funding: This research was supported by the National Natural Science Fund (31501213), the Science and Technology Key project of Henan Province (172102310244, 172102310696) and (182102110250, 182102110356), and the China Postdoctoral Science Foundation (2017M612399).

Conflicts of Interest: The authors declare no conflict of interest.

References

1. Tang, S.; Li, Z.; Chen, L. *Fault Detection in Analog and Mixed-Signal Circuits by Using Hilbert-Huang Transform and Coherence Analysis*; Elsevier: Amsterdam, The Netherlands, 2015.
2. Wang, Y.H.; Yan, Y.Z.; Signal, S. Wavelet-based feature extraction in fault diagnosis for biquad high-pass filter circuit. *Math. Probl. Eng.* **2016**, *2016*, 1–13. [CrossRef]
3. Zhang, C.L.; He, Y.G.; Yuan, LF. A Novel Approach for Diagnosis of Analog Circuit Fault by Using GMKL-SVM and PSO. *J. Electron. Test.-Theory Appl.* **2016**, *32*, 531–540. [CrossRef]
4. Long, Y.; Xiong, Y.J.; He, Y.G. A new switched current circuit fault diagnosis approach based on pseudorandom test and preprocess by using entropy and Haar wavelet transform. *Analog Integr. Circuits Signal Process.* **2017**, *91*, 445–461. [CrossRef]
5. Li, J.M. The Application of Dual-Tree Complex Wavelet Packet Transform in Fault Diagnosis. *Agro Food Ind. Hi-Tech* **2017**, *28*, 406–410.
6. Xie, X.; Li, X.; Bi, D.; Zhou, Q.; Xie, S.; Xie, Y. Analog Circuits Soft Fault Diagnosis Using Rényi's Entropy. *J. Electron. Test.* **2015**, *31*, 217–224. [CrossRef]
7. Long, T.; Jiang, S.; Luo, H.; Deng, C. Conditional entropy-based feature selection for fault detection in analog circuits. *Dyna* **2016**, *91*, 309–318. [CrossRef]
8. He, W.; He, Y.; Li, B.; Zhang, C. Analog Circuit Fault Diagnosis via Joint Cross-Wavelet Singular Entropy and Parametric t-SNE. *Entropy* **2018**, *20*, 604. [CrossRef]
9. Song, P.; He, Y.; Cui, W. Statistical property feature extraction based on FRFT for fault diagnosis of analog circuits. *Analog Integr. Circuits Signal Process.* **2016**, *87*, 427–436. [CrossRef]
10. Zhao, D.; He, Y. A novel binary bat algorithm with chaos and Doppler effect in echoes for analog fault diagnosis. *Analog Integr. Circuits Signal Process.* **2016**, *87*, 437–450. [CrossRef]
11. Prieto-Moreno, A.; Llanes-Santiago, O.; García-Moreno, E. Principal components selection for dimensionality reduction using discriminant information applied to fault diagnosis. *J. Process Control* **2015**, *33*, 14–24. [CrossRef]
12. Haddad, R.Z.; Strangas, E.G. On the Accuracy of Fault Detection and Separation in Permanent Magnet Synchronous Machines Using MCSA/MVSA and LDA. *IEEE Trans. Energy Convers.* **2016**, *31*, 924–934. [CrossRef]
13. Sugiyama, M. Dimensionality Reduction of Multimodal Labeled Data by Local Fisher Discriminant Analysis. *J. Mach. Learn. Res.* **2007**, *8*, 1027–1061.
14. Spina, R.; Upadhyaya, S. Linear circuit fault diagnosis using neuromorphic analyzers. *IEEE Trans. Circuits Syst. II Analog Digit. Signal Process.* **1997**, *44*, 188–196. [CrossRef]
15. Jia, W.; Zhao, D.; Shen, T.; Ding, S.; Zhao, Y.; Hu, C. An optimized classification algorithm by BP neural network based on PLS and HCA. *Appl. Intell.* **2015**, *43*, 1–16. [CrossRef]

16. Yuan, Z.; He, Y.; Yuan, L. Diagnostics Method for Analog Circuits Based on Improved KECA and Minimum Variance ELM. *IOP Conf. Ser.Mater. Sci. Eng.* **2017**. [CrossRef]

17. Yu, W.X.; Sui, Y.; Wang, J. The Faults Diagnostic Analysis for Analog Circuit Based on FA-TM-ELM. *J. Electron. Test.* **2016**, *32*, 1–7. [CrossRef]

18. Ma, Q.; He, Y.; Zhou, F. A new decision tree approach of support vector machine for analog circuit fault diagnosis. *Analog Integr. Circuits Signal Process.* **2016**, *88*, 455–463. [CrossRef]

19. Cui, Y.Q.; Shi, J.Y.; Wang, Z.L. Analog circuit fault diagnosis based on Quantum Clustering based Multi-valued Quantum Fuzzification Decision Tree (QC-MQFDT). *Measurement* **2016**, *93*, 421–434. [CrossRef]

20. Liu, Z.B.; Jia, Z.; Vong, C.M. Capturing High-Discriminative Fault Features for Electronics-Rich Analog System via Deep Learning. *IEEE Trans. Ind. Inform.* **2017**, *13*, 1213–1226. [CrossRef]

21. Zhuang, L.; Zhou, Z.; Gao, S.; Yin, J.; Lin, Z.; Ma, Y. Label Information Guided Graph Construction for Semi-Supervised Learning. *IEEE Trans. Image Process.* **2017**, *26*, 4182–4192. [CrossRef]

22. Zhou, X.; Prasad, S. Active and Semisupervised Learning with Morphological Component Analysis for Hyperspectral Image Classification. *IEEE Geosci. Remote Sens. Lett.* **2017**, *26*, 1–5. [CrossRef]

23. Guoming, S.; Houjun, W.; Hong, L. Analog circuit fault diagnosis using lifting wavelet transform and SVM. *J. Electron. Meas. Instrum.* **2010**, *24*, 17–22.

24. Qing, Y.; Feng, T.; Dazhi, W.; Dongsheng, W.; Anna, W. Real-time fault diagnosis approach based on lifting wavelet and recursive LSSVM. *Chin. J. Sci. Instrum.* **2011**, *32*, 596–602.

25. Pan, H.; Siu, W.C.; Law, N.F. A fast and low memory image coding algorithm based on lifting wavelet transform and modified SPIHT. *Signal Process. Image Commun.* **2008**, *23*, 146–161. [CrossRef]

26. Hou, X.; Yang, J.; Jiang, G.; Qian, X. Complex SAR Image Compression Based on Directional Lifting Wavelet Transform with High Clustering Capability. *IEEE Trans. Geosci. Remote Sens.* **2013**, *51*, 527–538. [CrossRef]

27. Roy, A.; Misra, A.P. Audio signal encryption using chaotic Hénon map and lifting wavelet transforms. *Eur. Phys. J. Plus* **2017**, *132*, 524. [CrossRef]

28. Chiang, L.H.; Kotanchek, M.E.; Kordon, A.K. Fault diagnosis based on Fisher discriminant analysis and support vector machines. *Comput. Chem. Eng.* **2004**, *28*, 1389–1401. [CrossRef]

29. Yin, Y.; Hao, Y.; Bai, Y.; Yu, H. A Gaussian-based kernel Fisher discriminant analysis for electronic nose data and applications in spirit and vinegar classification. *J. Food Meas. Charact.* **2017**, *11*, 24–32. [CrossRef]

30. Li, C.; Jiang, K.; Zhao, X.; Fan, P.; Wang, X.; Liu, C. Spectral identification of melon seeds variety based on k-nearest neighbor and Fisher discriminant analysis. In Proceedings of the AOPC 2017: Optical Spectroscopy and Imaging, Beijing, China, 4–6 June 2017.

31. Wang, Z.; Ruan, Q.; An, G. Facial expression recognition using sparse local Fisher discriminant analysis. *Neurocomputing* **2016**, *174*, 756–766. [CrossRef]

32. Yu, Q.; Wang, R.; Li, B.N.; Yang, X.; Yao, M. Robust Locality Preserving Projections With Cosine-Based Dissimilarity for Linear Dimensionality Reduction. *IEEE Access* **2017**, *5*, 2676–2684. [CrossRef]

33. Sugiyama, M.; Idé, T.; Nakajima, S.; Sese, J. Semi-supervised local Fisher discriminant analysis for dimensionality reduction. *Mach. Learn.* **2010**, *78*, 35. [CrossRef]

34. Wang, S.; Lu, J.; Gu, X.; Du, H.; Yang, J. Semi-supervised linear discriminant analysis for dimension reduction and classification. *Pattern Recognit.* **2016**, *57*, 179–189. [CrossRef]

35. Cheng, G.; Zhu, F.; Xiang, S.; Wang, Y.; Pan, C. Semisupervised Hyperspectral Image Classification via Discriminant Analysis and Robust Regression. *IEEE J. Sel. Top. Appl. Earth Obs. Remote Sens.* **2017**, *9*, 595–608. [CrossRef]

36. Blum, A.; Mitchell, T. Combining labeled and unlabeled data with co-training. In Proceedings of the Eleventh Annual Conference on Computational Learning Theory, Madison, WI, USA, 24–26 July 1998; pp. 92–100.

37. Zhao, J.H.; Wei-Hua, L.I. One of semi-supervised classification algorithm named Co-S3OM based on cooperative training. *Appl. Res. Comput.* **2013**, *30*, 3237–3239.

38. Díaz-Uriarte, R.; De Andres, S.A. Gene selection and classification of microarray data using random forest. *BMC Bioinform.* **2006**, *7*, 3. [CrossRef] [PubMed]

39. Belgiu, M.; Drăguţ, L. Random forest in remote sensing: A review of applications and future directions. *ISPRS J. Photogramm. Remote Sens.* **2016**, *114*, 24–31. [CrossRef]

40. Li, C.; Sanchez, R.V.; Zurita, G.; Cerrada, M.; Cabrera, D.; Vásquez, R.E. Gearbox fault diagnosis based on deep random forest fusion of acoustic and vibratory signals. *Mech. Syst. Signal Process.* **2016**, *76*, 283–293. [CrossRef]

41. Mellor, A.; Boukir, S.; Haywood, A.; Jones, S. Exploring issues of training data imbalance and mislabelling on random forest performance for large area land cover classification using the ensemble margin. *ISPRS J. Photogramm. Remote Sens.* **2015**, *105*, 155–168. [CrossRef]
42. Jiang, Y.; Wang, Y.; Luo, H. Fault diagnosis of analog circuit based on a second map SVDD. *Analog Integr. Circuits Signal Process.* **2015**, *85*, 395–404. [CrossRef]

symmetry

MDPI

Article

A Prediction Method for the Damping Effect of Ring Dampers Applied to Thin-Walled Gears Based on Energy Method

Yanrong Wang [1,2], Hang Ye [1,2], Xianghua Jiang [1,2] and Aimei Tian [3,*]

1   School of Energy and Power Engineering, Beihang University, Beijing 100191, China;
    yrwang@buaa.edu.cn (Y.W.); yeahhang155@gmail.com (H.Y.); jxh@buaa.edu.cn(X.J.)
2   Collaborative-Innovation Center for Advanced Aero-Engine, Beijing 100191, China
3   School of Astronautics, Beihang University, Beijing 100191, China
*   Correspondence: amtian@buaa.edu.cn

Received: 16 November 2018; Accepted: 30 November 2018; Published: 30 November 2018

Abstract: In turbomachinery applications, thin-walled gears are cyclic symmetric structures and often subject to dynamic meshing loading which may result in high cycle fatigue (HCF) of the thin-walled gear. To avoid HCF failure, ring dampers are designed for gears to increase damping and reduce resonance amplitude. Ring dampers are installed in the groove. They are held in contact with the groove by normal pressure generated by interference or centrifugal force. Vibration energy is attenuated (converted to heat) by frictional force on the contact interface when the relative motion between ring dampers and gears takes place. In this article, a numerical method for the prediction of friction damping in thin-walled gears with ring dampers is proposed. The nonlinear damping due to the friction is expressed as equivalent mechanical damping in the form of vibration stress dependence. This method avoids the forced response analysis of nonlinear structures, thereby significantly reducing the time required for calculation. The validity of this numerical method is examined by a comparison with literature data. The method is applied to a thin-walled gear with a ring damper and the effect of design parameters on friction damping is studied. It is shown that the rotating speed, geometric size of ring dampers and friction coefficient significantly influence the damping performance.

Keywords: thin-walled gear; ring damper; vibration; energy dissipation; friction damping

---

1. Introduction

Vibrations of gears are mainly caused by dynamic meshing loads. Resonance of the gear may occur if the excitation frequency is close to the resonance frequencies of the gear within its range of operating speeds. To avoid fatigue failure owing to high resonance stresses, the ideal solution is to redesign the gear to move its natural frequencies away from any potential external excitation. This method is called detuning [1]. However, for the thin-walled Gear, which is typically lightweight and operates at high rotating speed, detuning may not be feasible because each gear has multiple natural frequencies in coincidence with the mesh frequency within its operating range.

If detuning does not prevent resonance, then damping, as a passive control technique, is a feasible option to avoid high cycle fatigue failures. Friction dampers are effective approaches to provide damping in turbomachinery [2,3]. Friction dampers are substructures that remain in contact with the main structure through elastic deformation or centrifugal force. The vibration energy of the system is attenuated (converted to heat) by friction on the interface via the relative motion between the damper and primary structure [4].

Thin-walled structural components in aircraft gas turbine engines are easily excited to high vibration level. To reduce the vibrational stress of turbomachinery blades caused by the forced response from aerodynamic exciting sources and negative aerodynamic damping, i.e., flutter [5–8], many types of friction dampers have been studied and applied in actual structures. Among them, the under-platform damper has been extensively studied in detail [9–14]. This type of damper is installed under the platform or between neighboring blades. However, gears do not have suitable positions to install the under-platform damper. Therefore, ring dampers are used as damping devices for gears. In contrast with under-platform dampers, limited work has been carried out to investigate ring dampers. Lopez [15,16] used ring dampers on the train wheels to reduce the vibration emitted by freight traffic. The results revealed that increasing the mass of the ring damper is beneficial to vibration reduction. Laxalde [17] studied the damping strategy of ring dampers by using the dynamic Lagrangian frequency-time method to derive the forced response of blisks in the presence of ring dampers. The results showed that the size of the alternating stick-slip area determines the damping effectiveness of ring dampers. A nonlinear modal analysis method is proposed by Laxalde [18], and applied to analyze the effect of design parameters of ring dampers. Zucca [19] studied the effect of the key parameters (for example, mass and friction coefficient) of ring dampers on the vibration amplitude. The authors used the contact element to link the static and dynamic differential equations and calculated the forced response of the coupling system. Tang [20] proposed a novel reduced-order modeling method to solve the forced responses of the blisk–damper systems based on Craig–Bampton component mode synthesis. The authors studied the effect of geometric parameters of ring dampers on the blisk forced responses [21] by this method.

For ring dampers to be effective, they are typically located on the rim of the gear where large vibration amplitudes occur, as shown in Figure 1. Otherwise the energy dissipation due to friction will be reduced and even equal to zero, and the ring damper will be ineffective. Ring dampers are mostly effective only for the fundamental mode shapes of the gear [22]. These modes are characterized by a large amplitude at the rim of the gear. For thin-walled gears, friction damping is produced by the relative motion caused by the different extension deformations between ring dampers and gears along the tangential direction of the contact surface [23]. However, note that the circumferential deformation is caused by radial vibration. In other applications, for example train wheels, vibration energy is attenuated by the axial component of the vibration, and friction damping is produced by relative motion in the axial direction [15]. Zucca [22] analyzed the axial and circumferential relative motion of a bevel gear with a ring damper in different response conditions. The results show that although the radial and axial components of the vibration have the same order of magnitude, the ring damper worked mainly in the circumferential direction because the relative displacement along the circumferential direction is much larger than along the axial direction. No relative motion occurs in the radial direction due to the ring damper maintaining contact with the primary structure by centrifugal force.

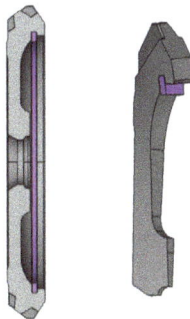

Figure 1. An example of a gear with a ring damper.

Although all of these papers show that vibration amplitude will decrease when ring dampers are used, limited work to investigate the nonlinear friction damping of thin-walled gears with ring dampers has been done. Most previous theoretical analyses have focused on the forced response of main structures in the presence of ring dampers. In contrast, the energy dissipation by ring dampers has been seldom studied. Niemotka [24] proposed a design method for split ring dampers to lower the vibration amplitude of annular air seals in gas turbine engines based on a quasi-static energy dissipation analysis.

The primary objective of this work is to construct a numerical model to predict the damping of ring dampers in thin-walled gears. In the model, the nonlinear friction damping is expressed as equivalent mechanical damping in the form of vibration stress dependent. Macro-slip is used in the friction model to calculate the energy dissipation. The validity of the proposed method is confirmed by a comparison with forced response analysis results. The secondary objective is to investigate the influence of rotating speed, temperature, parameters of ring dampers, and friction coefficient on the damping performance by means of method proposed in this paper.

The rest of this paper is arranged as follows. The theoretical background, including the equation of motion and modal analysis, is introduced in Section 2. Theoretical derivation of equivalent damping ratio of the ring damper is shown in Section 3. Method validation and parameter analysis are performed on a thin-walled gear in Section 4, followed by conclusions in Section 5.

2. Vibration Analysis of The Gear-Ring Damper System

2.1. The Equations of Motion

The equations of motion in time domain of the gear-ring damper system can be written as

$$M\ddot{X} + C\dot{X} + KX + F_{nl}(X, \dot{X}, t) = F(t) \tag{1}$$

where M, C, and K are the mass, damping, and stiffness matrices of the gear, respectively, and X is the vector of the displacements. $F(t)$ is the vector of the external excitation force. $F_{nl}(X, \dot{X}, t)$ is the vector of the nonlinear forces generated by the ring damper and depends on the vibration displacement and vibration velocity of the system. $F_{nl}(X, \dot{X}, t)$ can be given by the equivalent damping and stiffness matrices as [25]

$$F_{nl}(X, \dot{X}, t) = C_{eq}\dot{X} + K_{eq}X \tag{2}$$

The equivalent damping matrix C_{eq} and the equivalent stiffness matrix K_{eq} depend on the motion of the gear.

The displacement vector X is a function of time and can be expressed as a linear combination of the natural modes of the un-damped system.

$$X(t) = \Phi q(t) \tag{3}$$

Thus, Equation (1) can be rewritten as

$$M\Phi\ddot{q}(t) + C\Phi\dot{q}(t) + K\Phi q(t) + C_{eq}\Phi\dot{q}(t) + K_{eq}\Phi q(t) = F(t) \tag{4}$$

where Φ is the mass-normalized eigenvector matrix of the gear.

Premultiplying Equation (4) throughout by Φ^T:

$$I\ddot{q}(t) + Z\dot{q}(t) + \Lambda q(t) + Z_{eq}\dot{q}(t) + \Lambda_{eq}q(t) = Q(t) \tag{5}$$

where

$$I = \Phi^T M\Phi, \; Z = \Phi^T C\Phi, \; \Lambda = \Phi^T K\Phi, \; Z_{eq} = \Phi^T C_{eq}\Phi, \; \Lambda_{eq} = \Phi^T \Lambda_{eq}, \; Q(t) = \Phi^T F(t) \tag{6}$$

because I denotes the unity matrix and Z, Λ, Z_{eq}, and Λ_{eq} are all diagonal. In the vicinity of the jth natural frequency, Equation (5) can be rewritten as

$$\ddot{q}_j(t) + 2(\zeta_j + \zeta_{j,eq})\omega_j\dot{q}_j(t) + (k_j + k_{j,eq})q_j(t) = Q_j(t), \quad \text{with } j = 1, 2, \cdots, n \tag{7}$$

where ζ_j and $\zeta_{j,eq}$ are the modal damping ratio and the equivalent damping ratio caused by the ring damper for the jth mode, respectively; k_j and $k_{j,eq}$ are the modal stiffness and equivalent stiffness for the jth mode, respectively; and $k_j = \omega_j^2$; ω_j is the jth natural frequency of the undamped system.

The n equations represented in Equation (7) can be uncoupled from all other equations. Therefore, the forced response of the jth mode can be calculated if the relationship between the equivalent damping and the equivalent stiffness and response amplitude can be pre-calculated.

In general, the mass of the ring damper is much smaller than the mass of the main structure. Let the weight penalty be defined as

$$\beta = \frac{\text{mass of the ring damper}}{\text{mass of the gear}} \tag{8}$$

In this study, the weight penalty is less than 5%. Note that the magnitudes of M and K are much larger than the magnitude of $F_{nl}(X, \dot{X}, t)$, thus $k_{j,eq}$ is much smaller than k_j. Generally, for the ring damper, $k_{j,eq}$ is two orders of magnitude lower than k_j. In other words, the ring damper does not affect the shape of the vibration mode; rather, it affects only the vibration amplitude. Moreover, the influence of the damper on the resonance frequency of the primary structure can be neglected. However, the equivalent damping matrix is of the same order of magnitude or even larger with respect to the damping matrix because the structural damping is usually small (For steel, the damping ratio is $1\sim5\times10^{-4}$). The results of other scholars [2,3,20,25–27] also showed that the influence of the ring damper on the frequency is negligible. With or without ring dampers, the frequency variation is less than 1%. Thus, the damper ring reduces the resonant amplitude of the gear, primarily by providing damping, rather than changing the stiffness of the gear system.

2.2. Modal Analysis

Modal analysis was performed with the FEM software ANSYS 14.5. The gear and ring damper finite element models are shown in Figure 2. The gear is a cyclic symmetry structure, comprising z fundamental sectors (Figure 2a). The ring damper is machined to be C-shaped for ease of installation. There is a split in the axial direction, as shown in Figure 2b.

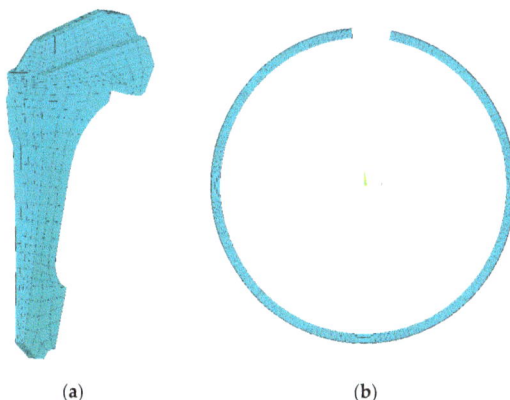

(a)   (b)

Figure 2. Finite element model: (a) The gear (fundamental sector); (b) the ring damper.

Typical gear resonance failure in practice [1] is shown in Figure 3. The mode shapes (Figure 4) that lead to gears failure have the following features:

1. The modal amplitude has an integer number of harmonic distributions along the circumferential direction.
2. The nodal line passes through the center of rotation, and the vibration amplitude of the nodal line is zero.
3. For thin-walled gears, the gear rim vibrates mainly in the radial direction.

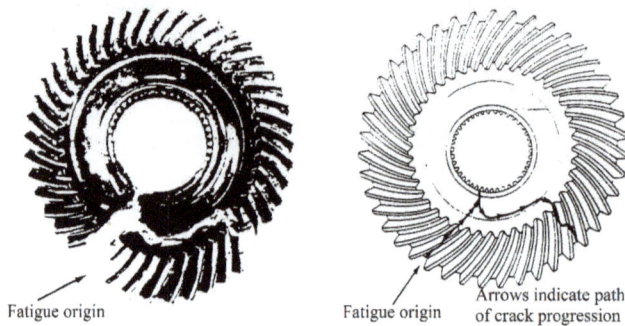

Figure 3. Typical gear resonance failure [1].

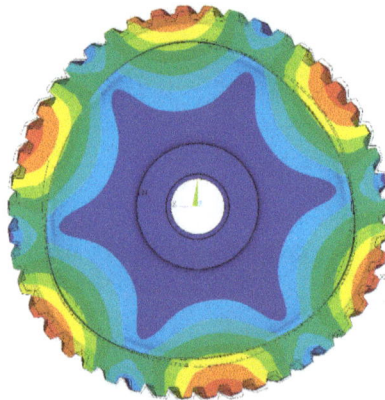

Figure 4. Mode shape of the gear with 3 nodal diameters.

Therefore, in this study, we focused on nodal diameter vibration. For N nodal diameters (ND), the radial displacement of the groove of the gear can be assumed as

$$w = B \cos(N\theta) \tag{9}$$

where B is the maximum amplitude of the groove of the gear, N is the number of nodal diameters, θ is circumferential angle.

3. Theoretical Model of Equivalent Damping Ratio of The Ring Damper

3.1. Energy Dissipated by Frictional Force

In this paper, the motion of the gear is assumed to be small amplitude vibrations, i.e., only elastic deformation is considered and in the same mode shape, and the vibration stress is proportional to the vibration amplitude. The following energy dissipation analysis is based on the method proposed by Alford [28–31] and Niemotka [24].

Generally, deflections of the structure at a resonance are very small compared to its size; otherwise, the structure will suffer fatigue failure in a short time. For small deformations, the strain-curvature relation is

$$\varepsilon = \kappa y \tag{10}$$

where ε, κ, and y are strain, curvature, and the distance from the neutral line, respectively.

The preceding equation shows that the circumferential strains are proportional to the curvature and are linearly related with the distance y from the neutral line. Here tensile strain is defined as positive and compressive strain is defined as negative.

The curvature can be expressed by the bending moment:

$$\kappa = \frac{M}{EI} \tag{11}$$

where M, E, and I are bending moment, Young's modulus, and moment of inertia of ring dampers, respectively. Equation (11) is known as the moment-curvature equation. When the radius of curvature of a ring is sufficiently large compared to its radial height, the relationship between the bending moment M and radial displacement w can be expressed as [32]

$$\frac{M}{EI} = \frac{1}{R^2}\left[w + \frac{d^2w}{d\theta^2}\right] \tag{12}$$

By substituting Equation (9) into Equation (12), the following relationship is obtained:

$$\frac{M}{EI} = \frac{1}{R^2}B(1 - N^2)\cos(N\theta) \tag{13}$$

At a distance y from the mean radius R, the bending strain is:

$$\varepsilon_y = \frac{y}{R^2}B(1 - N^2)\cos(N\theta) \tag{14}$$

For the gear, the strain on the contact surface of the gear is tensile on the groove interface; in contrast, it is compressive for the ring damper on the contact surface and vice versa, as shown in Figure 5.

$$\varepsilon_g = -\frac{c_g}{R_g^2}B(1 - N^2)\cos(N\theta) \tag{15}$$

$$\varepsilon_d = \frac{c_d}{R_d^2}B(1 - N^2)\cos(N\theta) \tag{16}$$

where c and R are the half of the radial thickness and the radius. Subscript g and d represent gear and damper respectively.

When there is no relative motion on the contact surface, the contact state is the stick state. The relationship between strain caused by friction and bending strain is

$$\varepsilon_f = \varepsilon_g - \varepsilon_d = -\left(\frac{c_d}{R_d^2} + \frac{c_g}{R_g^2}\right)B(1 - N^2)\cos(N\theta) \tag{17}$$

The strain caused by friction in the ring damper also can be calculated by dividing the frictional force by the product of the damper cross-sectional area and its Young's modulus. F_f is defined as the frictional force per unit length, where F_f is a function of circumferential angle θ.

$$\varepsilon_f = \frac{R_d}{A_d E}\int F_f d\theta \tag{18}$$

By substituting Equation (17) into Equation (18), F_f can be written as

$$F_f = \frac{A_d E}{R_d} \frac{d\varepsilon_f}{d\theta} = -\frac{BA_d E}{R_d}\left(\frac{c_d}{R_d^2} + \frac{c_g}{R_g^2}\right)N(1-N^2)\sin(N\theta) \tag{19}$$

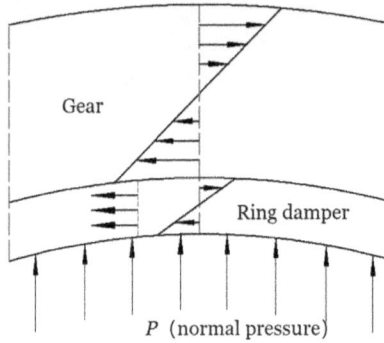

Figure 5. Local behavior in the contact region.

When no slipping occurs on the entire contact surface, $F_{f\,max}$ will appears at $\theta = \pi/2N$. And $F_{f\,max} = \frac{BA_d E}{R_d}\left(\frac{c_d}{R_d^2} + \frac{c_g}{R_g^2}\right)N(1-N^2)$. When tangential force is greater than the maximum static friction, slipping occurs at $\theta < \pi/2N$, and over the zone $\theta_0 < \theta < \pi/2N$, $F_{f\,max} = \mu P$. Where μ is friction coefficient, and P is normal pressure on the contact surface.

At $\theta_0 = \theta$,

$$F_{f\,max} = \mu P = -\frac{BA_d E}{R_d}\left(\frac{c_d}{R_d^2} + \frac{c_g}{R_g^2}\right)N(1-N^2)\sin(N\theta_0) \tag{20}$$

Thus,

$$N\theta_0 = \arcsin\frac{\mu P}{\frac{BEA_d}{R_d}\left(\frac{c_g}{R_g^2} + \frac{c_d}{R_d^2}\right)N(N^2-1)} \tag{21}$$

where θ_0 represents the angle where slippage starts, which is called the critical slip angle.

When normal pressure P is constant, over a vibration cycle, the condition that no slipping occurs on the entire contact surface is the maximum vibration amplitude of the gear B is less than the critical vibration amplitude B_c.

$$B_c = \frac{\mu P R_d}{E\left(\frac{c_g}{R_g^2} + \frac{c_d}{R_d^2}\right)} \frac{1}{N(N^2-1)} \tag{22}$$

In the sliding zone, the frictional force is equal to sliding frictional force μP. Therefore, the strain caused by friction can be written as

$$\varepsilon_f = -\frac{\mu P R_d}{E A_d}\left(\theta - \frac{\pi}{2N}\right) \tag{23}$$

where R_d and A_d are respectively radius and the cross-sectional area of the ring damper.

The relative displacement on the contact surface can be obtained by integrating the strain. Note that displacement is 0 at the beginning of the sliding zone.

$$\begin{cases} s(\theta) = 0, & 0 \le \theta \le \theta_0 \\ s(\theta) = \int_{\theta_0}^{\theta}(\varepsilon_g - \varepsilon_d - \varepsilon_f)R_f d\theta, & \theta_0 \le \theta \le \frac{\pi}{2N} \end{cases} \tag{24}$$

Therefore,

$$
\begin{aligned}
s(\theta) &= \int_{\theta_0}^{\theta} (\varepsilon_g - \varepsilon_d - \varepsilon_f) R_d d\theta \\
&= \frac{\mu P R_d R_f}{A_d E} \left[-\frac{1}{N^2} \frac{R_d}{R_g} \left(\frac{\sin(N\theta)}{\sin(N\theta_0)} - 1 \right) + \frac{1}{2} \left[(\frac{\pi}{2N} - \theta)^2 - (\frac{\pi}{2N} - \theta_0)^2 \right] \right]
\end{aligned}
\tag{25}
$$

The energy dissipated by the ring damper in a complete vibration cycle, ΔW, can be obtained by integrating the product of the frictional force F_f and the relative displacement $s(\theta)$ in the slip region.

$$
\Delta W = 16N \int_{\theta_0}^{\frac{\pi}{2N}} F_f \Delta s(\theta) R_f d\theta = 16 \frac{(\mu P)^2 R_f^3}{N^2 E A_r} \left\{ \left[\cot(N\theta_0) + N\theta_0 - \frac{\pi}{2} \right] - \frac{1}{3} (\frac{\pi}{2} - N\theta_0)^3 \right\}
\tag{26}
$$

Note that ΔW depends on the critical slip angle θ_0. According to Equation (21), θ_0 is a nonlinear function of B. Therefore, ΔW is a function of B.

3.2. Equivalent Damping Ratio

The loss coefficient η or damping ratio ζ is commonly used to indicate the damping capacity of engineering structures. The loss coefficient η is defined as the ratio of the energy dissipated per radian and the total vibration energy [33]:

$$
\eta = \frac{\Delta W / 2\pi}{W} \simeq 2\zeta
\tag{27}
$$

For small damping, the total vibration energy of the system W approximately equal to the maximum kinetic energy [33]. Thus, the total vibration energy for the jth normal mode can be expressed as

$$
W = \frac{1}{2} [\dot{X}]^T [M][\dot{X}] = \frac{1}{2} \omega_j^2 q_j^2
\tag{28}
$$

Thus, the equivalent structural damping ratio $k_{j,eq}$ in Equation (7) can be rewritten as

$$
\zeta_{j,eq} = \frac{\Delta W}{W} = \frac{16 \frac{(\mu P)^2 R_f^3}{N^2 E A_r} \left\{ \left[\cot(N\theta_0) + N\theta_0 - \frac{\pi}{2} \right] - \frac{1}{3} (\frac{\pi}{2} - N\theta_0)^3 \right\}}{\omega_j^2 q_j^2 / 2}
\tag{29}
$$

4. Application and Discussion

To validate the method shown in this article, the numerical simulation is applied to a real thin-walled gear made of 4310 steel (Young's modulus $E = 207$ GPa and density $\rho = 7.84 \times 10^3$ kg/m³). The mass of the gear is 425 g. Figure 4 shows the mode shape of the model with 3 ND. The corresponding natural frequency is 3758 Hz. For reasons of confidentiality, some of results are given in a normalized form.

4.1. Method Validation

The influence of the normal pressure on damping effect is compared with the results from the forced response analysis based on the harmonic balance method in [34], as shown in Figure 6. The results obtained by the two analysis methods are highly consistent. However, the method shown in this article does not need to calculate the equation of motion in the frequency domain or time domain, so it has faster calculation speed. Since the numerical method shown in this article is independent of excitation and inherent mechanical damping, the excitation and mechanical damping are given in accordance with [34]. Since the normal pressure is not directly given in [34], the normal pressure in this section is a relative value (defined as normalized normal pressure P').

At $P' = 0$, the frictional force at the contact surfaces is 0, and the ring damper can freely slide relative to the gear. The energy dissipated by frictional is 0, and the ring dampers is ineffective. An increment of P' leads to the vibration to decrease down to a minimum value, corresponding to the optimum normalized normal pressure (about 0.45). A further increment of P' causes the vibration to increase again. When P' is large enough (about 1.65), the vibration amplitude increases to the amplitude at $P' = 0$. In this case, no relative motion takes place on the contact surface of the two structures Thus, the ring damper ceased to be effective. It is worth mentioning that two different analysis methods show that when the normal pressure is greater than about 3.7 times of the optimal normal pressure, the ring damper ceased to be effective, which will be further explained in the following parameter sensitivity analyses.

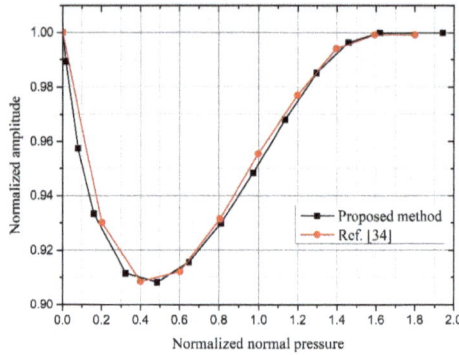

Figure 6. Validation of the proposed method by [34] results: Resonance amplitude by normalized normal pressure.

For a given normal pressure (or rotating speed), when the vibration amplitude B is small, the ring damper is full-stick, and there is no slip, as shown in Figure 7. When B increases to the critical vibration amplitude B_c, sliding appears in $\theta_0 = \pi/2N$. When B increases, the critical slip angle decreases and the slip area increases. When the vibration amplitude is large enough, the critical slip angle approaches 0, and the ring damper is approximately full-slip. In this case, the energy dissipation caused by the ring damper is approximately linear with the vibration amplitude, as shown in Figure 8.

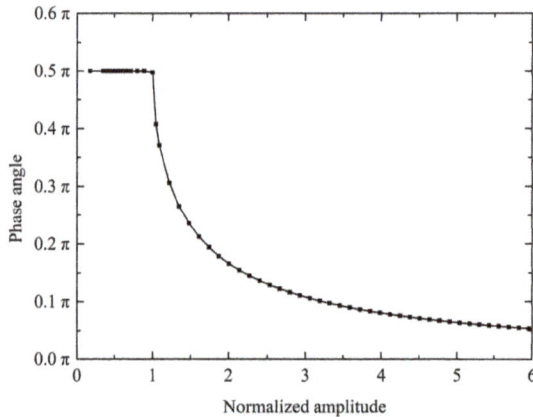

Figure 7. The critical angle versus the normalized amplitude.

The normalized frictional force and the contact state between the gear and the ring damper along the circumferential direction are shown in Figure 9. In Figure 9a, when the vibration amplitude B is less than B_c, the contact state is stick. Thus, no relative motion occurs on the contact surface. Frictional force is a function of θ, and the maximum frictional force appears at the position of the nodal line. When $B = B_c$, slip appears at the position of the nodal line, as shown in Figure 9b. When $B > B_c$, the slip region expands to both sides as B increases, as shown in Figure 9c. When $B \gg B_c$, the slip region increases slowly as the vibration amplitude increases. In this case, the contact status is approximately full-slip, as shown in Figure 9d. This observation is highly consistent with other studies [9,20], according to those studies, when the excitation frequency is far from the natural frequency, the response amplitude is small and the contact status is stick. When the excitation frequency gradually approaches the natural frequency, relative slip appears on the contact surface and the slip region gradually increases.

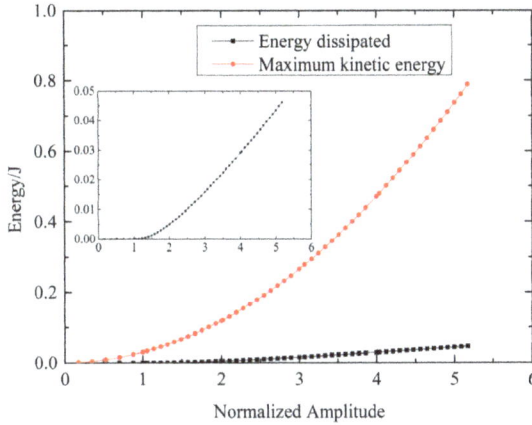

Figure 8. Energy dissipated per cycle by the ring damper and maximum kinetic energy of the system versus normalized amplitude.

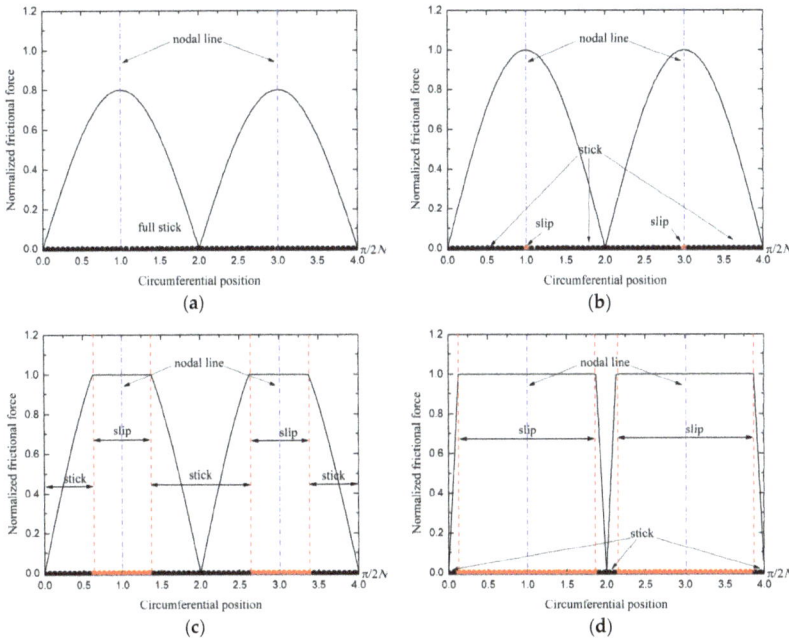

Figure 9. Normalized frictional force and contact state: (**a**) $B < B_c$; (**b**) $B = B_c$; (**c**) $B > B_c$; (**d**) $B \gg B_c$.

4.2. Effect of Ring Damper Parameters

4.2.1. Effect of Rotating Speed or Normal Pressure

The normal pressure on the contact surface depends on the rotating speed of the system, and the normal pressure is proportional to the square of the rotational speed. Thus, in this article, only the effect of rotating speed is shown.

Figure 10 shows the effect of rotating speed on the equivalent damping performance. In Figure 10a, a decrement of the rotating speed causes the contact surface to slide more easily at a given resonance stress, resulting in a lower critical vibration stress. Also, the vibration stress corresponding to the maximum damping ratio decreases as the rotating speed decreases. In Figure 10b, for a given resonance stress, when the rotating speed is greater than about 1.9 times of the optimal rotating speed, the contact surface is full-stick, where the optimum rotating speed is defined as the rotating speed corresponding to the maximum damping ratio.

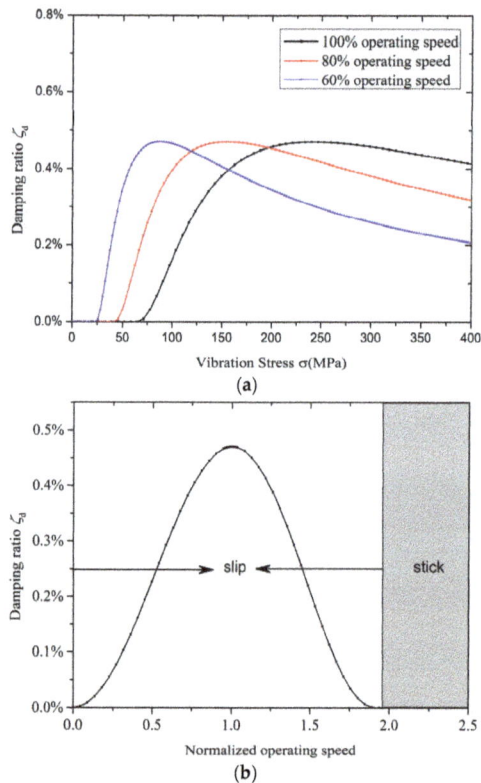

Figure 10. Effect of the rotating speed: (**a**) Friction damping at various rotating speed; (**b**) friction damping for normalized rotating speed (for a given vibration stress).

4.2.2. Effect of Temperature

Figure 11 shows the effect of temperature on the damping performance. The effect of temperature is negligible. This, of course, is because the change in Young's modulus E is small during the operating temperature range. The stiffness of the ring damper is almost unchanged. This also indicates that the ring damper can work at high temperatures and with good temperature adaptability.

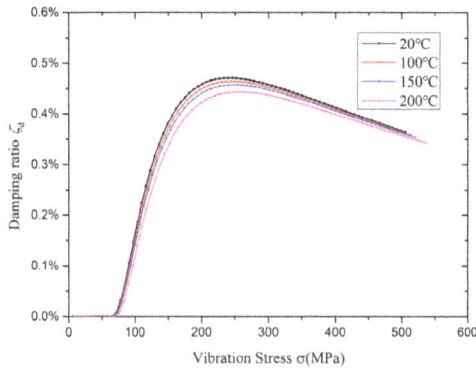

Figure 11. Effect of temperature.

4.2.3. Effect of the Ring Damper Density

The effect of the density is investigated according to its effect on the normal pressure acting on the contact surface. The normal direction is defined as along the radial direction of the gear.

The effect of the ring damper density on the damping performance is shown in Figure 12. The critical vibration stress increases with an increase of the ring damper density. If the density is too large, then the ring damper ceases to be effective due to the contact surface tends to stick. In this case no energy is dissipated by frictional force.

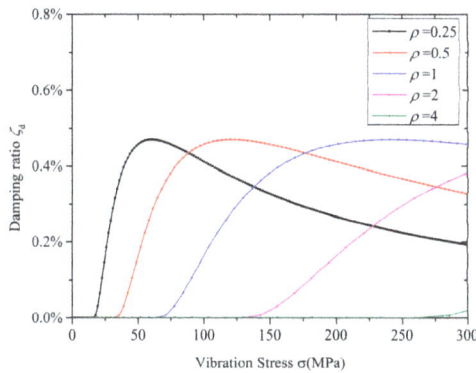

Figure 12. Effect of the ring damper density.

4.2.4. Effect of the Friction Coefficient

As shown in Figure 13a, the effect of the friction coefficient μ on the damping performance is similar to density. Increasing μ results in an increase in critical vibration stress. Moreover, in this case, the contact surface tends to be full-stick due to an increase in μ. In contrast, a decrease in μ results in the contact surface tending to be full-slip. However, due to $F_{f\,max} = \mu P$, the maximum frictional force on the contact surface $F_{f\,max}$ decreased with a decrease in μ. For a given vibration stress, there is an optimum density that maximizes frictional damping. When the density is greater than 3.7 times the optimal density, the ring damper will cease to be effective again, as shown in Figure 13b.

4.2.5. Effect of the Cross-Sectional Area of the Ring Damper

The cross-sectional area is equal to the product of the radial thickness and the axial thickness of the ring damper. The effect of the radial thickness is shown in Figure 14. The critical vibration stress decreases and the peak damping ratio increased with an increase in the radial thickness. Increasing the radial thickness can significantly improve the damping performance. The effect of axial thickness is shown in Figure 15. The critical vibration stress is not affected by the axial thickness. However, the peak damping ratios increase with an increase in the axial thickness. In the premise that the mass of ring dampers is much smaller than the mass of gears, the equivalent damping ratio is approximately linear with the axial thickness.

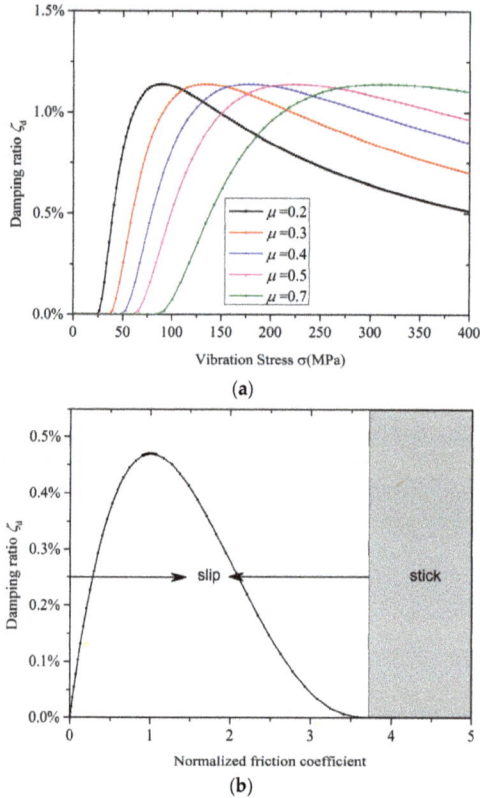

(a)

(b)

Figure 13. Effect of the friction coefficient:(a) Friction damping at various friction coefficient; (b) friction damping for normalized friction coefficient (for a given vibration stress).

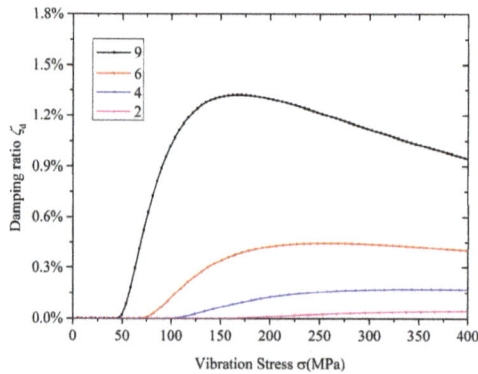

Figure 14. Effect of the radial thickness of the ring damper.

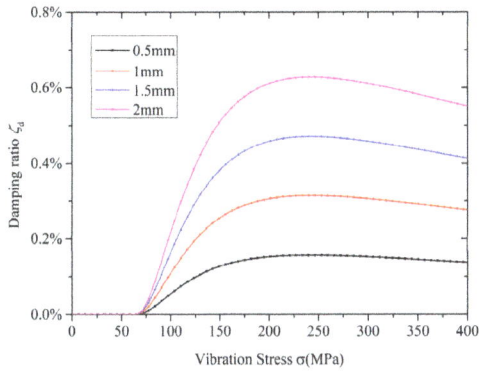

Figure 15. Effect of the axial thickness of the ring damper.

Therefore, for a given cross-sectional area, the ring damper with a large ratio of radial thickness to axial thickness has a better damping effect.

5. Conclusions

In this article, a theoretical study of ring dampers for thin-walled gears has been shown. A numerical method to predict the damping performance of ring dampers is proposed. In the proposed method, the energy dissipated by the ring damper is calculated through a quasi-static process then it is expressed as the equivalent mechanical damping function that depends on vibration stress. The validity of the model is confirmed by a comparison with forced response analysis results. Compared with forced response analysis, the method shown in this article only needs once modal analysis of the primary structure. The proposed method avoids computation of the periodical response of the nonlinear structure. Therefore, minimal computation is required to obtain the damping performance, which greatly improves the efficiency of ring dampers design.

The damping performance of the ring damper depends on the vibration amplitude of the gear B and the damper parameters. When B is less than the critical vibration amplitude B_c, the ring damper is ineffective. When B is greater than B_c, the ring damper can provide friction damping. By increasing B, slip first appears at position of the nodal line, and the slip region expands to both sides as B increases. At approximately 3.7 times the critical vibration amplitude, the efficiency of the damper is theoretically maximized.

For a given amplitude, there is optimum speed, density and friction coefficient to maximize damping. Excessively increasing or decreasing the rotating speed, the ring damper density and the friction coefficient will cause the contact surface to be full-stick or full-slide. In both cases, the ring damper does not provide frictional damping. For a given mass of ring dampers, different damping performances may be observed if the density and the ratio of radial thickness to axial thickness are different.

The proposed method works well when the mass of the ring damper is significantly less than the mass of the primary structure. The ring damper can provide substantial damping and only weakly affects the mode shape of the system. This methodology is suitable for specific applications such as gears or blisks with ring dampers.

Author Contributions: Conceptualization and Methodology, Y.W. and H.Y.; Software development, parameter analysis, and Writing—Original Draft Preparation, H.Y.; Review and Editing, X.J.; Funding acquisition, A.M.

Funding: This work was supported by the National Nature Science Foundation of China (No. 51475022).

Acknowledgments: The authors would like to thank all of the reviewers for their constructive comments. In addition, Hang Ye especially wishes to thank Yiheng Zhang for her continuous support.

Conflicts of Interest: The authors declare no conflict of interest.

Notation

B	vibration amplitude	Subscript g	gear
B_c	critical vibration amplitude	Subscript d	ring damper
C	damping matrices of the gear	Subscript eq	equivalent
c	half-width of the gear rim or the ring damper	W	total energy of the system
E	Young's modulus	w	radial displacement of the groove
$F(t)$	external periodic force	X	displacement vector
$F_{nl}(X, \dot{X}, t)$	nonlinear frictional force	z	number of teeth of the gear
F_f	frictional force per unit length	ε	strain
I	sectional moment of inertia	η	loss coefficient
K	stiffness matrices of the gear	κ	curvature
M	mass matrices of the gear	μ	friction coefficient
M	bending moment	θ	circumferential angle
N	number of nodal diameters	θ_0	critical slip angle
P	normal pressure	ρ	density of the ring damper
P'	normalized normal pressure	ζ	damping ratio
R	radius	ζ_{eq}	equivalent damping ratio provided by the ring damper
s	relative displacement	ΔW	energy dissipated per cycle by the ring damper

References

1. Drago, R.J.; Brown, F.W. The analytical and experimental evaluation of resonant response in high-speed, lightweight, highly loaded gearing. *J. Mech. Des.* **1981**, *103*, 346–356. [CrossRef]
2. Griffin, J.H. Friction damping of resonant stresses in gas turbine engine airfoils. *J. Eng. Power* **1980**, *102*, 329–333. [CrossRef]
3. Cameron, T.M.; Griffin, J.H.; Kielb, R.E. An integrated approach for friction damper design. *J. Vib. Acoust.* **1990**, *112*, 175–182. [CrossRef]
4. Ferri, A.A. Friction damping and isolation systems. *J. Vib. Acoust.* **1995**, *117*, 196–206. [CrossRef]
5. Wang, Y.; Fu, Z.; Jiang, X.; Tian, A. Mistuning effects on aero-elastic stability of axial compressor rotor blades. *J. Eng. Gas Turbines Power* **2015**, *137*, 102504. [CrossRef]
6. Fu, Z.; Wang, Y.; Jiang, X.; Wei, D. Tip Clearance effects on aero-elastic stability of axial compressor blades. *J. Eng. Gas Turbines Power* **2015**, *137*, 012501. [CrossRef]
7. Zhang, X.; Wang, Y.; Xu, K. Mechanisms and key parameters for compressor blade stall flutter. *J. Turbomach.* **2013**, *135*, 024501. [CrossRef]
8. Zhang, X.; Wang, Y.; Xu, K. Flutter prediction in turbomachinery with energy method. *J. Aerosp. Eng.* **2011**, *225*, 995–1002. [CrossRef]
9. Koh, K.H.; Griffin, J.H.; Filippi, S.; Akay, A. Characterization of turbine blade friction dampers. *J. Eng. Gas Turbines Power* **2005**, *127*, 856–862. [CrossRef]
10. Petrov, E.P.; Ewins, D.J. Advanced modeling of underplatform friction dampers for analysis of bladed disk vibration. *J. Turbomach.* **2007**, *129*, 143–150. [CrossRef]
11. Botto, D.; Gastadi, C.; Gola, M.M.; Umer, M. An experimental investigation of the dynamics of a blade with two under-platform dampers. *J. Eng. Gas Turbines Power* **2017**, *140*, 032504. [CrossRef]
12. Botto, D.; Umer, M. A novel test rig to investigate under-platform damper dynamics. *Mech. Syst. Signal. Pr.* **2018**, *100*. [CrossRef]
13. Pesaresia, L.; Sallesa, L.; Jonesb, A.; Greenb, J.S.; Schwingshackla, C.W. Modelling the nonlinear behaviour of an underplatform damper test rig for turbine applications. *Mech. Syst. Signal. Pr.* **2017**. [CrossRef]
14. Kaneko, Y. Vibration response analysis of mistuned bladed disk with under-platform damper: Effect of variation of contact condition on vibration characteristics. In Proceedings of the ASME Turbo Expo 2017: Turbomachinery Technical Conference and Exposition, Charlotte, NC, USA, 26–30 June 2017.
15. Lopez, I.; Busturia, J.M.; Nijmeijer, H. Energy dissipation of a friction damper. *J. Sound Vib.* **2004**, *278*, 539–561. [CrossRef]

16. Lopez, I.; Nijmeijer, H. Prediction and validation of the energy dissipation of a friction damper. *J. Sound Vib.* **2009**, *328*, 396–410. [CrossRef]

17. Laxalde, D.; Thouverez, F.; Lombard, J.P. Forced response analysis of integrally bladed disks with friction ring dampers. *J. Vib. Acoust.* **2010**, *132*, 011013. [CrossRef]

18. Laxalde, D.; Salles, L.; Blanc, L.; Thouverez, F. Non-linear modal analysis for bladed disks with friction contact interfaces. In Proceedings of the ASME Turbo Expo 2008: Power for Land, Sea, and Air. American Society of Mechanical Engineers, Berlin, Germany, 9–13 June 2008.

19. Zucca, S.; Firrone, C.M.; Facchini, M. A method for the design of ring dampers for gears in aeronautical applications. *J. Mech. Des.* **2012**, *134*, 091003. [CrossRef]

20. Tang, W.; Epureanu, B.I. Geometric optimization of dry friction ring dampers. *Int. J. Nonlin. Mech.* **2018**. [CrossRef]

21. Tang, W.; Epureanu, B.I. Nonlinear dynamics of mistuned bladed disks with ring dampers, International. *Int. J. Nonlin. Mech.* **2017**, *97*, 30–40. [CrossRef]

22. Firrone, C.M.; Zucca, S. Passive control of vibration of thin-walled gears: Advanced modelling of ring dampers. *Nonlinear Dyn.* **2014**, *76*, 263–280. [CrossRef]

23. Buyukataman, K. A theoretical study on the vibration damping of aircraft gearbox gears. In Proceedings of the 27th Joint Propulsion Conference, Sacramento, CA, USA, 24–26 June 1991.

24. Niemotka, M.A.; Ziegert, J.C. Optimal design of split ring dampers for gas turbine engines. *J. Eng. Gas Turbines Power* **1995**, *117*, 569–575. [CrossRef]

25. Baek, S.; Epureanu, B. Reduced-order modeling of bladed disks with friction ring dampers. *J. Vib. Acoust.* **2017**, *139*, 061011. [CrossRef]

26. Tangpong, X.W.; Wickert, J.A.; Akay, A. Finite element model for hysteretic friction damping of traveling wave vibration in axisymmetric structures. *J. Vib. Acoust.* **2008**, *130*, 011005. [CrossRef]

27. Tangpong, X.W.; Wickert, J.A.; Akay, A. Distributed friction damping of travelling wave vibration in rods. *Philos. Trans. R. Soc. A* **2008**, *366*, 811–827. [CrossRef] [PubMed]

28. Alford, J.S. Protection of labyrinth seals from flexural vibration. *J. Eng. Power* **1964**, *86*, 141–147. [CrossRef]

29. Alford, J.S. Protecting turbomachinery from self-excited rotor whirl. *J. Eng. Power* **1965**, *87*, 333–343. [CrossRef]

30. Alford, J.S. Protecting turbomachinery from unstable and oscillatory flows. *J. Eng. Power* **1967**, *89*, 513–527. [CrossRef]

31. Alford, J.S. Nature, causes, and prevention of labyrinth air seal failures. *J. Aircraft* **1975**, *12*, 313–318. [CrossRef]

32. Weaver, J.W.; Timoshenko, S.P.; Young, D.H. *Vibration Problems in Engineering*, 5th ed.; Wiley-Interscience: Hoboken, NJ, USA, 1990; ISBN 9780471632283.

33. Rao, S.S. *Mechanical Vibrations*, 5th ed.; Prentice Hall: New York, NY, USA, 1995; ISBN 978-0-13-212819-3.

34. Tang, W.; Baek, S.; Epureanu, B.I. Reduced-order models for blisks with small and large mistuning and friction dampers. *J. Eng. Gas Turbines Power* **2017**, *139*, 012507. [CrossRef]

symmetry

MDPI

Article

A Novel Tool for Supervised Segmentation Using 3D Slicer

Daniel Chalupa * and Jan Mikulka

Department of Theoretical and Experimental Electrical Engineering, Brno University of Technology,
Technická 3082/12, 616 00 Brno, Czech Republic; mikulka@feec.vutbr.cz
* Correspondence: daniel.chalupa@vut.cz

Received: 15 October 2018; Accepted: 7 November 2018; Published: 12 November 2018

Abstract: The rather impressive extension library of medical image-processing platform 3D Slicer lacks a wide range of machine-learning toolboxes. The authors have developed such a toolbox that incorporates commonly used machine-learning libraries. The extension uses a simple graphical user interface that allows the user to preprocess data, train a classifier, and use that classifier in common medical image-classification tasks, such as tumor staging or various anatomical segmentations without a deeper knowledge of the inner workings of the classifiers. A series of experiments were carried out to showcase the capabilities of the extension and quantify the symmetry between the physical characteristics of pathological tissues and the parameters of a classifying model. These experiments also include an analysis of the impact of training vector size and feature selection on the sensitivity and specificity of all included classifiers. The results indicate that training vector size can be minimized for all classifiers. Using the data from the Brain Tumor Segmentation Challenge, Random Forest appears to have the widest range of parameters that produce sufficiently accurate segmentations, while optimal Support Vector Machines' training parameters are concentrated in a narrow feature space.

Keywords: 3D slicer; classification; extension; random forest; segmentation; sensitivity analysis; support vector machine; tumor

---

1. Introduction

3D Slicer [1] is a free open-source platform for medical image visualization and processing. Its main functionality comes from the Extension Library, which consists of various modules that allow specific analyses of the input data, such as filtering, artefact suppression, or surface reconstruction. There is a lack of machine-learning extensions except for the open-source DeepInfer [2]. This deep-learning deployment kit uses 3D convolutional neural networks to detect and localize the target tissue. The development team demonstrated the use of this kit on the prostate segmentation problem for image-guided therapy. Researchers and practitioners are able to select a publicly available task-oriented network through the module without the need to design or train it.

To enable the use of other machine-learning techniques, we developed the Supervised Segmentation Toolbox as an extensible machine-learning platform for 3D Slicer. Currently, Support Vector Machine (SVM) and Random Forest (RF) classifiers are included. These classifiers are well-researched and often used in image-processing tasks, as demonstrated in References [3–8].

SVMs [9] train by maximizing the distance between marginal samples (also referred to as support vectors) and a discriminative hyperplane by maximizing f in the equation:

$$f(\alpha_1 \cdots \alpha_n) = \sum \alpha_i - \frac{1}{2} \sum_i \sum_j \alpha_i \alpha_j y_i y_j \vec{x_i} \cdot \vec{x_j}, \tag{1}$$

where y is defined as +1 for a class A sample and −1 for a class B sample, α is a Lagrangian multiplier, and $\vec{x}$ is the feature vector of the individual sample. On real data, this is often too strict because of noisy samples that might cross their class boundary. This is solved by using a combination of techniques known as the kernel trick and soft margining. The kernel trick uses a kernel function ϕ to remap the original feature space to a higher-dimensional one by replacing the dot product in Equation (1) with $\phi(x_i) \cdot \phi(x_j)$. This allows linear separation of the data as required by the SVM. An example of the kernel function is the radial basis function used in this study. Incorporating a soft margin allows some samples to cross their class boundary. Examples of such soft-margin SVMs are the C-SVM [10] and N-SVM [11]. The C parameter of the C-SVM modifies the influence of each support vector on the final discriminatory hyperplane. The larger the C, the closer the soft-margin C-SVM is to a hard-margin SVM. The N parameter of the N-SVM defines a minimum number of support vectors and, consequently, an upper bound of the guaranteed maximum percentage of misclassifications. Further modifications to the SVM can also be done, such as using fuzzy-data points for the training dataset, as demonstrated in References [5,6].

RF uses a voting system on the results of the individual decision trees. Each decision tree is created on the basis of a bootstrapped sample of the training data [12].

2. Materials and Methods

The data used in this study describe Low Grade Glioma (LGG) obtained by MRI. The dataset consists of four images of one patient (LG_0001, Figure 1): T1- and T2-weighted, contrast-enhanced T1C, and Fluid Attenuated Inversion Recovery (FL). These data are part of the training set featured in the Brain Tumor Segmentation (BraTS) 2012 Challenge.

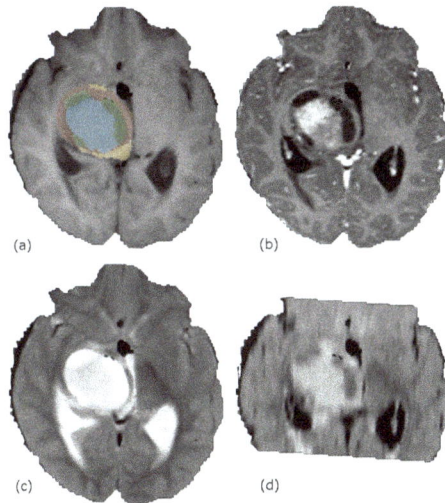

Figure 1. Slice 88 of the LG_0001 data from the 2012 Brain Tumor Segmentation (BraTS) Challenge. The dataset consists of a (**a**) T1-weighted image with labels as overlay; (**b**) postcontrast T1-weighted image; (**c**) T2-weighted image; and (**d**) Fluid Attenuated Inversion Recovery (FL) image.

The free and open-source Supervised Segmentation Toolbox extension [13] of the 3D Slicer was used throughout this study. The extension allows the user to train a range of classifiers using labeled data. The user is also able to perform a grid search to select the optimal parameters for the classifier. To achieve this, the extension uses either an already available function or a cross-validation algorithm developed by the author of the extension, depending on the classifier library used. Currently, N-SVM and C-SVM from the dlib library [14] and C-SVM and Random Forest from Shark-ml library [15] are incorporated. The extension takes care of the parallelizable parts of the training and classification subtasks, thus significantly reducing computation times. A preprocessing algorithm selection is also a part of the extension. This allows for artefact correction or feature extraction. The extension workflow is depicted in Figure 2.

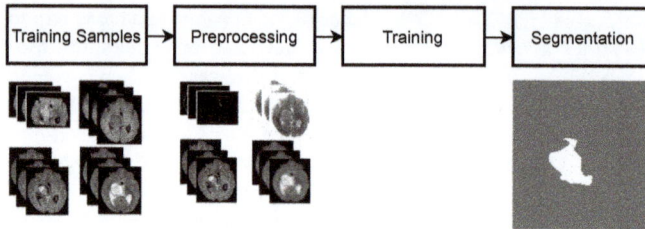

Figure 2. Supervised segmentation extension workflow.

3. Results and Discussion

A series of tests were performed in order to provide a sense of the speed and accuracy of the provided classifiers. Sensitivity and specificity metrics were used to evaluate the results. A classifier that had a larger sum of specificity and sensitivity (or their respective means, when cross-validation was used) was considered a better classifier. During the first test run, each type of classifier was trained and evaluated using the single patient image set. Optimal training parameters of the classifiers were obtained using a grid-search approach. The results are presented in Figures 3–5. The γ parameter is common in both SVM classifiers and influences the variance of the radial basis kernel. A large γ means that more data points will look similar, thus preventing overfitting. Using the aforementioned dataset, the results indicate a relative insensitivity of the classification accuracy on this parameter. For the given dataset, C values of the C-SVM larger than 1 seem optimal. The best results of the N-SVM classifier are obtained with N around 10% or lower combined with a high-variance radial basis function. Optimal RF training parameters were: a small node size of under 5, number of trees higher than 800, no out-of-bag samples, and no random attributes.

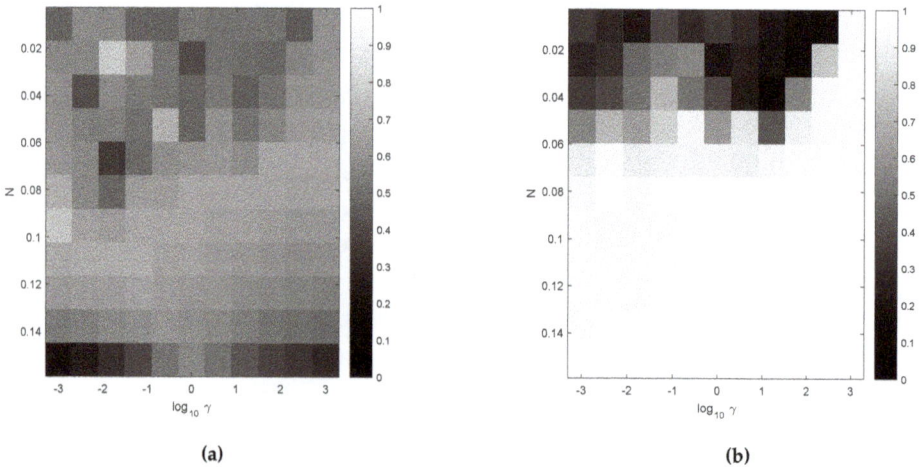

(a)

(b)

Figure 3. N-Support Vector Machine (SVM)-classifier (**a**) sensitivity and (**b**) specificity using different parameter pairs.

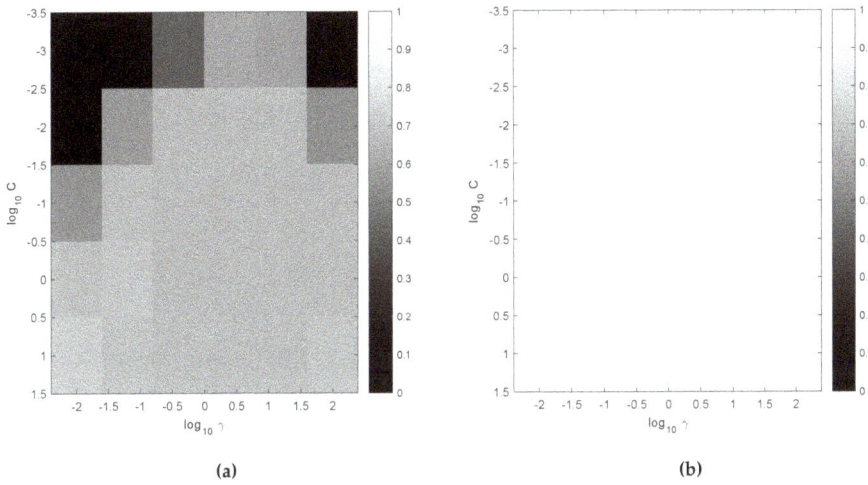

(a)   (b)

Figure 4. C-SVM-classifier (**a**) sensitivity and (**b**) specificity using different parameter pairs.

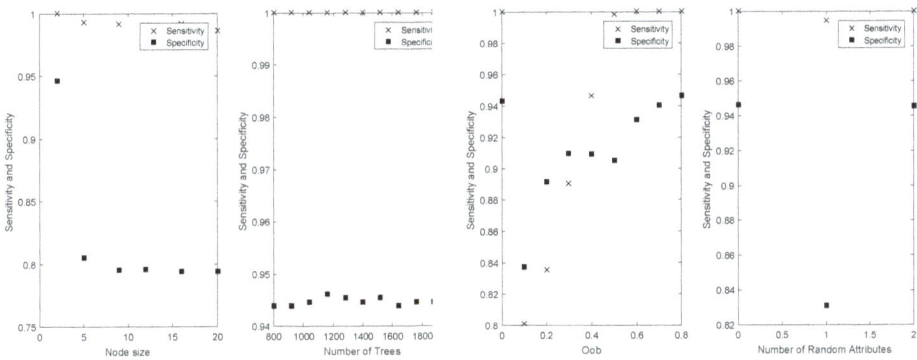

Figure 5. Random Forest classifier sensitivity and specificity using different parameters. Left to right: Different node size, number of trees, OOB and number of random attributes.

The second test run consisted of using a different number of slices around the center of the tumor to reveal the impact of the size of the training set on the specificity and sensitivity of all classifiers (Figures 6a and 7). The results indicate that reducing the number of unique training samples has a negligible effect on the subsequent classification accuracy. RF shows slightly better classification accuracy improvement when using a larger training vector. Using a reduced training dataset influences training process length and might result in a simpler classifier, which is easier to interpret and has shorter classification computation times. The classification time is a limiting factor of using these methods in real-time applications.

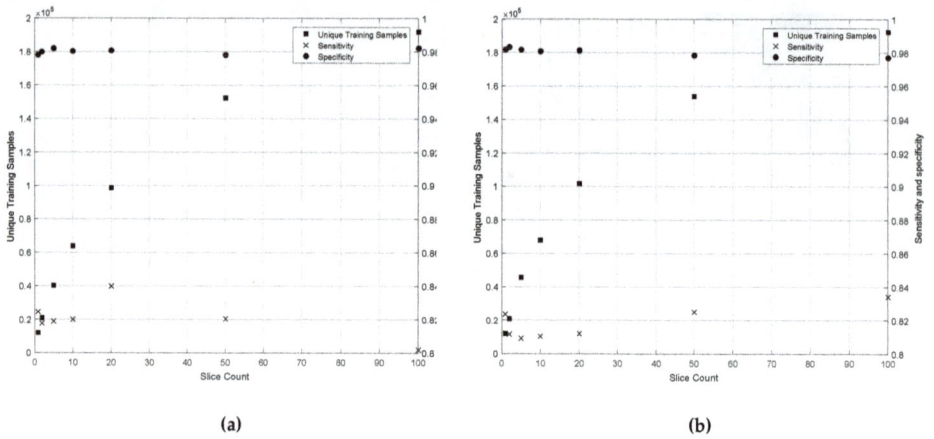

(a)   (b)

Figure 6. (a) N-SVM and (b) C-SVM sensitivity and specificity using different training vector sizes.

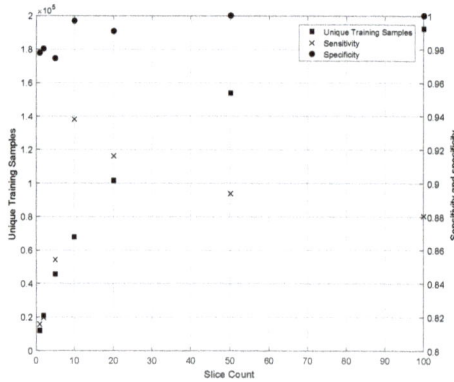

Figure 7. RF-classifier sensitivity and specificity using different training vector sizes.

The effect of different image types on classifier accuracy was examined in the last test run (Figures 8a and 9). Slices 88 of the LG_0001 images were used as a source of training samples. Sufficient sensitivity and specificity were obtained by only using T1- and T2-weighted images. Furthermore, all classifiers benefited from the addition of a postcontrast T1-weighted image. The RF classifier achieved best overall results with the use of FL, and postcontrast T1- and T2-weighted images.

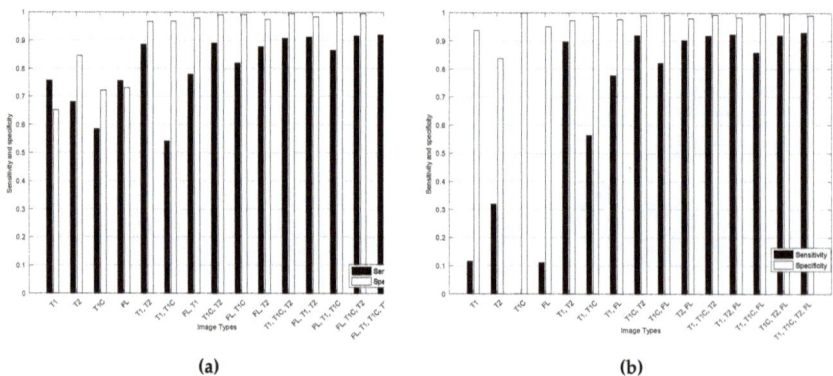

(a)   (b)

Figure 8. Classifier sensitivity using different input images. N-SVM (**a**), C-SVM (**b**).

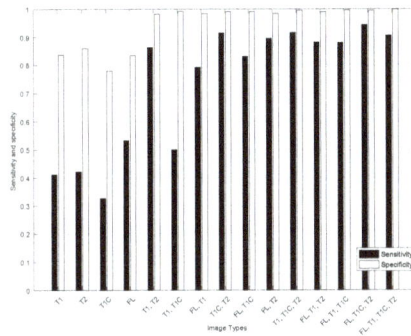

Figure 9. RF-Classifier sensitivity using different input images.

The following standardized procedure was designed in order to compare classifier performance. Training samples were extracted from the whole volume of the unmodified T1C and T2 images. Then, sensitivity and specificity were obtained using fivefold cross-validation. The best performing parameters and results are reported in Table 1. Segmentations are shown in Figure 10. Classification results can be further improved by using preprocessed data instead of raw data, and by means of postprocessing to remove the outlying voxels and inlying holes as demonstrated in Reference [16].

Table 1. Classifier comparison and best-performing parameters.

	Parameters	Sensitivity	Specificity	Acc.	Prec.	DICE	Jaccard
C-SVM	$\gamma = 1.0, C = 1.0$	0.72	0.98	0.99	0.96	0.80	0.66
N-SVM	$\gamma = 10.0^{-3}, N = 0.1$	0.77	0.97	0.99	0.96	0.82	0.70
RF	0 % OOB, 0 random attributes, 1200 trees, node size 2	1.00	0.95	1.00	0.96	0.94	0.89

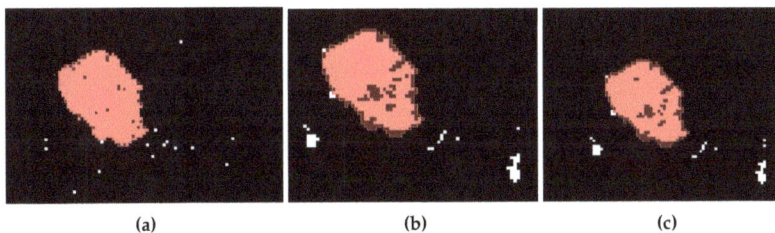

(a)   (b)   (c)

Figure 10. (a) RF, (b) C-SVM, and (c) N-SVM classification results of the slice 88 (white) and ground truth (red).

Lastly, the performance of the RF classifier trained on all tumor cores of the 20 real high-grade glioma volumes using the 3D Slicer extension were compared to similar studies performed on the BraTS dataset. The values were obtained as a mean of fivefold cross-validation. This comparison is shown in Table 2. The other DICE values are from Reference [17].

The means to combine the results of different classifiers to further expand the usability of the Supervised Segmentation Toolbox extension were added. Currently, logical AND and OR and a majority voting system are implemented. An addition of a Multiple Classifier System (MCS) is currently considered. A review of the advantages of MCS is provided by Wozniak et al. [18]. Termenon and Graña [19] used a two-stage MCS where the second classifier was trained on low-confidence data obtained by training and analysis of the first classifier. In the future, the authors expect implementing additional classifiers as well. Adding a Relevance Vector Machine (RVM), for example, might bring an improvement over SVM [20].

Table 2. RF classifier comparison with similar studies. The classifier was trained using all 20 of the real high-grade glioma volumes, and the DICE value is a mean of fivefold cross-validation.

Paper	Approach	DICE
This paper	RF	0.43
Geremia [21]	Spatial decision forests with intrinsic hierarchy	0.32
Kapás [16]	RF	0.58
Bauer [22]	Integrated hierarchical RF	0.48
Zikic [23]	Context-sensitive features with a decision tree ensemble	0.47
Festa [24]	RF using neighborhood and local context features	0.50

4. Conclusions

The Supervised Segmentation Toolbox extension was presented as an addition to the 3D Slicer extension library. This extension allows the user to train and use three types of classifiers, with more to be added in the future. The usability of the extension was demonstrated on a brain-tumor segmentation use case. The effects of the training parameters of all classifiers on the final sensitivity and specificity of the classification were considered to provide an insight into usable parameter selection for future studies. A low γ in combination with softer margin terms resulted in a better performing classifier commonly for both SVM classifiers. This might be largely due to a limited training sample, and a broader dataset should be analyzed in order to generalize the results. The RF classifier performed best using no added randomization, a relatively large tree count, and a small node size. The possibility of reducing training vector size in order to reduce model complexity and decrease classification time is verified. A 20-fold increase of the number of unique training samples resulted, at best, in a 2% increase of specificity. All combinations of input images are considered as a training input for all classifiers, and the significance of adding more types of images is discussed. A combination of T1C and T2 images performed sufficiently for all classifiers. The addition of the FL image brought a slight improvement in sensitivity. Lastly, best-performing parameter combinations were listed and the corresponding results were compared. The RF classifier had the largest sensitivity and worst specificity, and C-SVM performed oppositely. The significance of these two metrics largely depends on the type of task for which the classifiers are used. All sensitivity and specificity data were obtained directly using the 3D Slicer extension.

Supplementary Materials: The authors publicly release the source code of the Supervised Segmentation Toolbox. The source code can be found on GitHub [13]. The BraTS 2012 dataset is available on the challenge's website.

Author Contributions: J.M. conceived and designed the experiments; D.C. performed the experiments, programmed the extension, analyzed the data, and wrote the paper.

Funding: This work was supported by the Ministry of Health of the Czech Republic, grant no. NV18-08-00459.

Conflicts of Interest: The authors declare no conflict of interest.

Abbreviations

The following abbreviations are used in this manuscript:

BraTS	Brain-Tumor Segmentation
FL	Fluid-Attenuated Inversion Recovery
LGG	Low-Grade Glioma
MCS	Multiple Classifier System
OOB	Out Of Box
RVM	Relevance Vector Machine
RF	Random Forest
SVM	Support Vector Machine

References

1. Fedorov, A.; Beichel, R.; Kalpathy-Cramer, J.; Finet, J.; Fillion-Robin, J.C.; Pujol, S.; Bauer, C.; Jennings, D.; Fennessy, F.; Sonka, M.; et al. 3D Slicer as an image computing platform for the Quantitative Imaging Network. *Magn. Reson. Imaging* **2012**, *30*, 1323–1341. [CrossRef] [PubMed]
2. Mehrtash, A.; Sedghi, A.; Ghafoorian, M.; Taghipour, M.; Tempany, C.M.; Wells, W.M.; Kapur, T.; Mousavi, P.; Abolmaesumi, P.; Fedorov, A. Classification of Clinical Significance of MRI Prostate Findings Using 3D Convolutional Neural Networks. *Proc. SPIE Int. Soc. Optical Eng.* **2017**, *10134*.
3. Zhang, Y.; Dong, Z.; Wang, S.; Ji, G.; Yang, J. Preclinical diagnosis of magnetic resonance (MR) brain images via discrete wavelet packet transform with Tsallis entropy and generalized eigenvalue proximal support vector machine (GEPSVM). *Entropy* **2015**, *17*, 1795–1813. [CrossRef]
4. Zhang, Y.; Wang, S.; Phillips, P.; Dong, Z.; Ji, G.; Yang, J. Detection of Alzheimer's disease and mild cognitive impairment based on structural volumetric MR images using 3D-DWT and WTA-KSVM trained by PSOTVAC. *Biomed. Signal Process. Control* **2015**, *21*, 58–73. [CrossRef]
5. Wang, S.; Li, Y.; Shao, Y.; Cattani, C.; Zhang, Y.; Du, S. Detection of dendritic spines using wavelet packet entropy and fuzzy support vector machine. *CNS Neurol. Dis.-Drug Targets* **2017**, *16*, 116–121. [CrossRef] [PubMed]
6. Zhang, Y.; Yang, Z.; Lu, H.; Zhou, X.; Phillips, P.; Liu, Q.; Wang, S. Facial emotion recognition based on biorthogonal wavelet entropy, fuzzy support vector machine, and stratified cross validation. *IEEE Access* **2016**, *4*, 8375–8385. [CrossRef]
7. Zheng, Q.; Wu, Y.; Fan, Y. Integrating semi-supervised and supervised learning methods for label fusion in multi-atlas based image segmentation. *Front. Neuroinform.* **2018**, *12*, 69. [CrossRef] [PubMed]
8. Amiri, S.; Mahjoub, M.A.; Rekik, I. Tree-based Ensemble Classifier Learning for Automatic Brain Glioma Segmentation. *Neurocomputing* **2018**, *313*, 135–142. [CrossRef]
9. Cortes, C.; Vapnik, V. Support-vector networks. *Mach. Learn.* **1995**, *20*, 273–297. [CrossRef]
10. Chang, C.C.; Lin, C.J. LIBSVM: A library for support vector machines. *ACM Trans. Intell. Syst. Technol.* **2011**, *2*, 1–27. [CrossRef]
11. Schölkopf, B.; Smola, A.J.; Williamson, R.C.; Bartlett, P.L. New support vector algorithms. *Neural Comput.* **2000**, *12*, 1207–1245. [CrossRef] [PubMed]
12. Ho, T.K. Random decision forests. In Proceedings of the 3rd International Conference on Document Analysis and Recognition, Montreal, QC, Canada, 14–16 August 1995; pp. 278–282.
13. Chalupa, D. Supervised Segmentation Toolbox for 3D Slicer. Source Code Available Under the GNU General Public License v3.0. Available online: https://github.com/chalupaDaniel/slicerSupervisedSegmentation (accessed on 14 October 2018).
14. King, D.E. Dlib-ml: A machine learning toolkit. *J. Mach. Learn. Res.* **2009**, *10*, 1755–1758.
15. Igel, C.; Heidrich-Meisner, V.; Glasmachers, T. Shark. *J. Mach. Learn. Res.* **2008**, *9*, 993–996.
16. Kapás, Z.; Lefkovits, L.; Szilágyi, L. Automatic Detection and Segmentation of Brain Tumor Using Random Forest Approach. In *Modeling Decisions for Artificial Intelligence*; Springer: New York, NY, USA, 2016; pp. 301–312.
17. Menze, B.H.; Jakab, A.; Bauer, S.; Kalpathy-Cramer, J.; Farahani, K.; Kirby, J.; Burren, Y.; Porz, N.; Slotboom, J.; Wiest, R.; et al. The Multimodal Brain Tumor Image Segmentation Benchmark (BRATS). *IEEE Trans. Med. Imaging* **2015**, *34*, 1993–2024. [CrossRef] [PubMed]
18. Woźniak, M.; Graña, M.; Corchado, E. A survey of multiple classifier systems as hybrid systems. *Inf. Fusion* **2014**, *16*, 3–17. [CrossRef]
19. Termenon, M.; Graña, M. A two stage sequential ensemble applied to the classification of Alzheimer's disease based on MRI features. *Neural Process. Lett.* **2012**, *35*, 1–12. [CrossRef]
20. Tipping, M.E. Sparse Bayesian learning and the relevance vector machine. *J. Mach. Learn. Res.* **2001**, *1*, 211–244.
21. Geremia, E.; Menze, B.H.; Ayache, N. Spatial Decision Forests for Glioma Segmentation in Multi-Channel MR Images. *MICCAI Chall. Multimodal Brain Tumor Segmentation* **2012**, *34*.
22. Bauer, S.; Nolte, L.P.; Reyes, M. Fully automatic segmentation of brain tumor images using support vector machine classification in combination with hierarchical conditional random field regularization. In *Medical Image Computing and Computer-Assisted Intervention—MICCAI*; Springer: Berlin/Heidelberg, Germany, 2011.

23. Criminisi, A.; Zikic, D.; Glocker, B.; Shotton, J. Context-sensitive classification forests for segmentation of brain tumor tissues. *Proc MICCAI-BraTS* **2012**, *1*, 1–9.

24. Festa, J.; Pereira, S.; Mariz, J.; Sousa, N.; Silva, C. Automatic brain tumor segmentation of multi-sequence MR images using random decision forests. *Proc. NCI-MICCAI BRATS* **2013**, *1*, 23–26.

symmetry

MDPI

Article

Accessibility Evaluation of High Order Urban Hospitals for the Elderly: A Case Study of First-Level Hospitals in Xi'an, China

Ke Ruan and Qi Zhang *

School of Civil Engineering, Xi'an University of Architecture and Technology, Xi'an 710055, China;
rebecca111ruanke@sina.com
* Correspondence: zhangqi-xauat@163.com; Tel.: +86-135-7253-3872

Received: 11 September 2018; Accepted: 8 October 2018; Published: 12 October 2018

Abstract: With reference to the hospitalizing trips made by the elderly, the impedance of these trips that require the use of public transportation, is introduced. An evaluation model that can accurately detect the accessibility of high order urban hospitals (HOUHs) for the elderly is established with the help of Geographic Information System (GIS) technology. Furthermore, the established model is employed to detect the accessibility of first-level hospitals in Xi'an City. Results showed that the traffic connection between hospitals and their service objects is an important factor for the feasibility and effectiveness of an accessibility evaluation. It is suggested that special evaluations of the accessibility of hospitals for the elderly are needed to achieve the human-oriented goal of urban traffic planning. The well-served spatial pattern of hospitalizing accessibility for the elderly in Xi'an City has been established in recent years because of the strategies for public transit metropolis. The accessibility constraints can be divided into three types: The imprisonment, the antagonism and the running-in, for which the corresponding countermeasures to settle the low accessibility of hospitals will be taken by the planning administration. Attention is paid to specific population groups during their hospitalizing trips in the accessibility research, which is beneficial for enabling the improvement of the current traditional method which is mainly based on travel facilities.

Keywords: aged; high order urban hospitals (HOUHs); accessibility; evaluation model; trip impedance based on public transportation; urban traffic planning

---

1. Introduction

The concept of accessibility, which is the opportunity for nodes in network traffic to interact with one another, was firstly proposed by Hansen in 1959 [1]. From then on, a lot of attention has been paid to researching traffic accessibility in many scientific and professional fields such as economic geography, urban and rural planning, and transportation engineering [2–4]. Although there are differences in the theoretical perspectives of different subjects, there is a general definition for traffic accessibility. That is, the degree of convenience of travelling from one place to another by means of a specific mode of transportation.

With reference to spatial scale and research objects, research on traffic accessibility can be divided into two categories: Regional traffic accessibility, and the accessibility of urban public service facilities [5]. Comparatively speaking, research on the spatial pattern and the temporal and spatial evolution of regional transportation accessibility has made significant progress in China and other countries [6,7]; however, relevant research on the transportation accessibility of urban public service facilities started relatively late in China [8]. As urbanization in China is speeding up, it has become an important and urgent task to evaluate the rationality of the spatial layout of various urban public service facilities or the planning of urban transportation systems based on the measure of traffic accessibility, to promote the optimization of urban development.

The traffic accessibility of hospitals, one of the most fundamental service facilities in the city, is an important aspect to consider when measuring the fairness of the allocation of medical service facilities in the city. In other countries, Higgs and Gould [9] made evaluations on the traffic accessibility of hospitals by selecting factors such as the number of beds, the number of doctors, transportation supply and population density. Joseph [10] put forward a potential model when studying hospital accessibility in Wellington Formation of Noble Country, Oklahoma, which was used to characterize the space competition effect on supply and demand. In China, Wang and Zhang [11] used the Thiessen Polygon Method to initially determine the nearest service area of each residential area in Pudong New District in Shanghai, with the support of a GIS platform. Following this they put the actual layout of the hospitals into these service areas to obtain the accessibility of each hospital. Song [12] and others revealed the difference in the spatial accessibility of the Hospital of Rudong County in Jiangsu province through the appropriate selection of travel friction coefficients, and the potential model was supplemented with GIS technology. Hou and Jiang [13] discussed the temporal and spatial distribution characteristics of accessibility between residential districts and hospitals in Changchun city by using the ArcGIS Network Analysis method, and the shortest accessible time and public transportation service frequency were set as evaluation indexes for the characteristics of traffic conditions during off-peak and peak periods.

Currently, eminent progress has been made in Chinese research on hospital traffic accessibility. However, few special studies on hospitalizing accessibility for the elderly focus on the theoretical framework construction and analysis model. With the gradual aging of the Chinese society, the elderly urban population is getting larger [14], and its health condition is declining every day. As a result, the elderly are bound to become the most urban travel population with the greatest demand for medical services. Meanwhile, the elderly population is still situated at a weak position among other users of the urban transportation system, as their declining physical strength and energy determines their gradual inability to drive both motor and non-motor vehicles. Thus, the elderly have less of a choice when choosing their mode of transportation in comparison with other urban populations.

With the aim of paying attention to the disadvantaged elderly groups that are affected by both urban transportation and the construction of a people-oriented urban medical service system, the elderly population was set as a research object for hospitalizing accessibility, and was used to explore the traffic mode preference and travel time characteristics of the elderly. In the secondary development of the Arc-GIS platform and in the specific case of Xi'an City, an effective evaluation method was built for the traffic accessibility of high order urban hospitals (HOUHs) in the city, with the aim of serving the elderly.

This paper tries to establish a new evaluation method for hospitalizing accessibility by adding the trip influences of specific population groups to the traditional model, based on travel facilities. The case study of 45 blocks and 17 HOUHs in Xi'an City shows that the well-served spatial pattern of hospitalizing accessibility for the elderly has been formed. On the other hand, accessibility constraints can still be found for the following reasons: To ensure that targeted countermeasures with regards to urban planning and administration, are taken. However, little consideration is given to population groups other than the elderly (e.g., children and the disabled) in this study, while there is also a lack of sub-classifications of the elderly. Both should be considered in future studies.

2. Background and Concepts

2.1. Potential Model

The potential model is a classic model that uses the laws of gravitation as reference points to study the spatial interaction of social and economic activities. The general connotation is that the energy produced in the system by Object A to Object B is directly proportional to the scale of activity of Object A at its own position, and is inversely proportional to the trip impedance (e.g., distance or time) between the locations of Object A and Object B, while the potential of any point in the system is equal to the sum of the potential produced by each object to that point [15]. The formulas for applying the potential model to study hospitalizing accessibility are:

$$A_{ti} = \sum_{j=1}^{n} \frac{M_j}{D_{ij}^{\beta} V_j}, \ V_j = \sum_{k=1}^{m} \frac{P_k}{D_{kj}^{\beta}}. \tag{1}$$

In Formula (1) A_{ti} is the cumulative attraction value for all the hospitals in the research area to the research unit i; M_j stands for the service ability of the hospital, which is mostly expressed by the number of health personnel or beds in the hospital; D_{kj} stands for trip impedance between the research unit i (such as residential area or block) and the hospital (j), which can be described by the travel distance or travel time between the origin and the destination; P_k indicates the number of populations for research unit k; V_j stands for the influencing factor of the population size; β denotes the travel friction coefficient; and n and m, respectively, represent the number of hospitals and research units. The greater the value of A_{ti} is, the better the spatial accessibility will be. Therefore, as trip impedance (D_{kj}) increases, A_{ti} will decrease accordingly.

2.2. Trip Impedance

How to measure trip impedance is the key to building a traffic accessibility evaluation method based on a potential model. When measuring regional accessibility [16,17], due to the large research scale, the research object can usually be abstracted as a particle, and the linear distance between point pairs can be used to approximately calculate the degree of trip impedance. The difference is that travel paths are usually blocked by various buildings and occur on many tortuous urban roads on the microscopic interior scale of the city. Therefore, some researchers [18,19] have proposed a method to define trip impedance by searching the shortest path of urban roads. In addition, some researchers [20,21] have noticed that factors such as city road grades and road conditions may affect the traffic speed of the road itself, which may in turn influence the traffic accessibility measurement. Thus, it is suggested that the travel time obtained by stacking the travel distance and the travel speed, should be the fundamental basis for determining trip impedance.

There are two main characteristics for the urban elderly traveling to HOUHs [22]. Firstly, they usually take vehicles that do not require autonomous operation. In the case of major diseases, the elderly need to take public transportation to HOUHs for medical treatment. In other cases, they will walk to low-grade hospitals or clinics for treatment when they have ailments. Secondly, the travel time distribution does not coincide with the morning and evening peaks of urban vehicles. With the widespread promotion of expert online appointment registrations in HOUHs, the elderly are more inclined to decide their appointment time for medical treatment hospitals accord to their daily schedules.

Under the constraints of the above travel characteristics, not only must the simple representation method for measuring the linear distance be eliminated, but also the idea of measuring the shortest path or the travel time on urban roads cannot be adopted to determine trip impedance because public transport cannot completely follow the shortest route to enable the elderly to go to HOUHs. On account of safety, buses are also required to pass on various grades of urban roads at stable, medium, and low speed. Only when there is traffic congestion during rush hour in the morning or evening, will there be an obvious reduction in speed. For this sake, the distribution of public transportation lines is used as the basis for studying trip impedance, and a traffic accessibility evaluation model is set up for HOUHs which serve the elderly.

2.3. The Improved Potential Model Based on the Trip Impedance of Public Transportation

The calculation formula of trip impedance expressed by public transportation lines is:

$$D_{bus} = \frac{1}{m} \sum_{i=1}^{m} \frac{a_i}{(b_i/L)}. \tag{2}$$

In Formula (2) a_i indicates the length of the i-th bus line from the target hospital to the target research unit, wherein the bigger a_i is, the longer the length between the hospitals to the target research unit is, and the stronger road resistance will be; b_i stands for the length of the bus line inside the target research unit, wherein the bigger b_i is, the longer the length of the bus line inside the target research unit will be and the weaker the road resistance will be; L indicates the total length of the roads that the public transportation lines can pass through in the target research unit; b_i/L indicates the overall usage of the roads that public transportation lines pass through in the target research unit; and m indicates the number of bus lines connecting the target hospital to the target research unit.

If there is no direct bus line to the hospital in the target research unit, the elderly in the research unit will need to transfer to other research units for public bus lines when seeking medical treatment. At this time, the trip impedance of the current target research unit will be converted from the trip impedance of other research units. The calculation formula is:

$$D_{bus} = \begin{cases} T_1 \cdot D'_{bus} \\ T_2 \cdot D''_{bus} \end{cases}. \tag{3}$$

In Formula (3), when the transfer condition coefficient $T_1 = 1.5$, it indicates that the bus line cannot reach the target research unit directly from the target hospital, but that it can reach the adjacent research unit of the target research unit. The trip impedance of the target research unit is converted from the road resistance (D'_{bus}) of the research unit with the largest number of bus lines to the adjacent research unit. When the transfer condition coefficient $T_2 = 2$, it indicates that there is no bus route from either the target hospital to the target research unit or to neighboring research units. The trip impedance of the target research unit is converted from the road resistance (D''_{bus}) of the research unit that is the nearest to its geometric center, where there is a bus route to the target hospital. The values of D'_{bus} and D''_{bus} are still calculated according to Formula (2).

According to the analysis and improvement of trip impedance, the potential model for the traffic accessibility evaluation based on trip impedance is:

$$A_{ci} = \sum_{j=1}^{n} \frac{M_j}{D_{bus-ij} V_j}, \quad V_j = \sum_{k=1}^{m} \frac{P_k}{D_{bus-ij}}. \tag{4}$$

In Formula (4) A_{ci} indicates the traffic accessibility of the HOUHs, which serve the elderly population in research unit I; M_j indicates the number of beds in high order urban hospitals (j); D_{bus-ij} indicates the bus trip impedance between research unit i and the high order urban hospital; P_k indicates the number of elderly population in research unit k; V_j indicates the influence of the coefficient on the size of the population; and n and m, respectively, represent the number of HOUHs and research units.

3. Materials and Methods

3.1. Determination of Research Areas and Research Units

Xi'an is an important regional central city in Western China, and the capital of Shaanxi Province. Using the literature [23], the third ring road within the main urban area in Xi'an is chosen as the research area. When the selecting research units, some information that was needed, such as the address and the number of beds of 22 first-level hospitals in the research area, were obtained from the official website of Xi'an Municipal Health Bureau [24]. After removing some specialized hospitals that were not suitable for the elderly and merging the hospitals that were in close proximity to each other, the regional distribution of 17 HOUHs occurred as shown in Figure 1. On this basis, considering the availability of elderly population data, the 45 sub-district administrative divisions within the research area were taken as the basic research unit of accessibility analysis-blocks. The geometric center of each block was used to represent the most populated center of the block (Figure 1).

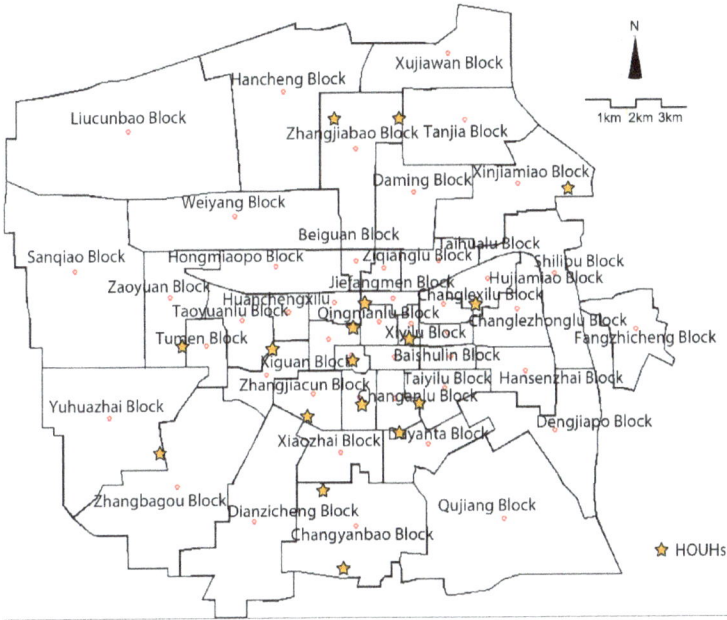

Figure 1. Location of high order urban hospitals (HOUHs) and the research blocks in Xi'an City.

3.2. Data Acquisition of the Elderly, Urban Roads and Bus Trip Impedance

The number of elderly people is obtained from the population data, wherein the number of people aged 60 and above in each block unit of Xi'an City in 2010 was registered. Using the road network map that was built and which was under construction in Xi'an in 2010 (provided by Xi'an City Planning and Design Research Institute), the urban road distribution data of the research area was established on the ArcGIS platform. Furthermore, using the public transportation route operation map provided by the Xi'an Public Transportation Network [25], the public transportation route collection was searched, ranging from bus stops found within 300 m around the location of each first-level hospital. Following this, the public transportation operation route included in the collection was loaded into ArcGIS, where the urban road distribution map of the research area had already been established in order to obtain the primary trip impedance data of the public transportation route between the hospital and the block unit.

4. Results and Discussion

4.1. The Traffic Accessibility is Significantly Attenuated from the Center of the City Based on Linear DistanceTrip Impedance

The hospital location data set up in Section 2.2 were imported into the common potential model of Formula (1) to measure the traffic accessibility level of HOUHs based on linear distance trip impedance. This was used as a basis and a reference point for further analysis. Linear distance impedance was calculated by using ArcGIS engine and using c# language under the net environment. Then, ArcGIS desktop products were used to solve elements related to accessibility, and visually expressed them. The results are shown in Figure 2.

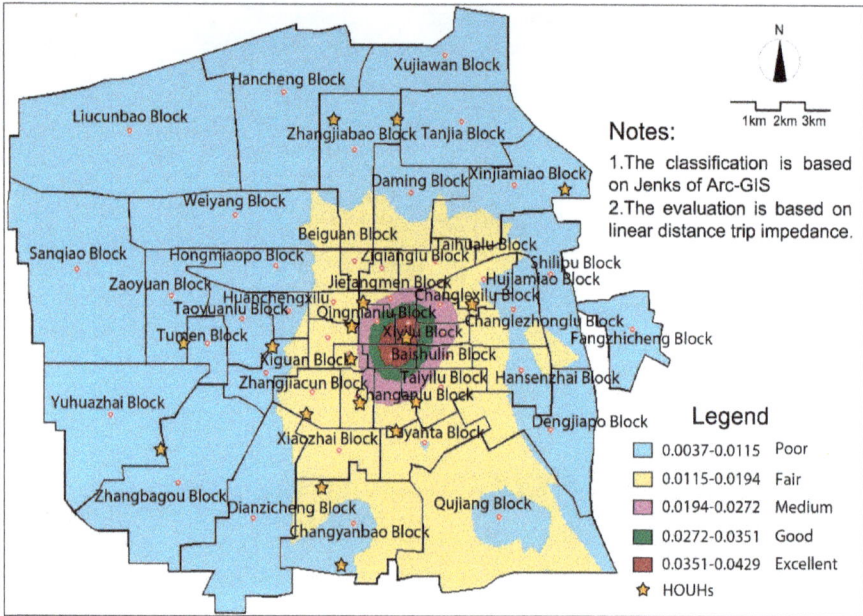

Figure 2. Accessibility distribution of high order urban hospitals under the linear impedance in Xi'an City, 2017.

Under the condition that trip impedance is characterized by a straight-line distance, the common potential model reveals that the traffic accessibility of HOUHs in Xi'an is significantly attenuated from the center of the city. Research units with intermediate and higher accessibility levels in the inner urban area become smaller, and overall traffic accessibility of the peripheral urban research units is poor. However, the blocks in the southern part of the city have slightly higher accessibility.

The reason for this phenomenon is that Xi'an has formed a typical single-core urban spatial structure in its urbanization [26], which has led to the concentration of HOUHs in mature areas to undergo long-term urbanization development. Once a hospital is far away from the core areas in Xi'an, accessibility decreases rapidly. In addition, it should be noted that the planning and construction of HOUHs usually lags behind the development of cities. In newly urbanized areas, although land use planning includes land use arrangements for hospitals, most of the newly established hospitals cannot become HOUHs in a short period of time and have to experience long-term construction. Therefore, solely increasing the projected location of hospitals alone cannot completely solve the unfavorable situation of the low accessibility of HOUHs in cities. Thus, it is still necessary to study advantageous countermeasures using other factors.

4.2. "North–South Connectivity and Eastward Extension" Is Formed Based on Bus Trip Impedance

In order to accurately measure the traffic level of hospitalizing accessibility for the elderly, a program was created to determine bus trip impedance parameters by Microsoft Visual Studio 2018, a secondary development platform based on GIS. Using this, various related parameters were imported into the improved potential model of Formula (4), after which the ArcGIS software platform was used to solve and visually express traffic accessibility. The results are shown in Figure 3.

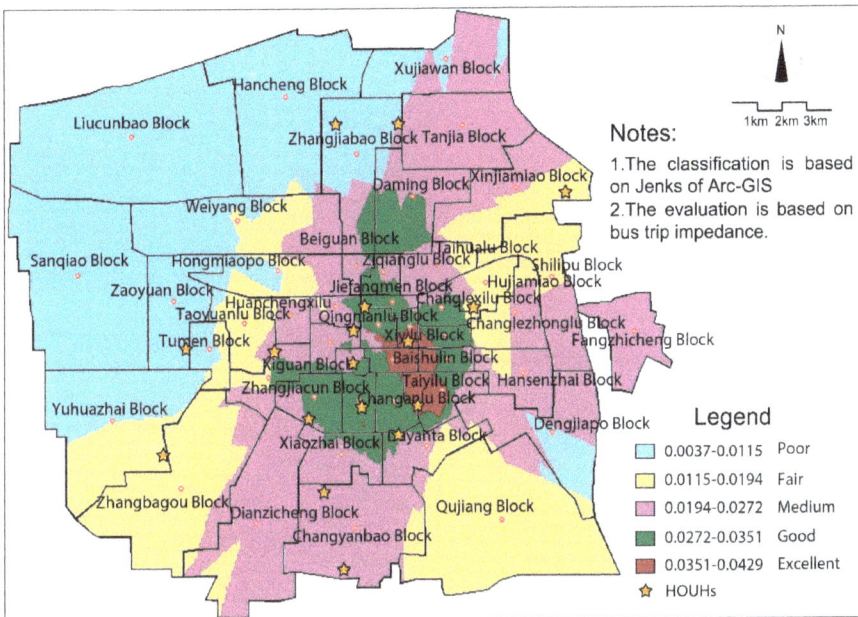

Figure 3. Accessibility distribution of high order urban hospitals due to the impedance of public transportation in Xi'an City, 2017.

Under the condition that public transportation lines are used to represent trip impedance, the distribution pattern of traffic accessibility to HOUHs no longer presents circle-shaped characteristics, which can be reflected by the measurement results of the improved potential model. Instead, research units with medium or high accessibility levels generally form a new spatial pattern of "North–South Connectivity and Eastward Extension" (Figure 3). The formation of this pattern has benefited from a recent and vigorously promoted strategy—"Public Transportation Metropolis", which was set up by Xi'an. Due to the increasing number of public transportation services, it is convenient for the elderly to take public transportation to HOUHs for medical treatment using the urban transportation system in Xi'an. Although the location has not changed, the transportation accessibility of HOUHs for the elderly has still greatly improved.

In the empirical study of accessibility which is based on bus trip impedance, some blocks' accessibility is still below the medium level. Due to urban planning and construction, the causes are classified as low accessibility.

The first type of accessibility is the imprisonment. This research unit is in fact a non-developable area (e.g., a humanity or nature relics reserve) in which infrastructure construction such as roads and so on, are restricted. As a result, there are no conditions for large-scale operations of public transport vehicles, nor are there any locations of HOUHs.

The second category is the antagonism. Most of the research units of this type are newly developed areas in the city. Various municipal infrastructure constructions including urban roads, should still be improved, otherwise this may hinder the future allocation of medical service facilities, and the operational planning of public transportation lines.

The third category is the running-in. This unit is a new urban area with certain developments, in which internal municipal infrastructure construction is relatively complete. However, the hospitals and public transportation services in this area have not yet matched the level of development and infrastructure construction in urban areas, and still need to be improved.

Based on the analysis of the causes and types of low-value performance of traffic accessibility in the HOUHs above, the block units with a poor accessibility evaluation in Xi'an (shown in Figure 3) can be classified, after which valid countermeasures can be put forward, as shown in Table 1.

Table 1. Patterns of accessibility constraints of high order urban hospitals and their corresponding countermeasures.

Type	Blocks Contained	Accessibility Improvement Countermeasures
Imprison	Weiyang, Liucunbao, Hancheng	• Control and relocate elderly residents
Antagonist	Sanqiao, Zaoyuan, Tumen, Yuhuazhai, Xinjiamiao, Shilipu	• Speed up urban planning and construction of road network, and plan public traffic lines at the same time
Running-in	Qujiang, Zhangbagou	• Optimize the coverage and connectivity of public traffic lines • Upgrade service level of conditional hospitals

5. Conclusions and Prospect

As the elderly, a weak group in the urban transportation system, is considered, a traffic accessibility evaluation method for HOUHs is proposed, based on the principle characteristics of their hospitalizing trip. In this theory, bus trip impedance is set as the core improvement content, and the first-level hospitals in Xi'an are used as examples to carry out the corresponding empirical research. Research shows that the location of the hospital determines the level of accessibility, however, the traffic conditions connecting the hospital to its patients are also important aspects that cannot be ignored in the accessibility evaluation. Due to the lack of travel approaches for medical treatment and the elderly's high dependence on hospitals, it is important to effectively detect and improve the public transportation links between the blocks in which the elderly live and HOUHs, in order to implement a "people-oriented" principle in urban transportation planning. Through the empirical study of Xi'an City, various factors which restricted the rise of hospitalizing accessibility were resolved, and categorizing urban blocks based on these factors has become an important basis for adopting planning methods to deal with the problem of low accessibility.

Compared with other research [27], more historical and cultural textures of Xi'an were employed to evaluate and analyze the accessibility calculation results in this paper, including the control and relocation to the site of Chang'an of the Han Dynasty according to world culture heritage protection. Meanwhile, the methods were designed in theory for the trip characteristics of special population groups, and also to fill in the missing data of accessibility measurements for metropolitan hospitals in Western China. Moreover, the traditional accessibility studies which were only based on travel facilities, was improved through paying attention to the influence of specific population groups and their modes of travel. Thus, current knowledge of traffic accessibility was refined through writing this paper.

However, little consideration was given to population groups (e.g., children and the disabled) other than the elderly when choosing the HOUHs for medical treatment. Additionally, walking to adjacent primary medical institutions may be more apt for the elderly when they develop common ailments. Therefore, with an increasing number of specific population groups, exploring the traffic accessibility measurement method for community health institutions which serve the elderly, should be carried out in future studies.

Author Contributions: K.R. is responsible for model development and data analysis. Q.Z. is responsible for framework design and conclusions.

Funding: This research received no external funding.

Conflicts of Interest: The authors declare no conflicts of interest.

References

1. Hansen, W.G. How accessibility shapes land use. *J. Am. Plan. Assoc.* **1959**, *25*, 73–76. [CrossRef]
2. Yang, J.W.; Zhou, Y.X. Accessibility: Concepts, metrics and applications. *Geogr. Territ. Res.* **1999**, *15*, 61–66.
3. Liu, X.T.; Gu, C.L. A spatial analysis of the modal accessibility gap in Nanjing metropolitan area. *Urban Plan. Forum* **2010**, *2*, 49–56.
4. Lu, H.P.; Wang, J.F.; Zhang, Y.B. Models and application of transport accessibility in urban transport planning. *J. Tsinghua Univ. (Sci. Technol.)* **2009**, *49*, 781–785.
5. Jiang, H.B.; Zhang, W.Z.; Qi, Y. Research progress on accessibility to regional transportation infrastructure. *Prog. Geogr.* **2013**, *32*, 807–817.
6. Zhang, L.; Lu, Y.L. Assessment on regional accessibility based on land transportation network: A case study of the Yangtze River Delta. *Acta Geogr. Sin.* **2006**, *61*, 1235–1246.
7. Holl, A. Twenty years of accessibility improvements: The case of the Spanish motorway building programme. *J. Transp. Geogr.* **2007**, *15*, 286–297. [CrossRef]
8. Li, P.H.; Lu, Y.Q. Metropolitan accessibility: Literature review and research progress in the west. *Urban Probl.* **2005**, *23*, 69–74.
9. Higgs, G.; Gould, M. Is there a role for GIS in the new NHS. *Health Place* **2001**, *7*, 247–259. [CrossRef]
10. Beckemeyer, R.J.; Hall, J.D. Permopanorpa inaequalis Tillyard, 1926 (Insecta: Holometabola: Panorpida: Permopanorpidae): A fossil mecopteroid newly reported for the Lower Permian Wellington Formation of Noble Country, Oklahoma. *Trans. Kansas Acad. Sci.* **2007**, *110*, 23–29. [CrossRef]
11. Wang, Y.F.; Zhang, C. GIS and gravity polygon based service area analysis of public facility: A case study of hospitals in Pudong New District. *Econ. Geogr.* **2005**, *25*, 800–803.
12. Song, Z.N.; Chen, W.; Che, Q.J.; Zhang, L. Measurement of spatial accessibility to health care facilities and defining health professional shortage areas based on improved potential modal: A case study of Rudong county in Jiangsu province. *Sci. Geogr. Sin.* **2010**, *30*, 213–219.
13. Hou, S.Y.; Jiang, H.T. An analysis on accessibility of hospitals in Changchun based on urban public transportation. *Geogr. Res.* **2014**, *33*, 915–925.
14. Li, L.J.; Wang, D.W.; Wang, Y.Q.; Li, J.J.; Sun, T.L. Analysis on the visit of elderly in urban communities and its influencing factors: A Case of Qingdao Southern. *Soft Sci. Health* **2014**, *28*, 95–97.
15. Song, Z.N.; Chen, W. Evaluation of spatial accessibility of medical facilities based on potential model price method. *Prog. Geogr.* **2009**, *28*, 848–854.
16. Lee, K.; Lee, H.Y. A new algorithm for graph-theoretic nodal accessibility measurement. *Geogr. Anal.* **1998**, *30*, 1–14. [CrossRef]
17. Hou, Q.; Li, S.M. Transport infrastructure development and changing spatial accessibility in the Greater Pearl River Delta, China, 1990–2020. *J. Transp. Geogr.* **2011**, *19*, 1350–1360. [CrossRef]
18. Cao, X.S.; Huang, X.Y.; Dong, Z. Public transport accessibility based on GIS resident travel characteristics. *J. South China Norm. Univ. (Natl. Sci. Ed.)* **2013**, *45*, 98–105.
19. Zhang, S.J.; Li, Y.S.; Zhang, Y.T. Based on GIS and minimum traffic impedance Bus travel best path algorithm. *J. Geomat. Sci. Technol.* **2008**, *25*, 359–362, 371.
20. Benenson, I.; Martens, K.; Rofe, Y.; Kwartler, A. Public transport versus private car GIS-based estimation of accessibility applies to the Tel Aviv metropolitan area. *Ann. Reg. Sci.* **2011**, *47*, 499–515. [CrossRef]
21. Deng, Y.; Cai, J.M.; Yang, Z.S.; Wang, H. Measuring time accessibility and its spatial characteristics in the urban areas of Beijing. *Acta Geogr. Sin.* **2012**, *67*, 169–178.
22. Shi, R.; Zhang, Q.; Tang, K. Study on Hospitalizing Trip Modes of the Aged in Cities. *Urban Roads Bridges Flood Control* **2015**, 12–15. [CrossRef]
23. Zhang, Q.; Li, T.S.; Han, X.; Wang, H. Research and demonstration on location selection of urban construction supervision enterprises. *Geogr. Res.* **2013**, *32*, 2121–2132.
24. Xi'an Municipal Health Bureau. Available online: http://xawjw.xa.gov.cn/ptl/index.html (accessed on 15 April 2018).
25. Xi'an Public Transportation Network. Available online: http://www1.xbus.cn/ (accessed on 15 April 2018).

26. Xue, D.Q.; Shi, N.; Gong, X.X. Study on spatial distribution and agglomeration model of producer services in Xi'an. *Sci. Geogr. Sin.* **2011**, *31*, 1195–1201.

27. Yu, J.X.; Zhang, X.Q. The spatial accessibility to health care facilities based on service radius: A case study of main areas in Xi'an city. *J. Shaanxi Norm. Univ. (Natl. Sci. Ed.)* **2017**, 78–84. [CrossRef]

symmetry

MDPI

Article

Path Planning for the Mobile Robot: A Review

Han-ye Zhang *, Wei-ming Lin and Ai-xia Chen

School of Mechanical & Materials Engineering, Jiujiang University, Jiujiang 332005, Jiangxi Province, China;
linwm123666@163.com (W.-m.L.); ChenAixia82@126.com (A.-x.C.)
* Correspondence: tzgjzhy@163.com

Received: 19 August 2018; Accepted: 22 September 2018; Published: 1 October 2018

Abstract: Good path planning technology of mobile robot can not only save a lot of time, but also reduce the wear and capital investment of mobile robot. Several methodologies have been proposed and reported in the literature for the path planning of mobile robot. Although these methodologies do not guarantee an optimal solution, they have been successfully applied in their works. The purpose of this paper is to review the modeling, optimization criteria and solution algorithms for the path planning of mobile robot. The survey shows GA (genetic algorithm), PSO (particle swarm optimization algorithm), APF (artificial potential field), and ACO (ant colony optimization algorithm) are the most used approaches to solve the path planning of mobile robot. Finally, future research is discussed which could provide reference for the path planning of mobile robot.

Keywords: path planning; mobile robot; environmental modeling; optimization criteria; path search

1. Introduction

Over the past decades, mobile robots have been successfully applied in different areas such as military, industry and security environments to execute crucial unmanned missions [1]. Path planning [2] is one of the most fundamental problems that have to be resolved before the mobile robots can navigate and explore autonomously in complex environments. Beginning with mid-1960s, the path planning has attracted interests from a lot of scholars. The path planning problem can be described in the following [3]: given a robot and its working environment, the mobile robots searches for an optimal or suboptimal path from the initial state to the target state according to a certain performance criteria. Good path planning technology of mobile robot can not only save a lot of time, but also reduce the wear and capital investment of mobile robot. Because the path planning of mobile robot has important application value, it has become a hot research topic both at home and abroad.

Generally speaking, the path planning can be divided into two categories: the global path planning and the local path planning (seen in Figure 1), according to whether all the information of the environment isaccessible or not. For the global path planning, all the information of the environment is known to the robot before starting. In contrast, for the local path planning, almost all the information of the environment is unknown to the robot before starting [4]. The path planning of mobile robot is retrieved by the database of Engineering Village, where the method of data retrieval is by Title: path planning & mobile robot & theme, for example, Title: path planning & mobile robot & genetic algorithm. Then, the results are summarized in Figure 2.

The remainder of the paper is as follows: Sections 2 and 3 provide the review of global path planning and local path planning, respectively. Section 4 concludes the paper.

Symmetry **2018**, *10*, 450

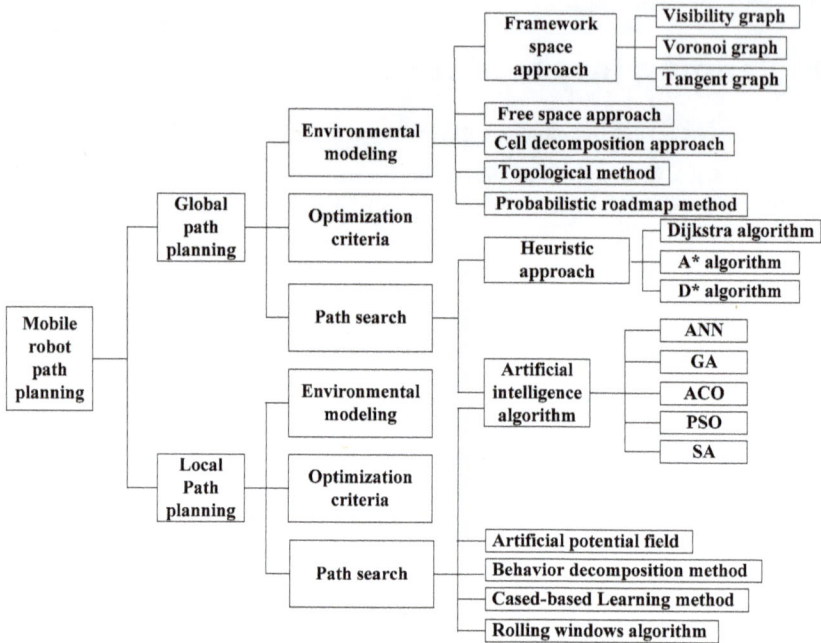

Figure 1. Classification of path planning.

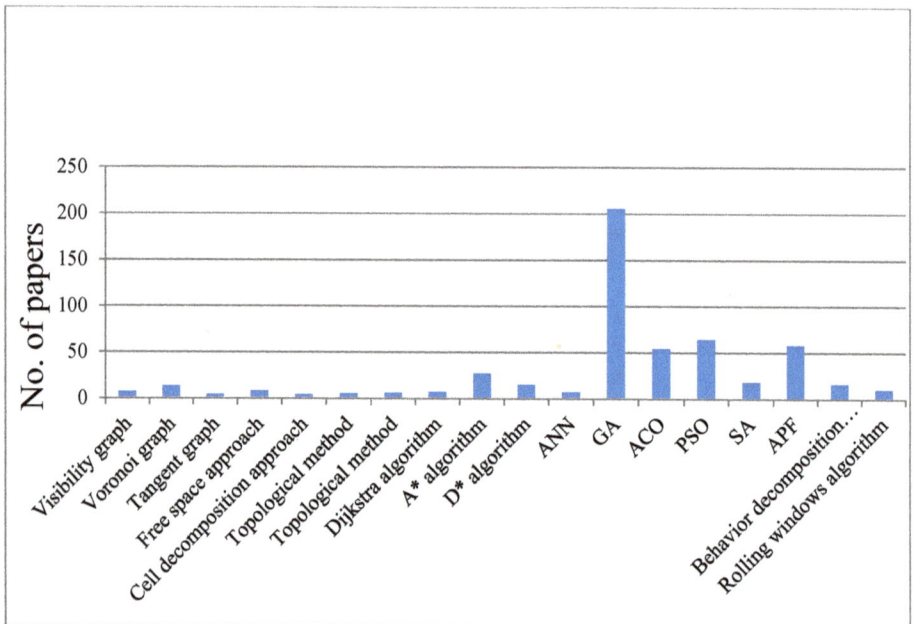

Figure 2. The number of papers retrieved by the database of Engineering Village.

2. Review of Global Path Planning

In the process of global path planning of the mobile robot, the following steps should be followed in the general case. (1) Environmental modeling. The environmental modeling is built according to the known map information: the actual environment for the mobile robot to perform task is converted to the map feature information which can storage conveniently. (2) Optimization criteria. (3) Path search algorithm. The path search algorithm is adopted to find a collision free path between the starting point and the target point in the state space which must satisfy a set of optimization criteria such as path length, smoothness, safety degree, etc. The principle of mobile robot global path planning is shown in Figure 3.

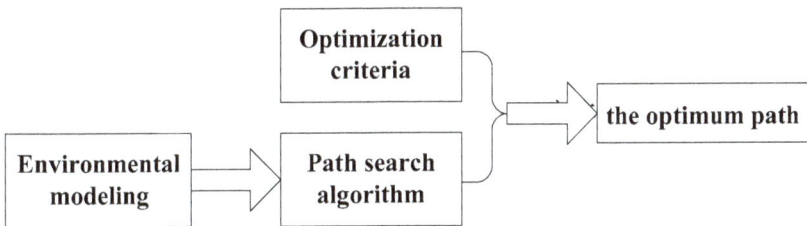

Figure 3. The principle of mobile robot global/local path planning.

2.1. Environmental Modeling

Before the mobile robot global path planning, a suitable environmental model will help to better understand the environmental variables, reduce unnecessary planning and greatly reduce the number of computations. Common methods of environmental modeling have framework space approach, free space approach, cell decomposition approach, topological method and probabilistic roadmap method.

2.1.1. Framework Space Approach

In order to simplify the problem, the mobile robot is usually reduced to a point, the obstacles around the mobile robot are scaled, the mobile robot can move freely in the obstacle space without colliding with obstacles and boundary. The framework space approach includes visibility graph, voronoi graph and tangent graph.

Visibility Graph

A polygon is used to represent an obstacle in the visibility graph (seen in Figure 4) method, and each endpoint is connected with all of its visible vertices to form a final map. In the range of the polygon, a vertex is connected to its total adjacent points, so the mobile robot can move along the polygon edge. Search the set of these lines and select an optimum path from the starting point to the end point. This method can successfully solve the small size problem in two-dimensional space and the path is optimal, but its time complexity is $O(N^2)$ [5]. However, with the increase of the problem's complexity, the efficiency of the visibility graph will be greatly reduced. If the visibility graph is used to the three-dimensional or a higher dimension space, then it is a NP-hard problem. At the same time, the mobile robot has a certain size and shape, all paths pass the end of obstacles, so the obtained path planning is likely to have a collision.

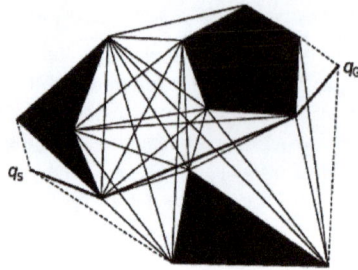

Figure 4. Visibility graph.

Voronoi Graph

The voronoi graph [6] (seen in Figure 5) is the trajectory of points that are equidistant from the nearest two or more barrier boundaries including the workspace boundary. The set of vertices is formed from points that are equidistant from three or more barrier boundaries, while the set of edges is formed from points that are equidistant from exactly two barrier boundaries. The merit of voronoi graph is of fast calculation speed and the drawback is of more mutational site.

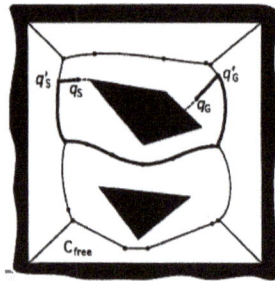

Figure 5. Voronoi graph.

Tangent Graph

In the tangent graph [7,8] (seen in Figure 6), the nodes represent tangent points on barrier boundaries, and the edges represent conflict-free common tangents of the obstacles or convex boundary segments between the tangent points. The tangent graph requires $O(K^2)$ memory, where K represents the number of convex segments of the barrier boundaries. The disadvantage is that if the position error is generated in the control process, the possibility of robot collision obstacles will be very high.

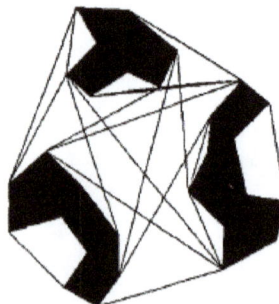

Figure 6. Tangent graph.

2.1.2. Free Space Approach

Based on the concept of free link, Habib et al. [9] developed a new technique to construct the obtainable free space between obstacles within the robotic environment in terms of free convex region. Then, a new graph named MAKLINK is built to provide the generation of a conflict-free path. The graph is built by use of the midpoints of common free links between free convex region as the passing points. These points represent nodes, and the connection between the points within each convex region represent arcs in the graph. A conflict-free path can be effectively generated by use of the MAKLINK graph. The complexity of searching for a conflict-free path is drastically reduced by minimizing the graphic size to be searched about the number of nodes and arcs connecting them. The advantage of free space approach is that it is more flexible, and easy to maintain the network diagram. In addition, it can flexibly change the starting point and target point of the robot. However, in an obstacle-intensive environment, the free space approach may fails and it can't obtain the optimal path.

2.1.3. Cell Decomposition Approach

The method decomposes the workspace of the mobile robot into a number of simple regions, and each region is generally called a cell. These grids form a connected graph and a path is searched from the initial grid to the target grid. In general, the path is represented by the ordinal number of the cell. The method is divided into two types: exact cell decomposition and approximate cell decomposition. The idea of exact cell decomposition is as follows. The free space is divided into n non-overlapping units. The space after the combination of these n units is exactly the same as the original free space. In approximate cell decomposition, all of the grids are in a predetermined shape (e.g., rectangular). The whole environment is divided into a number of larger rectangles, each rectangle is continuous. If any big rectangle contains obstacles or boundary, then it is divided into 4 small rectangular, all the larger grid are executed this operation, the operation is repeated until it reaches the solution boundaries. This structure is called quadtree shown in Figure 7 [10].

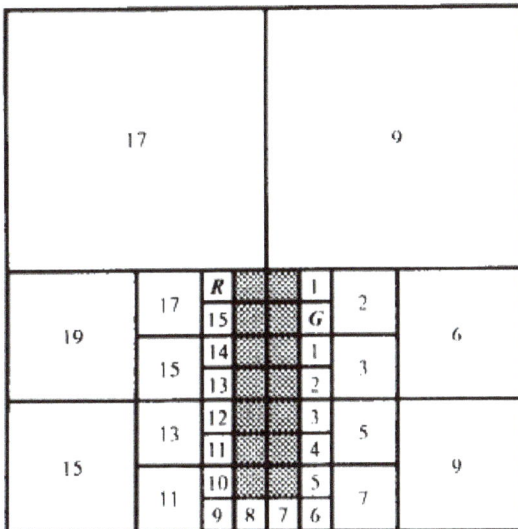

Figure 7. Standard quadtree-based approach.

2.1.4. Topological Method

The topological method is method of reducing dimensions, and the path planning problem in high dimensional geometry space is transformed into the discriminant problem of connectivity in low dimension. When the topology network is established, the robot planning path is obtained from the starting point to the target point. Compared to the cell decomposition approach, this method only needs less model building time and less storage space, the complexity of the topological method only depends on the number of obstacles, the topological method can achieve fast path planning. Topology method is suitable for the environment with obvious characteristics and sparse obstacles; otherwise, it is difficult to carry out reliable navigation control. Another drawback is that topology method of environment information is not easy to maintain, when the number of obstacle is increased or decreased, the network is hard to modify, because the process of establishing the topology network itself is quite complex [11].

2.1.5. Probabilistic Roadmap Method

Kavraki et al. [12] proposed the probabilistic roadmap method in 1994. Some scholars continued to research it [13,14]. The main idea of probabilistic roadmap method is as follows. Based on random sample, an undirected roadmap graph $R = (N, E)$ is built, where N is the nodes of obtained milestones by random sampling, E is the edge connecting these nodes. Given the starting-point s and finishing-point f, the probabilistic roadmap method is looking for two nodes s' and f' satisfying s and s' are directly connected, f and f' are directly connected. A path planning is obtained by searching the edge sequence which are directly connected to s' and f' in the undirected roadmap graph.

2.2. Optimization Criteria

Generally speaking, there are many factors that must be considered in the optimization criteria for planning a mobile path. Three commonly used optimization criteria are listed in the following.

2.2.1. Path Length

The path length D [15,16] is defined as

$$D = \sum_{j=0}^{n-1} \sqrt{(x_{j+1} - x_j)^2 + (y_{j+1} - y_j)^2}$$

where, x_j and y_j are the values of the X coordinate axis and Y coordinate axis of the nodes j, respectively.

2.2.2. Smoothness

The smoothness S [15] is defined as

$$S = \alpha \cdot \left(1 - \frac{DA_l}{N_f - 1}\right) + \beta \cdot \left(1 - \frac{S_{min}}{N_f}\right)$$

where, α and β are weighted coefficients, DA_l is the number of angle of deflection larger than the desired variable, N_f is the total number of path segments, S_{min} is the number of segments with the smallest number of path segments in the path.

2.2.3. Safety Degree

The safety degree SD [15] is defined as

$$SD = \sum_{j=1}^{n-1} C_j = \begin{cases} 0, d_j \geq \lambda \\ \sum_{j=1}^{n-1} e^{\lambda - d_j}, d_j < \lambda \end{cases}$$

where, d_i is the minimal distance between the i-th segment and its nearest obstacle, and λ is the threshold of the safety degree.

2.3. Path Search Algorithm

Generally speaking, the path search algorithm for the global path planning can beclassified into two categories: heuristic approach and artificial intelligence algorithm.

2.3.1. Heuristic Approach

Dijkstra Algorithm

The Dijkstra algorithm is proposed by E.W. Dijkstra in 1959 [17]. It is a typical shortest path algorithm for solving the shortest path problem in a directed graph. Its main feature is that the starting point is as the center to be extended to the end point.

Each edge of the graph is formed to an ordered element pair by the two vertices. The value of the edges are described by the weight function. The algorithm maintains two vertex sets named A and B. The initial set A is empty. Each time a vertex in B is moved to A, and the selected vertex ensures the sum of all the edge weight from the starting point to the point is minimized. Because the algorithm needs to traverse more nodes, so the efficiency is not high.

A^* Algorithm

Hart et al. [18] proposed A^* algorithm in 1968.

The A^* algorithm is developed on the basis of the Dijkstra algorithm. Starting from a specific node, the weighted value of the current child nodes are updated, and the child node which has the smallest weighted value is used to update the current node until all nodes are traversed. The key of A^* algorithm is to establish the evaluation function $f(n)$, $f(n) = g(n) + h(n)$, where $g(n)$ represents the actual cost from the initial node to the node n, and $h(n)$ represents the estimated cost of the optimal path from node n to the target node in the state space. The Euclidean distance between the two nodes is usually taken as the value of $h(n)$. When the value of $g(n)$ is constant, the value of $f(n)$ is mainly affected by the value of $h(n)$. When the node is close to the target node, the value of $h(n)$ is small, the value of $f(n)$ is relatively small. As a result, it guarantees the search for the shortest path always proceeds in the direction of the target point. The A^* algorithm considers the position information of the mobile robot's target point and searches along the target point. Compared with the Dijkstra algorithm, the path search efficiency of the A^* algorithm is higher.

D^* Algorithm

A^* algorithm is mainly used for the global search of the static environment. However, the path planning of mobile robots in practical application is gradually aware of the environmental information, and it is dynamic. Stentz [19] proposed D^* algorithm in 1994. It is mainly used for robot to explore the path. The problem space of the D^* algorithm is expressed as a series of state, and the states represent the direction of the robot's position. The principle of D^* algorithm is basically the same as that of D^* algorithm, the cost of arc used to ensure the direction of the search. In addition, some scholars have researched the D^* algorithm such as the field D^* algorithm [20] and Theta* algorithm [21,22].

2.3.2. Artificial Intelligence Algorithm

ANN

Path planning is a kind of mapping from the perceptual space to the space of behavior, and the artificial neural network (ANN) can express the mapping relationship.

The neural network is used to describe the constraints among the environment, and the energy is defined as the function of the path point. The level of the energy depends on the location of the path point, and the robot moves towards the direction of diminished energy. A path with the smallest total energy is obtained at last. Although this path has no obstacles, it is not the shortest or optimal path. Martin et al. [23] used ANN to solve the robotic path planning problem and discussed how neural networks may contribute to increase the performance of robotic path planners. Mulder et al. [24] constructed an interactive and competitive ANN tosettle the path planning problems. Combined ANN and Q-learning, Li et al. [25] proposed a hybrid method for solving the robotic path planning. The results show the hybrid method was better than either of the two methods. Raza et al. [26] used evolutionary ANN to solve path planning in RoboCup soccer. Contreras-Gonzalez et al. [27] proposed a back-propagation ANN for solving the path planning. The working environment of the mobile robot is random and it is difficult to describe by mathematical formula. It is hard to establish a neural network topology to describe the moving environment. In addition, the complex and large structure makes the weight setting of the neural network to be difficult.

GA

The genetic algorithm (GA) is proposed by Holland in 1975. In the GA, all the possible solutions of the problem are encoded to chromosomes, and all the chromosomes form an initial population. Several basic operations are constructed: crossover, mutation and selection. Initial population is generated, then the fitness value of each individual is calculated by the objectives. The individuals which are selected for crossover operation, mutation operation and selection operation are determined by the fitness value. The flowchart of GA is shown in Figure 8. Min et al. [28] used GA to settle the path planning for mobile manipulator. Liu et al. [29] presented a GA with two-layer encoding to settle the path planning. This kind of encoding can improve the expressing ability of codes. The heart of the two-layer encoding is to decrease the complexity of exploration through the middle-layer codes. Pehlivanoglu et al. [30] proposed vibrational genetic algorithm for path planning. Xu et al. [31] presented adaptive GA to solve the path planning of unmanned aerial vehicle. Example simulation shows that the new algorithm satisfies the requirements in the computation efficiency and the precision of the solution. Tsai et al. [32] proposed PEGA (parallel elite genetic algorithm) for autonomous robot navigation. The results show the PEGA is effective. Tuncer et al. [33] proposed an improved GA for mobile robots' dynamic path planning. Qu et al. [34] proposed an improved GA with co-evolutionary strategy to solve the global path planning for multiple mobile robots. The simulations show the method is efficient. Fei et al. [35] proposed tailored GA for mobile robot's optimal path planning. Shorakaei et al. [36] used a parallel GA for unmanned aerial vehicles' optimal cooperative path planning. The effectiveness of the method was shown by several simulations. The advantage of GA is that it is simple, robust, and has strong search capability and high search efficiency. However, it is prone to premature convergence. When it approaches the optimal solution of the problem, the convergence speed of the algorithm will decrease. It is usually used in the global path planning.

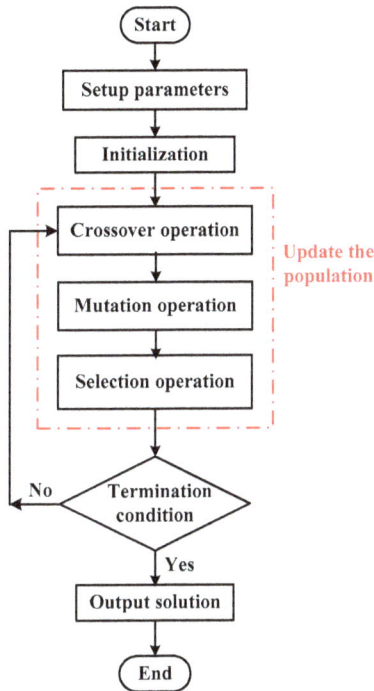

Figure 8. The flowchart of GA.

ACO

The ACO (Ant colony optimization algorithm) is proposed by Marco Dorigo in 1992.

The basic principle of the ACO is each ant will release a secretion on the path it walked as a reference and will also perceive the secretions released by other ants while it is searching for food. This secretion is usually called pheromone. Under the action of pheromones, the ant colony can communicate with each other and choose paths. When the pheromone on a path is more than other paths, the ant colony will spontaneously move to this path, and release more secretions during the movement, so that the concentration of the pheromone becomes higher to attract the latter ants which forms a mechanism of positive feedback. After a period of time, the concentration of pheromone on the shorter path is getting higher and higher, then the ants that choose it are gradually increasing, while the pheromones on other paths are gradually reduced until there is no. Finally the whole ant colony is concentrated in the optimal path. The process of ant foraging is similar to the path planning of robots. As long as there are enough ants in the nest, these ants will find the shortest path from the nest to the food to avoid obstacles. The principle of ant colony searching for food is shown in Figure 9. Wen et al. [37] modified ACO to optimize the global path. When only the pheromone was used to search the optimum path, the ACO converges easily. Wang et al. [38] used ACO to research on global path planning. Simulation results show the ACO algorithm is suitable for global path planning. Zhu et al. [39] proposed an improved ACO for the path planning of mobile robot. The results show the algorithm can not only increase the performance of path planning, but also the algorithm is effective. Zhao et al. [40] improved ACO to solve path planning of mobile robot. Simulation results show the improved algorithm converges quickly even in complex environment. Gao et al. [41] proposed an improved ACO for mobile robot's three-dimensional path planning. The results show that it was an effective approach. You et al. [42] proposed a chaotic ant colony system to solve the path planning of mobile robot. Simulation results show that the approach is not only more effective than the traditional ant colony system, but also improves the global search capabilities. The ACO has not only the global search ability of the population, but also has synergy between individuals. It can find a better path, even if the complete information of the environment is not known. However, in the early stage of the algorithm, the convergence speed is slow and it takes a lot of computation time. It is prone to prematurity.

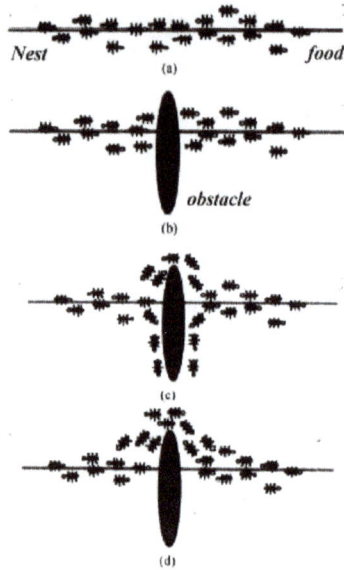

Figure 9. The principle of ant colony searching for food (ant colony searching for food from (**a**) to (**d**)).

PSO

Inspired by the regularity of the bird cluster activity, Eberhart and Kennedy proposed the PSO (particle swarm optimization) algorithm in 1995. It starts from a random solution. It finds the optimal solution through iteration. It evaluates the quality of the solution through fitness value, and it finds the global optimal by comparing the currently searched optimal value at last. This algorithm is used to solve the robotic path planning with the advantages of easy implementation, high precision and fast convergence. Zhang et al. [43] presented an improved PSO for a mobile robot's path planning. Simulation results show the method is effective. Based on multi objective PSO, Gong et al. [44] proposed a global path planning method. The effectiveness of the algorithm is verified by simulation. An improved chaos PSO was proposed to solve the path planning for unmanned aerial vehicle [45]. The results show the proposed algorithm was superior to the traditional PSO, especially in the three-dimensional environment. A fitness-scaling adaptive Chaotic PSO approach was presented to solve the path planning of UCAVs [46]. Based on PSO, Liu et al. [47] introduced some key technologies for path planning in radiation environment. The probability and effectiveness of the method is verified by the experiment. Yusof et al. [48] proposed a predetermined waypoints method. The results show the approach is promising. The algorithm is fast and efficient, but it is easy to fall into local optimum.

SA

The idea of simulated annealing (SA) algorithm was proposed by N. Metropolis et al. in 1953. It is a stochastic optimization algorithm based on the iterative solution strategy of Monte-Carlo. The SA starts from a certain higher initial temperature, and with the continuous decrease of temperature parameters, it randomly finds the global optimal solution of the objective function in the solution space combined with the feature of probability jump.

Martinez-Alfaro et al. [49] used SA to obtain an optimal conflict-free path for mobile robots or AGV in two-dimensional and three-dimensional environment. Vougioukas et al. [50] proposed an accelerated SA algorithm to resolve the path planning. Miao et al. [51] proposed a SA approach to obtain the optimal or near-optimal path quickly for a mobile robot in dynamic environments with static and dynamic obstacles. The effectiveness of the proposed approach was demonstrated. Chiu [52] used the SA to solve the path planning problem for mobile robots. The results show the method is effective. Hui et al. [53] developed an enhanced SA approach to solve the dynamic robot path planning. Behnck et al. [54] developed a modified SA algorithm to solve the path planning for SUAVs. The results show the modified SA is able to calculate paths matching POI and UAV types with an execution time. The algorithm has slow convergence speed, long execution time, and the performance relies on the initial value.

3. Review of Local Path Planning

The contents of environmental modeling and optimization criteria refer to the previous section of review of global path planning.

The path search algorithm for the local path planning can be divided into five categories: artificial potential field method, behavior decomposition method, cased-based Learning method, rolling windows algorithm, and artificial intelligence algorithm.

3.1. Artificial Potential Field

The idea of artificial potential field (APF) comes from the concept of potential field in physics, which regards the movement of objects as the result of two kinds of forces. The robot in the planning space is subjected to the gravitational force from the target point and repulsed by the obstacle. Under the action of the two forces, the robot moves toward the target point in the resultant force, and during the movement process it can effectively avoid the obstacles in the planning space and reach the target safely. The scheme of artificial potential field is shown in Figure 10.

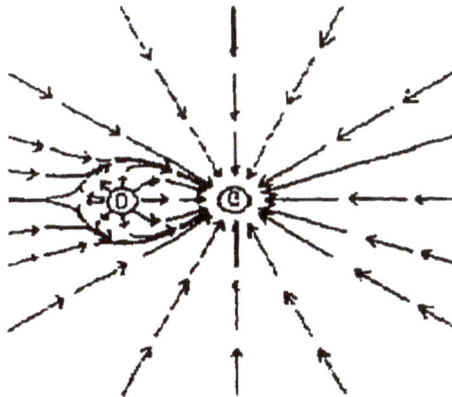

Figure 10. The scheme of artificial potential field.

Vadakkepat et al. [55] proposed a new approach called evolutionary APF (EAPF) for path planning of the real-time robot. The robustness and efficiency are verified by simulation. Min GP et al. [56] presented a virtual obstacle concept base on the APF to study the path planning of mobile robot. The results show the method is feasible and of small complexity. Cao et al. [57] proposed a modified APF approach for the path planning of mobile robot in a dynamic environment. Computer simulation and experiment demonstrated the effectiveness of the dynamic path-planning scheme. In order to solve the problems that path planning trapped in local minimum, Zhang et al. [58] proposed the evolutionary APF method. The feasibility and effectivity are verified by simulation. Zhou et al. [59] proposed an adaptive APF method for the path planning of robot obstacle avoidance. The results show the method can avoid falling into the local optimal solution. The merits of APF: (1) it is easy to operate and realize; (2) it can get a more secure path; (3) it needs a little map information and does not need a lot of computing in the planning process; (4) it can get a smoother trajectory. However, the APF ha some shortcomings, such as: (1) it is prone to shock before the obstacles; (2) if there is an obstacle near the target, the robot moves toward the target point, the smaller the distance between the robot and the target point, the greater the repulsion which will lead such a result the robot fail to reach the target point; (3) the path planning will trap into a locally optimal solution in some areas, etc.

3.2. Behavior Decomposition Method

Behavior decomposition method is a new trend in the path planning of mobile robot. Briefly speaking, it is to break down the navigation problem into a number of relatively independent navigation unit: behavioral primitives, such as collision avoidance, tracking, target guidance, etc. These behavior units are complete motion control unit with sensors and actuators, and have corresponding navigation. These behavior units coordinate with each other to complete the overall navigation tasks. Whitbrook et al. [60] integrated an idiotypic artificial immune system network with a reinforcement-learning-based control system for behavior planning control in robot navigation. Huq et al. [61] proposed a new approach combined fuzzy context dependent behavior modulation and motor schema. The results show it can obtain a conflict-free goal. Fernandez-Leon et al. [62] study the behaviors of scaling up in evolutive robotics. The results show it is efficient. Combined the improved beam curvature method (BCM) and the prediction model of collision, Shi et al. [63] presented a new method of local obstacle avoidance. The results show it can avoid moving obstacles in the dynamic environments. Toibero et al. [64] presented a switching control approach for the parking problem of non-holonomic mobile robot. The results show it is feasible.

3.3. Cased-Based Learning Method

Mobile robot needs to establish a proper case database before path planning. When the mobile robot encounters a new problem, it will search the information from the established case database. Based on the search results, it will compare and analyze to find a solution that is most similar to the new problem. Marefat et al. [65] developed a process planning system. The efficiency and the effectiveness of the approach is verified by experiment. Experience knowledge is accumulated from training and match practice, and is mainly used to match, as most knowledge is derived from experience. The way and formula are discussed and they succeed in applying to robotics soccer [66]. Case-based learning method was applied to the motion planning where volleyball robot's initial state was partially changed [67]. An intelligent typical case-based reasoning to path planning was put forward [68]. Combined with the knowledge about the road network, typical cases were defined and used to solve the problem. The experimental result showed that this path-planning algorithm can reduce the search space, speed up the progress of searching and satisfy people's preferences of running on the road they are familiar with.

3.4. Rolling Windows Algorithm

The mobile robot path planning based on the rolling window method uses the local environment information obtained by the mobile robot to establish a "window", and the path planning is realized by recursively calculating the "window" with its own surrounding information. At each step of the rolling calculation, sub-targets are obtained by heuristic method, then the obtained sub-targets are implemented real-time planning in the current rolling window. With the moving of the rolling window, the sub-targets are updated by the obtained information until the planning task is completed.

Combined rolling path planning and bug algorithm, Zhong et al. [69] presented a new approach to solve the path planning of mobile robot. The approach has environmental adaptability and good capability of obstacle avoidance by simulation. Combining the advantages of least-squares policy iteration (LSPI) and path planning based on rolling windows, a novel reactive navigation method based on LSPI, and rolling windows was presented [70]. The effectiveness and the adaptiveness are verified by simulation and experiment. Based on the dynamic window approach (DWA) for robot navigation, Chou et al. [71] proposed an approach named DWA*. The results show this approach has high performance by simulation and experiment. Because the local environment information of the mobile robot is measured in real time and is online planned in a rolling manner, it has good collision avoidance capability. However, the method may be trapped in a local optimization, it does not guarantee the obtained path is the optimal solution.

3.5. Artificial Intelligence Algorithm

The contents of artificial intelligence algorithm refers to the previous section of review of global path planning.

4. Conclusions

The path planning problem is an important research field of the mobile robot which has aroused the interest of many researchers both at home and abroad. Good path planning technology of mobile robot can not only save a lot of time, but also reduce the wear and capital investment of mobile robot. Different methodologies have been reviewed in this paper. The results shows GA, PSO, APF, and ACO are the most used four approaches to solve the path planning of mobile robot. Finally, future research is discussed which could provide reference for the path planning of mobile robot. Future research should include: (1) Each method can be suitable for different applications. As yet, there is no universal algorithm or method that can solve all above cases. The new path planning method should be researched, such as artificial immune algorithm [72], artificial bee colony [73,74], etc. Especially two or more algorithms are combined to improve the quality and efficiency of the solution. (2) Multi-sensor information should be inosculated into the path planning. Multi-sensor information fusion technology can overcome uncertainty and information incompleteness of the single sensor. It can more accurately and comprehensively understand and describe the environment and the measured object. (3) The task assignment, communication cooperation and path planning of multi-robot should be researched. (4) Path planning of mobile robots in high dimensional environment should be researched. (5) Air robot and underwater robot should be researched. (6) The combination of the robot bottom control and path planning algorithm should be researched.

Author Contributions: H.-y.Z. wrote the original manuscript; W.-m.L. and A.-x.C. polished the manuscript.

Funding: This research received no external funding.

Conflicts of Interest: The authors declare no conflicts of interest.

References

1. Tang, B.W.; Zhu, Z.X.; Luo, J.J. Hybridizing Particle Swarm Optimization and Differential Evolution for the Mobile Robot Global Path Planning. *Int. J. Adv. Robot. Syst.* **2016**, *13*, 1–17. [CrossRef]
2. Zhuang, Y.; Sun, Y.L.; Wang, W. Mobile robot hybrid path planning in an obstacle-cluttered environment based on steering control and improved distance propagating. *Int. J. Innov. Comput. Inf. Control* **2012**, *8*, 4095–4109.
3. Contreras-Cruz, M.A.; Ayala-Ramirez, V.; Hernandez-Belmonte, U.H. Mobile robot path planning using artificial bee colony and evolutionary programming. *Appl. Soft Comput.* **2015**, *30*, 319–328. [CrossRef]
4. Li, P.; Huang, X.H.; Wang, M. A novel hybrid method for mobile robot path planning in unknown dynamic environment based on hybrid DSm model grid map. *J. Exp. Theor. Artif. Intell.* **2011**, *23*, 5–22. [CrossRef]

5.   Goerzen, C.; Kong, Z.; Mettler, B. A survey of motion planning algorithms from the perspective of autonomous UAV guidance. *J. Intell. Robot. Syst. Theory Appl.* **2010**, *57*, 65–100. [CrossRef]
6.   Hoang, H.V.; Sang, H.A.; Tae, C.C. Dyna-Q-based vector direction for path planning problem of autonomous mobile robots in unknown environments. *Adv. Robot.* **2013**, *27*, 159–173.
7.   Liu, Y.; Arimoto, S. Proposal of tangent graph and extended tangent graph for path planning of mobile robots. In Proceedings of the 1991 IEEE International Conference on Robotics and Automation, Sacramento, CA, USA, 9–11 April 1991; pp. 312–317.
8.   Henrich, D. Fast motion planning by parallel processing—A review. *J. Intell. Robot. Syst. Theory Appl.* **1997**, *20*, 45–69. [CrossRef]
9.   Habib, M.K.; Asama, H. Efficient method to generate collision free paths for an autonomous mobile robot based on new free space structuring approach. In Proceedings of the IEEE/RSJ International Workshop on Intelligent Robots and Systems (IROS'91), Osaka, Japan, 3–5 November 1991; pp. 563–567.
10. Chen, D.Z.; Szczerba, R.J.; Uhran, J.J. Framed-quadtree approach for determining Euclidean shortest paths in a 2-D. environment. *IEEE Trans. Robot. Autom.* **1997**, *13*, 668–681. [CrossRef]
11. Liang, W.J. Research on Robot Path Planning for Dynamic Environment and Cooperation. Master's Thesis, Zhejiang University, Hangzhou, China, 2010.
12. Kavraki, L.; Latombe, J. Randomized preprocessing of configuration space for fast path planning. In Proceedings of the 1994 IEEE International Conference on Robotics and Automation, San Diego, CA, USA, 8–13 May 1994; pp. 2138–2145.
13. Kavraki, L.E.; Svestka, P.; Latombe, J.; Overmars, M.H. Probabilistic roadmaps for path planning in high-dimensional configuration spaces. *IEEE Trans. Robot. Autom.* **1996**, *12*, 566–580. [CrossRef]
14. Hsu, D.; Latombe, J.; Kurniawati, H. On the probabilistic foundations of probabilistic roadmap planning. *Int. J. Robot. Res.* **2006**, *25*, 627–643. [CrossRef]
15. Wang, X.Y.; Zhang, G.X.; Zhao, J.B.; Rong, H.; Ipate, F.; Lefticaru, R. A Modified Membrane-Inspired Algorithm Based on Particle Swarm Optimization for Mobile Robot Path Planning. *Int. J. Comput. Commun. Control* **2015**, *10*, 732–745. [CrossRef]
16. Hidalgo-Paniagua, A.; Vega-Rodriguez, M.A.; Ferruz, J.; Pavón, N. MOSFLA-MRPP, Multi-Objective Shuffled Frog-Leaping Algorithm applied to Mobile Robot Path Planning. *Eng. Appl. Artif. Intell.* **2015**, *44*, 123–136. [CrossRef]
17. Dijkstra, E. A note on two problems in connexion with graphs. *Numer. Math.* **1959**, *4*, 269–271. [CrossRef]
18. Hart, P.E.; Nilsson, N.J.; Raphael, B. A formal basis for the heuristic determination of minimum cost paths in graphs. *IEEE Trans. Syst. Sci. Cybern.* **1968**, *4*, 100–107. [CrossRef]
19. Stentz, A. Optimal and efficient path planning for partially-known environments. In Proceedings of the 1994 IEEE International Conference on Robotics and Automation, San Diego, CA, USA, 8–13 May 1994; pp. 3310–3317.
20. Ferguson, D.; Stentz, A. Using interpolation to improve path planning the field D* algorithm. *J. Field Robot.* **2006**, *23*, 79–101. [CrossRef]
21. Nash, A.; Daniel, K.; Koenig, S.; Felner, A. Theta*, Any-angle path planning on grids. *J. Artif. Intell. Res.* **2010**, *39*, 533–579.
22. Xiao, K.; Gao, C.; Hu, X.; Pan, H.W. Improved Theta*, Improved any-angle path planning on girds. *J. Comput. Inf. Syst.* **2014**, *10*, 8881–8890.
23. Martin, P.; Del Pobil, A.P. Application of artificial neural networks to the robot path planning problem. In Proceedings of the Ninth International Conference on Applications of Artificial Intelligence in Engineering, Pennsylvania, PA, USA, 19–21 July 1994; pp. 73–80.
24. Mulder, E.; Mastebroek, H.A.K. Construction of an interactive and competitive artificial neural network for the solution of path planning problems. In Proceedings of the European Symposium on Artificial Neural Networks, Bruges, Belgium, 22–24 April 1998; pp. 407–412.
25. Li, C.H.; Zhang, J.Y.; Li, Y.B. Application of artificial neural network based on Q-learning for mobile robot path planning. In Proceedings of the 2006 IEEE International Conference on Information Acquisition, Weihai, China, 20–23 August 2006; pp. 978–982.
26. Raza, S.; Haider, S. Path planning in RoboCup soccer simulation 3D using evolutionary artificial neural network. In Proceedings of the Advances in Swarm Intelligence 4th International Conference, Harbin, China, 12–15 June 2013; pp. 342–350.

27. Contreras-Gonzalez, A.; Hernandez-Vega, J.; Hernandez-Santos, C.; Palomares-Gorham, D.G. A method to verify a path planning by a back-propagation artificial neural network. In Proceedings of the 10th Latin American Workshop on Logic/Languages, Algorithms and New Methods of Reasoning, Puebla, Mexico, 15 August 2016; pp. 98–105.

28. Min, Z.; Ansari, N.; Hou, E.S.H. Mobile manipulator path planning by a genetic algorithm. *J. Robot. Syst.* **1994**, *11*, 143–153.

29. Liu, Y.F.; Qiu, Y.H. Robot path planning based on genetic algorithms with two-layer encoding. *Control Theory Appl.* **2000**, *17*, 429–432.

30. Pehlivanoglu, Y.V.; Baysal, O.; Hacioglu, A. Path planning for autonomous UAV via vibrational genetic algorithm. *Aircr. Eng. Aerosp. Technol.* **2007**, *79*, 352–359. [CrossRef]

31. Xu, Z.; Tang, S. UAV path planning based on adaptive genetic algorithm. *J. Syst. Simul.* **2008**, *20*, 5411–5414.

32. Tsai, C.; Huang, H.; Chan, C. Parallel elite genetic algorithm and its application to global path planning for autonomous robot navigation. *IEEE Trans. Ind. Electron.* **2011**, *58*, 4813–4821. [CrossRef]

33. Tuncer, A.; Yildirim, M. Dynamic path planning of mobile robots with improved genetic algorithm. *Comput. Electr. Eng.* **2012**, *38*, 1564–1572. [CrossRef]

34. Qu, H.; Xing, K.; Alexander, T. An improved genetic algorithm with co-evolutionary strategy for global path planning of multiple mobile robots. *Neurocomputing* **2013**, *120*, 509–517. [CrossRef]

35. Liu, F.; Liang, S.; Xian, X.D. Optimal Path Planning for Mobile Robot Using Tailored Genetic Algorithm. *IAES TELKOMNIKA Indones. J. Electr. Eng.* **2014**, *12*, 1–9. [CrossRef]

36. Shorakaei, H.; Vahdani, M.; Imani, B.; Gholami, A. Optimal cooperative path planning of unmanned aerial vehicles by a parallel genetic algorithm. *Robotica* **2016**, *34*, 823–836. [CrossRef]

37. Wen, Z.Q.; Cai, Z.X. Global path planning approach based on ant colony optimization algorithm. *J. Cent. South Univ. Technol.* **2006**, *13*, 707–712. [CrossRef]

38. Wang, H.J.; Xiong, W. Research on global path planning based on ant colony optimization for AUV. *J. Mar. Sci. Appl.* **2009**, *8*, 58–64. [CrossRef]

39. Zhu, X.; Han, Q.; Wang, Z. The application of an improved ant colony algorithm in mobile robot path planning. *Key Eng. Mater.* **2011**, *467–469*, 222–225. [CrossRef]

40. Zhao, J.; Fu, X. Improved ant colony optimization algorithm and its application on path planning of mobile robot. *J. Comput.* **2012**, *7*, 2055–2062. [CrossRef]

41. Gao, M.K.; Chen, Y.M.; Liu, Q.; Huang, C.; Li, Z.; Zhang, D. Three-dimensional path planning and guidance of leg vascular based on improved ant colony algorithm in augmented reality. *J. Med. Syst.* **2015**, *39*, 111–133. [CrossRef] [PubMed]

42. You, X.; Liu, K.; Liu, S. A chaotic ant colony system for path planning of mobile robot. *Int. J. Hybrid Inf. Technol.* **2016**, *9*, 329–338. [CrossRef]

43. Zhang, Q.R.; Li, S.H. Path planning based on improved particle swarm optimization for a mobile robot. *WSEAS Trans. Syst. Control* **2007**, *2*, 347–352.

44. Gong, D.W.; Zhang, J.H.; Zhang, Y. Multi-objective Particle Swarm Optimization for Robot Path Planning in Environment with Danger Sources. *J. Comput.* **2011**, *6*, 1554–1561. [CrossRef]

45. Cheng, Z.; Tang, Y.X.; Liu, Y.L. 3-D Path Planning for UAV Based on Chaos Particle Swarm Optimization. *Appl. Mech. Mater.* **2012**, *232*, 625–630. [CrossRef]

46. Zhang, Y.D.; Wu, L.N.; Wang, S.H. UCAV Path Planning by Fitness-scaling Adaptive Chaotic Particle Swarm Optimization. *Math. Probl. Eng.* **2013**, *2013*, 705238. [CrossRef]

47. Liu, Y.K.; Li, M.K.; Xie, C.L.; Peng, M.; Xie, F. Path-planning research in radioactive environment based on particle swarm algorithm. *Prog. Nucl. Energy* **2014**, *74*, 184–192. [CrossRef]

48. Yusof, T.S.T.; Toha, S.F.; Yusof, H. Path Planning for Visually Impaired People in an Unfamiliar Environment Using Particle Swarm Optimization. *Procedia Comput. Sci.* **2015**, *76*, 80–86. [CrossRef]

49. Martinez-Alfaro, H.; Flugrad, D.R. Collision-free path planning for mobile robots and/or AGVs using simulated annealing. In Proceedings of the 1994 IEEE International Conference on Systems, Man and Cybernetics, San Antonio, TX, USA, 2–5 October 1994.

50. Vougioukas, S.; Ippolito, M.; Cugini, U. Path planning based on accelerated simulated annealing. In Proceedings of the 1996 Research Workshop on ERNET—European Robotics Network, Darmstadt, Germany, 9–10 September 1996; pp. 259–268.

51. Miao, H.; Tian, Y.C. Robot path planning in dynamic environments using a simulated annealing based approach. In Proceedings of the 2008 10th International Conference on Control, Automation, Robotics and Vision, Hanoi, Vietnam, 17–20 December 2008; pp. 1253–1258.

52. Chou, M.C. Path-planning for an autonomous robot using a simulating annealing. *J. Inf. Optim. Sci.* **2011**, *32*, 297–314. [CrossRef]

53. Miao, H.; Tian, Y.C. Dynamic robot path planning using an enhanced simulated annealing approach. *Appl. Math. Comput.* **2013**, *222*, 420–437. [CrossRef]

54. Behnck, L.P.; Doering, D.; Pereira, C.E.; Rettberg, A. A modified simulated annealing algorithm for SUAVs path planning. *IFAC-PapersOnLine* **2015**, *48*, 63–68. [CrossRef]

55. Vadakkepat, P.; Tan, K.C.; Wang, M.L. Evolutionary Artificial Potential Fields and their application in real time robot path planning. In Proceedings of the 2000 Congress on Evolutionary Computation, LA Jolla, CA, USA, 16–19 July 2000; pp. 256–263.

56. Park, M.G.; Lee, M.C. Artificial potential field based path planning for mobile robots using a virtual obstacle concept. In Proceedings of the 2003 IEEE/ASME International Conference on Advanced Intelligent Mechatronics, Kobe, Japan, 20–24 July 2003; pp. 735–740.

57. Cao, Q.X.; Huang, Y.W.; Zhou, J.L. An evolutionary artificial potential field algorithm for dynamic path planning of mobile robot. In Proceedings of the 2006 IEEE/RSJ International Conference on Intelligent Robots and Systems, Beijing, China, 9–15 October 2006; pp. 3331–3336.

58. Zhang, Q.S.; Chen, D.D.; Chen, T. An obstacle avoidance method of soccer robot based on evolutionary artificial potential field. *Energy Procedia* **2012**, *16*, 1792–1798. [CrossRef]

59. Zhou, L.; Li, W. Adaptive Artificial Potential Field Approach for Obstacle Avoidance Path Planning. In Proceedings of the 7th International Symposium on Computational Intelligence and Design, Hangzhou, China, 13–14 December 2014; pp. 429–432.

60. Whitbrook, A.M.; Aickelin, U.; Garibaldi, J.M. Idiotypic immune networks in mobile-robot control. *IEEE Trans. Syst. Man Cybern. Part B Cybern.* **2007**, *37*, 1581–1598. [CrossRef]

61. Huq, R.; Mann, G.K.I.; Gosine, R.G. Mobile robot navigation using motor schema and fuzzy context dependent behavior modulation. *Appl. Soft Comput. J.* **2008**, *8*, 422–436. [CrossRef]

62. Fernandez-Leon, J.A.; Acosta, G.G.; Mayosky, M.A. Behavioral control through evolutionary neurocontrollers for autonomous mobile robot navigation. *Robot. Auton. Syst.* **2009**, *57*, 411–419. [CrossRef]

63. Shi, C.X.; Wang, Y.Q.; Yang, J.Y. A local obstacle avoidance method for mobile robots in partially known environment. *Robot. Auton. Syst.* **2010**, *58*, 425–434. [CrossRef]

64. Toibero, J.M.; Roberti, F.; Carelli, R.; Fiorini, P. Switching control approach for stable navigation of mobile robots in unknown environments. *Robot. Comput. Integr. Manuf.* **2011**, *27*, 558–568. [CrossRef]

65. Marefat, M.; Britanik, J. Case-based process planning using an object-oriented model representation. *Robot. Comput. Integr. Manuf.* **1997**, *13*, 229–251. [CrossRef]

66. Zhang, X.C.; Ji, G.; Shao, G.F.; Wei, J.; Li, Z.S. Research on soccer robot learning based on cases. *J. Harbin Inst. Technol.* **2004**, *36*, 905–907.

67. Zhang, P.Y.; Lu, T.S.; Song, L.B. The Case-Based Learning of Mot ion Planning and Its SVR Implementation for Volleyball Robot. *J. Shanghai Jiaotong Univ.* **2006**, *40*, 461–465.

68. Weng, M.; Du, Q.Y.; Qu, R.; Cai, Z. A Path Planning Algorithm Based on Typical Case Reasoning. *Geomat. Inf. Sci. Wuhan Univ.* **2008**, *33*, 1263–1266. [CrossRef]

69. Zhong, X.Y.; Peng, X.F.; Miao, M.L. Planning of robot paths through environment modelling and adaptive window. *J. Huazhong Univ. Sci. Technol.* **2010**, *38*, 107–111.

70. Liu, C.M.; Li, Z.B.; Huang, Z.H.; Zuo, L.; Wu, J.; Xu, X. A reactive navigation method of mobile robots based on LSPI and rolling windows. *J. Cent. South Univ. Sci. Technol.* **2013**, *44*, 970–977.

71. Chou, C.C.; Lian, F.L.; Wang, C.C. Characterizing Indoor Environment for Robot Navigation Using Velocity Space Approach with Region Analysis and Look-Ahead Verification. *IEEE Trans. Instrum. Meas.* **2011**, *60*, 442–451. [CrossRef]

72. Zhang, H.Y. An improved immune algorithm for simple assembly line balancing problem of type 1. *J. Algorithms Comput. Technol.* **2017**, *11*, 317–326. [CrossRef]

73. Khari, M.; Kumar, P.; Burgos, D.; Crespo, R.G. Optimized test suites for automated testing using different optimization techniques. *Soft Comput.* **2018**. [CrossRef]

74. Arora, S.; Singh, S. An Effective Hybrid Butterfly Optimization Algorithm with Artificial Bee Colony for Numerical Optimization. *Int. J. Interact. Multimedia Artif. Intell.* **2017**, *4*, 14–21. [CrossRef]

symmetry

MDPI

Article

Modeling the Service Network Design Problem in Railway Express Shipment Delivery

Siqi Liu, Boliang Lin *, Jianping Wu and Yinan Zhao

School of Traffic and Transportation, Beijing Jiaotong University, Beijing 100044, China;
15114206@bjtu.edu.cn (S.L.); 15114205@bjtu.edu.cn (J.W.); 17114216@bjtu.edu.cn (Y.Z.)
* Correspondence: bllin@bjtu.edu.cn; Tel.: +86-10-1568-2598

Received: 22 August 2018; Accepted: 2 September 2018; Published: 10 September 2018

Abstract: As air pollution becomes increasingly severe, express trains play a more important role in shifting road freight and reducing carbon emissions. Thus, the design of railway express shipment service networks has become a key issue, which needs to be addressed urgently both in theory and practice. The railway express shipment service network design problem (RESSNDP) not only involves the selection of train services and determination of service frequency, but it is also associated with shipment routing, which can be viewed as a service network design problem (SNDP) with railway characteristics. This paper proposes a non-linear integer programming model (INLP) which aims at finding a service network and shipment routing plan with minimum cost while satisfying the transportation time constraints of shipments, carrying capacity constraints of train services, flow conservation constraint and logical constraints among decision variables. In addition, a linearization technique was adopted to transform our model into a linear one to obtain a global optimal solution. To evaluate the effectiveness and efficiency of our approach, a small trial problem was solved by the state-of-the-art mathematical programming solver Gurobi 7.5.2.

Keywords: express shipment; service network design; linearization technique; railway network

---

1. Introduction

In recent years, air pollution has become a severe problem which has to be solved urgently across the world. Many countries have endeavored to reduce carbon emissions related to the transportation industry, which is one of the main sources of air pollution. As the average carbon emissions per workload of road transport is 3.36 to 4.64 times higher than that of railway transport [1], diverting road freight to rail is a wise option. In Europe, 30% of road freight over 300 km should shift to other modes such as rail or waterborne transport by 2030, and more than 50% by 2050 [2]. The Ministry of Transport of China also recommends that the railways and waterways should support more freight transportation. Since the shipments carried by road are generally high-value commodities, which are sensitive to delivery time instead of transportation cost, the express train, which features fast speed, guaranteed pickup and delivery time, plays an increasingly important role in competing with trucks and reducing carbon emissions. The railway express shipment service network design problem express shipment service network design problem (RESSNDP) can be viewed as a service network design problem service network design problem (SNDP) with characteristics of the railway industry. It not only involves the construction of service networks (selection of train services and determination of service frequency), but is also associated with the routing of shipments. Additionally, there is a trade-off between service quality (i.e., transportation capacity and delivery time) and total cost (including service providing cost, shipment routing cost and time cost). Theoretically, the combinations of train services grow exponentially with the number of stations, train speed levels, and stop strategies. Furthermore, each service network involves shipment routing problems which are also exponential explosion problems. Therefore, the RESSNDP has become a key issue that needs to be addressed both in theory and practice.

The problems that arise in railway transportation can be grouped into three levels: the strategic level, tactical level and operational level. Problems at the strategic level are characterized by lengthy time horizons and typically involve resource acquisition. Tactical level problems focus more on allocating resources across an infrastructure that is assumed to be fixed. Operational problems are defined as those that occur on a day-to-day basis. As the RESSNDP is involved with the allocation of express trains over the physical network, it is a tactical level problem, i.e., specific arrival/departure time of trains is not considered in this study.

The paper is organized as follows: Section 1 introduces the importance and complexity of the RESSNDP. Section 2 provides a brief survey of the literature devoted to the SNDP. In Section 3, we describe the RESSNDP in detail. In Section 4, we establish a non-linear integer programming model and transform it into a linear one by applying linearization techniques. A small-scale trial study is conducted in Section 5, and conclusions are presented in Section 6.

2. Literature Review

The RESSNDP can be abstracted as a SNDP, which is a classical problem in the field of traffic and transportation that many scholars have researched. Several studies that are closely related to the SNDP in the railway industry should be noted. A classical paper is that of Crainic et al. [3] who presented a general optimization model which took into account the interactions between routing freight traffic, scheduling train services and allocating classification work on a rail network. Barnhart et al. [4] formulated railroad blocking problems as a mixed integer programming (MIP) model with maximum degree and flow constraints on the nodes and proposed a heuristic Lagrangian relaxation approach to solve the problem. Campetella et al. [5] considered a SNDP in the Italian rail system, and proposed a model combining full and empty railcar management, service selection and frequency. Ahuja et al. [6] proposed a particularly detailed model for a railroad blocking problem arising in the United States and developed an algorithm using the very large-scale neighborhood search technique. Lin et al. [7] presented a formulation and solution for the train connection service problem in a Chinese railway network to determine the optimal freight train service, the frequency of service, and the distribution of classification workload among yards. Zhu et al. [8] addressed the scheduled service network design problem by proposing a model that integrated service selection and scheduling, car classification and blocking, train make-up, and the routing of time-dependent customer shipments based on a cyclic three-layer space-time network representation of the associated operations and decisions, and their relations and time dimensions. The literature mentioned above is mainly devoted to the selection and scheduling of services, as well as routing of shipments. In addition, a variety of techniques, such as Lagrangian relaxation, very large-scale neighborhood search, simulated annealing, etc., have been implemented to solve realistically-size problems.

There is a rich body of literature on express shipment SNDP. Barnhart and her research team [9,10] proposed a column generation-based method to address a real-life air cargo express delivery SNDP. However, it might be difficult to generalize the model to other freight transportation applications, especially to those without hub-and-spoke structures. Grunert et al. [11] and Smilowitz et al. [12] studied long-haul transportation in postal and package delivery systems, and established several models and solution approaches. Armacost et al. [13] proposed an approach for solving express shipment SNDPs by introducing composite variables, which were the combinations of aircraft routes, implicitly capturing package flows. Wang et al. [14] studied a dynamic SNDP for express shipment delivery and established a mixed integer programming model of multi-modal transportation. Yu et al. [15] constructed a bi-level model to optimize the flight transportation network of an express company. The upper-level program was aimed at designing the network and allocating the transportation capacity, and the lower-level program was intended to calculate the link flows in user equilibrium. Quesada et al. [16] developed a formulation in which the allocation of packages to hubs was a decision variable of the model. Also, the formulation was strengthened with three families of valid inequalities and the forcing constraints were reformulated for reducing the number of variables and constraints. Zhao et al. [17] built a hybrid hub-spoke express network decision model permitting direct shipment delivery, whose objective was to minimize the total transport costs of the network. Wang and He [18] investigated the resource planning optimization problem for the multi-modal express shipment network, and a multi-stage hybrid genetic algorithm was presented to solve the model. To our knowledge, the studies on RESSNDP started relatively late and there is limited related literature. Ceselli et al. [19] established three models, integer linear programming (ILP) models, for the express service network of Swiss federal railways using different methods, and designed corresponding solution methods that included a commercial solver approach, branch-and-cut approach, and column generation-based approach. A comparison between a few published studies on SNDP and this work is presented in Table 1.

Table 1. Comparison between a few published studies and this work.

Studies	Decision Variables			Constraints			Model Structure	Solution Approach
	Train Service	Service Frequency	Shipment Routing	Flow Conservation	Service Capacity	Transportation Time		
Barnhart et al. (2000)	√	×	√	√	√	×	MIP	A heuristic Lagrangian relaxation
Campetella et al. (2006)	×	√	√	√	√	×	INLP	A heuristic algorithm based on tabu search
Ahuja et al. (2007)	√	√	√	√	√	×	MIP	Very Large-scale Neighborhood Search
Ceselli et al. (2008)	√	×	√	√	√	√	ILP	Column Generation
Lin et al. (2012)	√	×	√	√	×	×	Bi-level	Simulated annealing
Zhu et al. (2014)	√	×	√	√	√	√	MIP	A matheuristic solution integrating exact and meta-heuristic principles
This work	√	√	√	√	√	√	ILP	Gurobi/Simulated annealing

As shown in Table 1, a majority of the models proposed in the abovementioned studies are integer/mixed-integer programming models, which are difficult to solve to optimality. Thus, a variety of heuristic algorithms are adopted to solve them. In addition, as much of the literature focuses on the railway SNDP of bulk cargo, transportation time and speed levels of freight trains are not considered. In fact, due to the high added-value of express shipments, cargo owners generally pay more attention to the transportation time instead of freight rate, and carriers usually provide express trains on different speed levels to meet the demands of cargo owners. Furthermore, as an express train might stop at a certain number of intermediate stations on its itinerary for the attachment/detachment of blocks (the railcars carrying the same shipment are defined as a block, which does not split until arrival at the destination), the stop strategy plays an important role in the construction of express service network. Thus, we are motivated to carry out this study, and the contribution of our study is as follows:

Firstly, a non-linear integer programming model is proposed which not only takes into account the constraints of transportation time, service capacity, and flow conservation, but also considers the speed levels and stop strategies of train services, the block attachment/detachment operations in intermediate stations, as well as the time cost of express shipments. Secondly, linearization techniques are employed to transform our model into a linear one, in order to obtain a global optimal solution. Thirdly, a small-scale trial study is conducted and a state-of-the-art commercial software Gurobi is adopted to solve our model.

3. Problem Description

The RESSNDP can be viewed as a SNDP with characteristics of the railway industry. It is not only involved with the construction of service network, but is also associated with the routing of shipments. To facilitate the description of the RESSNDP, we constructed a service network as depicted in Figure 1.

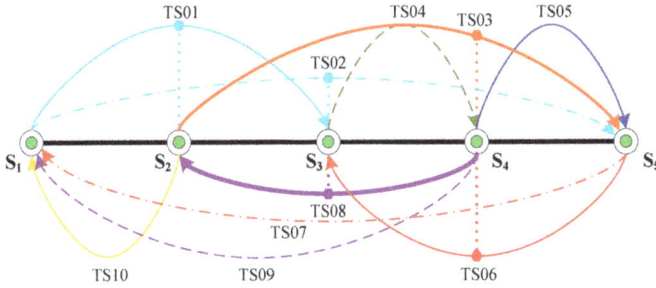

Figure 1. A simple railway service network.

As shown in Figure 1, there are 10 train services (TS01 to TS10), including six services at speed level I (denoted by the solid line, 80 km/h), two services at speed level II (dashed line, 120 km/h), and two services at speed level III (dot-dashed line, 160 km/h). The frequency (i.e., the number of trains dispatched per day) of TS03 and TS08 is two and three, respectively, while the frequency of other train services is set to one. Note that, TS01, TS02, TS03, TS06 and TS08 will stop (denoted by a hexagon) at station S_2, S_3, S_4, S_4 and S_3, respectively, for block attachment/detachment. Dispatching a train will incur a fixed cost and a variable cost which is involved with operation mileage. For example, the cost of providing service TS01 can be calculated by $(\alpha_1 + \theta_1 l_{S_1 \rightarrow S_3}) \times 1$, where α_1 is the fixed cost of dispatching an express train at speed level I, θ_1 is the transportation cost per kilometer of an express train at speed level I, $l_{S_1 \rightarrow S_3}$ is the distance between S_1 and S_3, and "1" is the number of trains dispatched from S_1 to S_3. Similarly, the cost of providing service TS02 can be calculated by $(\alpha_3 + \theta_3 l_{S_1 \rightarrow S_5}) \times 1$.

The second step is to assign the shipments to the service network we constructed. For example, if the shipment which originates at S_1 and is destined to S_5 is carried by TS02, it will be first sent to station S_2, then stop at S_2 for a couple of hours (about two hours in practice, mostly due to the attachment/detachment of other blocks, partly because of routine inspection), and finally delivered to the destination S_5. The routing cost of this shipment (S_1, S_5) can be calculated by:

$$\beta_3^{\text{Transport}} l_{S_1 \rightarrow S_5} f_{(S_1, S_5)} + \beta_{S_3}^{\text{Waiting}} f_{(S_1, S_5)} \tag{1}$$

where $\beta_3^{\text{Transport}}$ is the transportation cost of a loaded railcar per kilometer when carried by a train at speed level III, $f_{(S_1, S_5)}$ is the railcar volume of shipment (S_1, S_5), $\beta_{S_3}^{\text{Waiting}}$ is the waiting cost (including the expense of labor in inspecting the cargo damage and evaluating the performance of the train, and fuel consumption in the arrival and departure process at the intermediate station) of a railcar in average at station S_3. In fact, as high-value commodities are very sensitive to the transportation time, it is necessary to take into account the time cost of shipments. In this case, the time cost of shipment (S_1, S_5) can be expressed by:

$$\frac{\gamma}{\omega_3} l_{S_1 \rightarrow S_5} f_{(S_1, S_5)} + \gamma \sigma_{S_3}^{\text{Waiting}} f_{(S_1, S_5)} \tag{2}$$

where γ is the coefficient of converting car-hour cost into economic cost, ω_3 is the travel speed of train at level III, $\sigma_{S_3}^{\text{Waiting}}$ is the average waiting time of a railcar at station S_3. Therefore, the sum of routing cost and time cost of shipment (S_1, S_5) is equal to:

$$l_{S_1 \to S_5} f_{(S_1, S_5)} \left(\beta_3^{\text{Transport}} + \frac{\gamma}{\omega_3} \right) + f_{(S_1, S_5)} \left(\beta_{S_3}^{\text{Waiting}} + \gamma \sigma_{S_3}^{\text{Waiting}} \right) \tag{3}$$

In this way, the transport cost and waiting cost at S_3 of a railcar can be generalized by $\mu_3^{\text{Transport}}$ and $\mu_{S_3}^{\text{Waiting}}$, respectively, which are defined as follows:

$$\mu_3^{\text{Transport}} = \beta_3^{\text{Transport}} + \frac{\gamma}{\omega_3} \tag{4}$$

$$\mu_{S_3}^{\text{Waiting}} = \beta_{S_3}^{\text{Waiting}} + \gamma \sigma_{S_3}^{\text{Waiting}} \tag{5}$$

Similarly, if the shipment (S_1, S_5) is transport by TS01 and TS03, it will be first carried by TS01 from station S_1 to station S_2, then transferred (detached from a train and attached to another) to TS03 at S_2, and finally sent to the destination S_5 after a stop at S_4. It should be noted that transferring at a station might take about six hours. Throughout the whole process, the generalized routing cost (including routing cost and time cost) of the shipment will be:

$$\mu_1^{\text{Transport}} l_{S_1 \to S_2} f_{(S_1, S_5)} + \mu_{S_2}^{\text{Transfer}} f_{(S_1, S_5)} + \mu_1^{\text{Transport}} l_{S_2 \to S_4} f_{(S_1, S_5)} + \mu_{S_4}^{\text{Waiting}} f_{(S_1, S_5)} + \mu_1^{\text{Transport}} l_{S_4 \to S_5} f_{(S_1, S_5)} \tag{6}$$

where $\mu_{S_2}^{\text{Transfer}}$ is the generalized transfer cost at station S_2, which can be expressed by:

$$\mu_{S_2}^{\text{Transfer}} = \beta_{S_2}^{\text{Transfer}} + \gamma \sigma_{S_2}^{\text{Transfer}} \tag{7}$$

In fact, the combinations for constructing a service network grows exponentially with the number of stations, speed levels, and stop strategies. So, there are a huge number of combinations even in a small railway network. For instance, if there are three optional speed levels for each potential train service and the train might stop at a certain number of intermediate stations it passes through, the combinations of service network construction will be 64 ($2^3 \times 2^3$) for two stations, 2,985,984 ($2^{12} \times 3^6$) for three stations and over 2.1 quadrillion ($2^{18} \times 3^{12} \times 5^6$) for four stations. Furthermore, each service network is involved with a shipment routing problem which is also an exponential explosion problem. Thus, the RESSNDP is of huge complexity.

4. Mathematical Model

In this section, we first describe the assumptions proposed for our study, then introduce the notations used in this paper in Section 4.1, followed by model formulations (see Section 4.2), and finally, we adopt linearization techniques to linearize our model in Section 4.3.

Assumption 1. *Each shipment cannot be split until arriving at the destination. In other words, each shipment can only be delivered along one physical path, and can only choose one service chain (a sequence of train services). Although splitting a shipment and assigning the commodities to different paths and service chains might reduce the total operation cost, the no splitting rule is widely adopted by railway departments in practice for its simplicity in operations management.*

Assumption 2. *It is assumed that the physical path of each origin-destination (OD) demand (i.e., shipment) is specified in advance, which is a standard practice used in the Chinese railway system in freight transportation. As it might result in higher operational costs compared with joint optimization of the railcar itinerary and shipment delivery strategy, one solution to this problem is to optimize train paths and railcar reclassification plans simultaneously.*

Assumption 3. *For each potential train service, only one stop strategy can be selected. For example, for the OD pair (i, j), only one stop strategy can be selected for the express trains (at the same speed level, such as 80km/h)* *dispatched from i to j. It simplifies the problem to some extent by excluding a lot of optional stop strategies once a strategy is chosen.*

4.1. Notations

The notations used in this paper are listed in Table 2.

Table 2. Notations used in this paper.

Sets	Definition
V	Set of stations in a rail network.
H	Set of speed levels of train services.
W_{ij}	Set of stop strategies for express trains dispatched from i to j.
R_{ij}	Set of service arcs that might be provided from i to j, irrespective of the speed levels of the service.
Q_v	Set of service arcs leaving station v.
U_v	Set of service arcs entering station v.
N	Set of natural numbers.
Parameters	**Definition**
o, d, i, j, k, v	Index of stations, o, d, i, j, k, v.
h	Index of speed levels of services, $h \in H$.
r	Index of service arcs, $r \in R_{ij}$.
w	Index of stop strategies of services, $w \in W_{ij}$.
Γ_{ij}^w	The wth stop strategy for the train service provided from i to j, $\Gamma_{ij}^w = \left[\delta_{ij}^{wi}, \ldots, \delta_{ij}^{wk}, \ldots, \delta_{ij}^{wj} \right]$;
δ_{ij}^{wk}	Train stop parameter, it takes value one if an express train dispatched from i to j with the wth stop strategy stops at station k. Otherwise, it is zero. In addition, if this train service is provided, then $\delta_{ij}^{wi} = \delta_{ij}^{wj} = 1$; otherwise, $\delta_{ij}^{wi} = \delta_{ij}^{wj} = 0$.
τ_{ij}^{wr}	Service arc parameter, its value is one if the service arc r exists when the train service with the wth stop strategy is provided from i to j. Otherwise, it is zero.
f_{od}	The number of cars which origin at station o and are destined to d.
α_h	The fixed cost of dispatching an express train on a speed level of h.
θ_h	The variable cost of a train on speed level h per kilometer.
γ	The coefficient of converting car-hour cost into economic cost.
s_r	The starting station of arc r.
e_r	The ending station of arc r.
l_{ij}	The mileage from i to j.
l_r	The mileage of service arc r.
ω_h	The travel speed of an express train on speed level h.
m	The size of a train, i.e., the maximum number of railcars in a train.
$\beta_h^{\text{Transport}}$	The transport cost of a loaded railcar per kilometer when the speed level is h.
$\beta_k^{\text{Transfer}}$	The transfer cost of a railcar in average at station k.
β_k^{Waiting}	The waiting cost of a railcar in average at station k.
$\sigma_k^{\text{Transfer}}$	The average transfer delay of a railcar when transferring at station k.
$\sigma_k^{\text{Waiting}}$	The average waiting delay of a railcar when waiting at station k.
$\mu_h^{\text{Transport}}$	The generalized transport cost of a railcar per kilometer at a speed level of h.

Table 2. *Cont.*

Parameters	Definition
μ_k^{Transfer}	The generalized transfer cost of a railcar in average at station k.
μ_k^{Waiting}	The generalized waiting cost of a railcar in average at station k.
T_{od}^{Due}	The maximum time consumption for a shipment which is specified in advance.
M	A sufficiently large positive number.

Decision Variables	Definition
$y(i,j,w,h)$	Service selection variable; its value is one if the train service, whose speed level is h and stop strategy is w, is provided from i to j. Otherwise, it is zero.
$\eta(i,j,w,h)$	Service frequency variable; the number of trains dispatched from i to j per day, whose speed level is h and stop strategy is w. As express shipments are generally sensitive to the transportation time, at least one freight train will be dispatched for shipping the commodity in a day, if the shipment volume is positive. The value of this variable is an integer.
$x(o,d,i,j,r,h)$	Shipment routing variable; it takes a value of one if the shipment (o,d), i.e., the OD demand which originates at station o and are destined to d, is carried by service arc r at a speed level of h on its itinerary, where $r \in R_{ij}$. Otherwise, it is zero.

4.2. Formulations

In the delivery of express cargo, there are different types of cost that should be taken into consideration. The generalized transport cost (including transport cost and time cost) involved with the volume of loaded railcar and operation mileage can be expressed by:

$$C_{od}^{\text{Transport}} = \sum_{i \neq j \in V} \sum_{r \in R_{ij}} \sum_{h \in H} \mu_h^{\text{Transport}} l_r f_{od} x(o,d,i,j,r,h) \forall o \neq d \in V \tag{8}$$

As a block might be detached from one train and attached to another on its itinerary, the generalized block transfer cost should be considered. It can be expressed by:

$$C_{od}^{\text{Transfer}} = \sum_{i \neq j \in V} \sum_{h \in H} \sum_{r \in R_{ij}} \sum_{i' \neq j' \in V} \sum_{h' \in H} \sum_{r' \in R_{i'j'}, s_{r'} = e_r} \mu_{e_r}^{\text{Transfer}} f_{od} x(o,d,i,j,r,h) x(o,d,i',j',r',h')$$

$$- \sum_{i \neq j \in V} \sum_{h \in H} \sum_{r \neq r' \in R_{ij}, s_{r'} = e_r} \mu_{e_r}^{\text{Transfer}} f_{od} x(o,d,i,j,r,h) x(o,d,i,j,r',h) \tag{9}$$

$$\forall o \neq d \in V$$

Note that, block transfer will occur when the shipment is delivered by at least two trains with different origins, destinations or speed levels.

When a train stops at a certain station, the block that does not need to be transferred here may incur waiting cost, due to attachment/detachment of other blocks, routine inspection of cargo damage, train performance evaluation and other transit operations. The generalized waiting cost can be calculated as:

$$C_{od}^{\text{Waiting}} = \sum_{i \neq j \in V} \sum_{h \in H} \sum_{r \neq r' \in R_{ij}, s_{r'} = e_r} \mu_{e_r}^{\text{Waiting}} f_{od} x(o,d,i,j,r,h) x(o,d,i,j,r',h) \forall o \neq d \in V \tag{10}$$

In this paper, we define the sum of the generalized transport cost, transfer cost and waiting cost as the generalized shipment routing cost. In addition, when a train service is provided, a service providing cost will be incurred, consisting of a fixed cost associated with speed level and a variable cost involved with operation mileage, which can be expressed by:

$$C^{\text{Train}} = \sum_{i \neq j \in V} \sum_{w \in W_{ij}} \sum_{h \in H} (\alpha_h + \theta_h l_{ij}) \eta(i, j, w, h) \tag{11}$$

For an express train dispatched from i to j, the set of its stop strategy is denoted as W_{ij}. Among them, a special strategy should be noted, i.e., $\Gamma_{ij}^0 = [0, \ldots, 0, \ldots, 0]$. In this case, the express train will not stop at any station on its itinerary, including i to j. In other words, no service from i to j will be provided. Conversely, if an express train stops at every station it passes through on its itinerary, then the stop strategy might be $\Gamma_{ij}^{|W_{ij}|} = [1, \ldots, 1, \ldots, 1]$. The set of decision variables for a potential service on speed level h can be denoted as:

$$Y_{ij}^h = \{y(i, j, 0, h), \cdots y(i, j, w, h), \cdots, y(i, j, |W_{ij}|, h)\} \tag{12}$$

According to Assumption 3, the following constraint should be considered:

$$\sum_{w \in W_{ij}} y(i, j, w, h) = 1 \forall i \neq j \in V, h \in H \tag{13}$$

Therefore, the RESSNDP can be formulated as a non-linear integer programming model whose objective function and constraints are written as follows:

$$\min C^{\text{Train}} + \sum_{o \neq d \in V} \left[C_{od}^{\text{Transport}} + C_{od}^{\text{Transfer}} + C_{od}^{\text{Waiting}} \right] \tag{14}$$

$$\text{S.T.} \sum_{w \in W_{ij}} y(i, j, w, h) = 1 \forall i \neq j \in V, h \in H \tag{15}$$

$$\sum_{i \neq j \in V} \sum_{r \in Q_v \cap R_{ij}} \sum_{h \in H} x(o, d, i, j, r, h) - \sum_{i \neq j \in V} \sum_{r \in U_v \cap R_{ij}} \sum_{h \in H} x(o, d, i, j, r, h) = \begin{cases} 1 v = o \\ 0 v \neq o \neq d \forall v \in V, o \neq d \in V \\ -1 v = d \end{cases} \tag{16}$$

$$\sum_{i \neq j \in V} \sum_{r \in R_{ij}} \sum_{h \in H} \frac{l_r}{\omega_h} x(o, d, i, j, r, h) + \sum_{i \neq j \in V} \sum_{h \in H} \sum_{r \neq r' \in R_{ij}, s_{r'} = e_r} \sigma_{e_r}^{\text{Waiting}} x(o, d, i, j, r, h) x(o, d, i, j, r', h)$$

$$+ \sum_{i \neq j \in V} \sum_{h \in H} \sum_{r \in R_{ij}} \sum_{i' \neq j' \in V} \sum_{h' \in H} \sum_{r' \in R_{i'j'}, s_{r'} = e_r} \sigma_{e_r}^{\text{Transfer}} x(o, d, i, j, r, h) x(o, d, i', j', r', h') \qquad \forall o \neq d \in V \tag{17}$$

$$- \sum_{i \neq j \in V} \sum_{h \in H} \sum_{r \neq r' \in R_{ij}, s_{r'} = e_r} \sigma_{e_r}^{\text{Transfer}} x(o, d, i, j, r, h) x(o, d, i, j, r', h) \leq T_{od}^{\text{Due}}$$

$$\sum_{o \neq d \in V} f_{od} x(o, d, i, j, r, h) \leq m \sum_{w \in W_{ij}} \eta(i, j, w, h) \forall i \neq j \in V, r \in R_{ij}, h \in H \tag{18}$$

$$x(o, d, i, j, r, h) \leq \sum_{w \in W_{ij}} \tau_{ijh}^{wr} y(i, j, w, h) \forall o \neq d \in V, i \neq j \in V, r \in R_{ij}, h \in H \tag{19}$$

$$\eta(i, j, w, h) \leq M \sum_{r \in R_{ij}} x(o, d, i, j, r, h) o = i, d = j, \forall i \neq j \in V, w \in W_{ij}, h \in H \tag{20}$$

$$\sum_{h \in H} \sum_{r \in R_{ij}} \sum_{i' \neq j' \in V} \sum_{h' \in H} \sum_{r' \in R_{i'j'}, s_{r'} = e_r} x(o, d, i, j, r, h) x(o, d, i', j', r', h')$$

$$- \sum_{h \in H} \sum_{r \neq r' \in R_{ij}, s_{r'} = e_r} x(o, d, i, j, r, h) x(o, d, i, j, r', h) = 0 o = i, d = j, \forall i \neq j \in V \tag{21}$$

$$\eta(i, j, w, h) \leq M y(i, j, w, h) \forall i \neq j \in V, w \in W_{ij}, h \in H \tag{22}$$

$$y(i, j, w, h) \leq \eta(i, j, w, h) \forall i \neq j \in V, w \neq 0 \in W_{ij}, h \in H \tag{23}$$

$$\eta(i, j, w, h) \in N \forall i \neq j \in V, h \in H, w \in W_{ij} \tag{24}$$

$$y(i, j, w, h), x(o, d, i, j, r, h) \in \{0, 1\} \forall o \neq d \in V, i \neq j \in V, w \in W_{ij}, r \in R_{ij}, h \in H \tag{25}$$

The objective function is to minimize the sum of the service providing cost, shipment routing cost and time cost. The constraint (15) ensures that, for each OD pair (i, j), only one stop strategy can be selected for the train service provided from i to j at speed level h, $h \in H$. The constraint (16) is the flow conservation constraint which guarantees that all shipments can be delivered to their destination. The constraint (17) is the transportation time constraint which ensures that the sum of transport time, transfer time and waiting time of a shipment is less than the pre-specified due time. The constraint (18) is the capacity constraint which guarantees that the workload of a certain service arc is less than its carrying capacity, where m is the size of the train. The constraint (19) is the logical constraint for the service selection variable and shipment routing variable, which ensures that the service arc r would not be selected if it does not exist when stop strategy w is chosen for the train service. The constraints (20) and (21) together guarantee that a shipment will be assigned to the service whose origin and destination are the same as that of the shipment, if the train service is provided. In other words, if there is no shipment whose origin and destination are the same as that of a candidate train service, the service will not be provided. The constraints (22) and (23) are the logical constraints for the service selection variable and service frequency variable, which indicate that at least one train will be provided when a stop strategy (excluding the non-stop strategy, i.e., $w = 0$) is chosen. It should be noted that no train will be dispatched if stop strategy "0" is selected.

4.3. Linearization of the Model

As the INLP model is very difficult to be solved to optimality, linearization techniques are adopted to linearize the model we proposed. Apparently, the objective function (14), (17) and (21) have nonlinear terms, i.e., the product of binary variables: $x(o, d, i, j, r, h)x(o, d, i', j', r', h')$ and $x(o, d, i, j, r, h)x(o, d, i, j, r', h)$. To solve it, we first introduced the method of linearizing the product of a sequence of binary variables. Let us assume that $b = a_1 \times a_2 \times \cdots \times a_k$, where $a_1, a_2, \cdots, a_k$ are all binary variables. Then b can be equivalently expressed by the following linear inequality:

$$a_1 + a_2 + \cdots + a_k - (k - 1) \leq b \leq \frac{1}{k} \times (a_1 + a_2 + \cdots + a_k) \tag{26}$$

According to the formula $b = a_1 \times a_2 \times \cdots \times a_k$, it is clear that b is a binary variable. and when $a_1, a_2, \cdots, a_k$ all take the value of one, b is obviously equal to one. In contrast, if at least one variable of $a_1, a_2, \cdots, a_k$ takes the value of zero, then b is equal to zero. Similarly, on the basis of formula (26), when $a_1, a_2, \cdots, a_k$ all take the value of one, this inequality can be rewritten as $k - (k - 1) \leq b \leq \frac{1}{k} \times k$, i.e., $1 \leq b \leq 1$. Thus, b has to take the value of one. Conversely, if not all variables of $a_1, a_2, \cdots, a_k$ take the value of one, i.e., $a_1 + a_2 + \cdots + a_k < k$, then the formula (26) can be converted to $a_1 + a_2 + \cdots + a_k - (k - 1) \leq 0 \leq b \leq \frac{1}{k} \times (a_1 + a_2 + \cdots + a_k) < \frac{1}{k} \times k = 1$, i.e., $0 \leq b < 1$. Thus, b has to take the value of zero. To summarize, inequality (26) is equivalent to the equation $b = a_1 \times a_2 \times \cdots \times a_k$. Based on the method mentioned above, the transfer variable $z(o, d, i, j, r, h, i', j', r', h')$ and waiting variable $\zeta(o, d, i, j, r, r', h)$ are introduced, which can be expressed by:

$$z(o, d, i, j, r, h, i', j', r', h') = x(o, d, i, j, r, h)x(o, d, i', j', r', h')$$
$$e_r = s_{r'}, (i, j) \neq (i', j') \text{ or } h \neq h', \forall o \neq d \in V, i \neq j \in V, i' \neq j' \in V, r \in R_{ij}, r' \in R_{i'j'}, h \in H, h' \in H \tag{27}$$

$$\mu(o, d, i, j, r, r', h) = x(o, d, i, j, r, h)x(o, d, i, j, r', h)$$
$$e_r = s_{r'}, \forall o \neq d \in V, i \neq j \in V, r \in R_{ij}, r' \in R_{ij}, h \in H \tag{28}$$

The transfer variable takes the value of one if the shipment (o, d) is detached from a train and attached to another at station e_r. Otherwise, it is zero. Similarly, the waiting variable is equal to one if the shipment (o, d) is carried by the same train before and after the stop at a certain intermediate station e_r. Otherwise, it is zero. Moreover, several constraints should be added to the original model:

$$x(o,d,i,j,r,h) + x(o,d,i',j',r',h') - 1 \leq z(o,d,i,j,r,h,i',j',r',h')$$
$$e_r = s_{r'}, (i,j) \neq (i',j') \text{ or } h \neq h', \forall o \neq d \in V, i \neq j \in V, i' \neq j' \in V, r \in R_{ij}, r' \in R_{i'j'}, h \in H, h' \in H \tag{29}$$

$$z(o,d,i,j,r,h,i',j',r',h') \leq \frac{x(o,d,i,j,r,h) + x(o,d,i',j',r',h')}{2}$$
$$e_r = s_{r'}, (i,j) \neq (i',j') \text{ or } h \neq h', \forall o \neq d \in V, i \neq j \in V, i' \neq j' \in V, r \in R_{ij}, r' \in R_{i'j'}, h \in H, h' \in H \tag{30}$$

$$z(o,d,i,j,r,h,i',j',r',h') \in \{0,1\}$$
$$e_r = s_{r'}, (i,j) \neq (i',j') \text{ or } h \neq h', \forall o \neq d \in V, i \neq j \in V, i' \neq j' \in V, r \in R_{ij}, r' \in R_{i'j'}, h \in H, h' \in H \tag{31}$$

$$x(o,d,i,j,r,h) + x(o,d,i,j,r',h) - 1 \leq \zeta(o,d,i,j,r,r',h)$$
$$e_r = s_{r'}, \forall o \neq d \in V, i \neq j \in V, r \in R_{ij}, r' \in R_{i'j'}, h \in H \tag{32}$$

$$\zeta(o,d,i,j,r,r',h) \leq \frac{x(o,d,i,j,r,h) + x(o,d,i,j,r',h)}{2}$$
$$e_r = s_{r'}, \forall o \neq d \in V, i \neq j \in V, r \in R_{ij}, r' \in R_{i'j'}, h \in H \tag{33}$$

$$\zeta(o,d,i,j,r,r',h) \in \{0,1\} e_r = s_{r'}, \forall o \neq d \in V, i \neq j \in V, r \in R_{ij}, r' \in R_{i'j'}, h \in H \tag{34}$$

In this way, the original non-linear model can be converted to a linear one which can be expressed by:

$$\min C^{\text{Train}} + \sum_{o \neq d \in V} \left[C^{\text{Transport}}_{od} + C^{\text{Transfer}}_{od} + C^{\text{Waiting}}_{od} \right] \tag{35}$$

s.t. Constraints (15) to (25) and (29) to (34). Thus, we can directly use a standard optimization solver to solve the model.

5. Numerical Study

In this section, we carried out a small-scale trial study to evaluate the effectiveness and validity of our model and approach. The procedures are described in detail as follows: (1) A physical railway network consisting of five stations, denoted as S_1 through S_5, was constructed which is depicted by Figure 2. (2) The OD matrix, transportation time constraint of each shipment, mileage of each OD pair, transfer cost, transfer delay, waiting cost and waiting delay at each station were determined and are listed in Table 3. (3) A commercial software, Gurobi 7.5.2, was adopted to solve the model. (4) We obtained the stop strategy and frequency of express trains which are to be provided, as well as the service chain (the sequence of express trains carrying a certain shipment) of each shipment. (5) To evaluate the impact of some critical parameters on the total cost and number of trains dispatched, sensitivity analysis was carried out.

According to the operational practices of express trains in China, the travel speed of trains at three speed levels (I, II, III) were set to 80 km/h, 120 km/h and 160 km/h respectively. The fixed cost of dispatching an express train at three speed levels were set to 5000 CNY, 6000 CNY and 7000 CNY, respectively, while the variable cost of an express train at three speed levels per kilometer were set to 40 CNY/km, 50 CNY/km and 60 CNY/km, respectively. In addition, the unit transportation cost of a railcar at three speed levels were set to 5 CNY/km, 6 CNY/km and 7 CNY/km respectively. Moreover, the train size m was set to 25 cars.

The RESSNDP mentioned above was solved by Gurobi 7.5.2 on a 2.20 GHz Intel (R) Core (TM) i5-5200U CPU computer with 4.0 GB of RAM. After about 51 seconds of computation, the optimal solution was obtained. The total cost is 1,200,561.5 CNY (where the total service providing cost is equal to 435,690 CNY, the total generalized transport cost is 764,098.1 CNY, the total generalized transfer cost is equal to 433.8 CNY, the total generalized waiting cost is 339.6 CNY), and ten train services are supposed to be provided to carry the shipments in the network, including seven services at speed level I, two services at speed level II and one service at speed level III. The express services provided and their frequency are listed in Table 4 and the optimal service network is depicted by Figure 3.

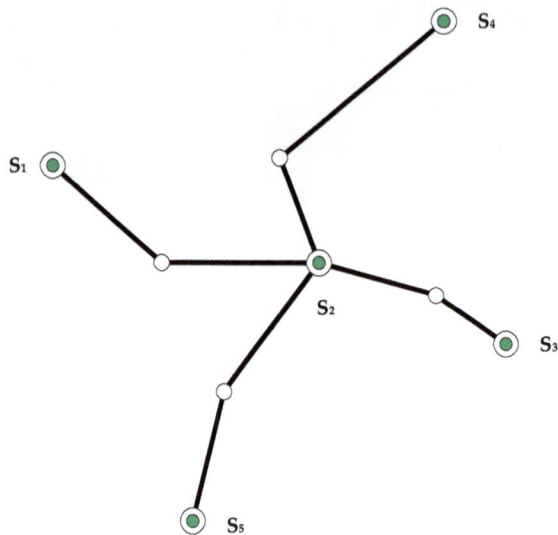

Figure 2. A railway sub-network.

Table 3. The input data.

Stations	S_1	S_2	S_3	S_4	S_5
Shipment Volume (Cars Per Day)					
S_1	—	11.7	9.5	10.1	2.3
S_2	12.6	—	14.7	17.7	8.1
S_3	8.7	12.9	—	8.8	10.7
S_4	7.8	15.4	9.7	—	11.3
S_5	4.5	13.1	7.2	7.6	—
Transportation Time Constraint (Hours)					
S_1	—	13.5	18.1	11.0	21.1
S_2	13.5	—	12.7	14.5	15.6
S_3	10.1	12.7	—	7.0	20.3
S_4	20.0	14.5	19.2	—	22.1
S_5	21.1	15.6	20.3	22.1	—
Mileage of Each OD Pair (Kilometers)					
S_1	—	439	811	960	1047
S_2	439	—	372	521	608
S_3	811	372	—	893	980
S_4	960	521	893	—	1129
S_5	1047	608	980	1129	—
Generalized Transfer cost (CNY/car)	20	18	22	19.2	20.8
Transfer Delay (Hours)	8	6	10	7.2	8.8
Generalized Waiting cost (CNY/car)	6	7.5	8	5	7
Waiting Delay (Hours)	2	2	3	1.5	2.5

Table 4. The express services provided and shipment routing plans.

No.	Train Service Origin to Destination	Speed Level	Intermediate Station	Stop Strategy	Service Frequency	Shipments Assigned to the Train Service
1	$S_1 \rightarrow S_3$	I	S_2	$[1,1,1]$	1	$(S_1, S_2), (S_1, S_3), (S_1, S_5), (S_2, S_3)$
2	$S_1 \rightarrow S_4$	II	S_2	$[1,1,1]$	1	$(S_1, S_4), (S_5, S_4)$
3	$S_2 \rightarrow S_4$	I	/	$[1,1]$	1	(S_2, S_4)
4	$S_3 \rightarrow S_1$	II	S_2	$[1,0,1]$	1	(S_3, S_1)
5	$S_3 \rightarrow S_4$	III	S_2	$[1,0,1]$	1	(S_3, S_4)
6	$S_3 \rightarrow S_5$	I	S_2	$[1,1,1]$	1	$(S_2, S_5), (S_3, S_2), (S_3, S_5)$
7	$S_4 \rightarrow S_1$	I	S_2	$[1,1,1]$	1	$(S_2, S_1), (S_4, S_1), (S_4, S_2), (S_5, S_1)$
8	$S_4 \rightarrow S_5$	I	S_2	$[1,1,1]$	1	$(S_1, S_5), (S_4, S_3), (S_4, S_5)$
9	$S_5 \rightarrow S_2$	I	/	$[1,1]$	1	(S_5, S_2)
10	$S_5 \rightarrow S_3$	I	S_2	$[1,1,1]$	1	$(S_4, S_3), (S_5, S_1), (S_5, S_3), (S_5, S_4)$

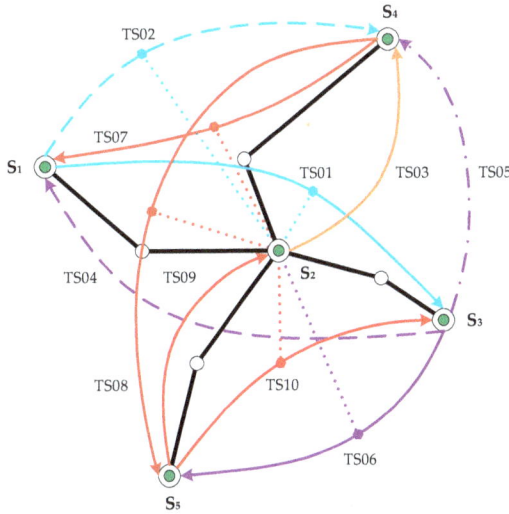

Figure 3. The optimal service network.

It should be noted that the shipment (S_1, S_5) will transfer from train service TS01 to TS08 at the intermediate station S_2; the shipment (S_4, S_3) will transfer from train service TS08 to train service TS10 at the intermediate station S_2; the shipment (S_5, S_1) will transfer from train service TS10 to train service TS07 at the intermediate station S_2; and the shipment (S_5, S_4) will transfer from train service TS10 to train service TS02 at the intermediate station S_2. Furthermore, the shipments $(S_1, S_3), (S_1, S_4), (S_3, S_5),$ $(S_4, S_1), (S_4, S_5), (S_5, S_3)$ will all wait at station S_2 for a couple of hours.

Sensitivity analysis was carried out to evaluate the impact of some critical parameters on total cost and number of trains dispatched, including the fixed cost and variable cost of dispatching an express train at a speed level of I, i.e., α_1 and θ_1; the generalized transportation cost of a loaded railcar at a speed level of I per kilometer, i.e., $\beta_1^{\text{Transport}}$; train size m. The standard value of these parameters and their step size are listed in Table 5.

Table 5. The value of some critical parameters in the trials.

Parameters	Standard Value	Step Size	Value in Trials
α_1	5000	500	3500, 4000, 4500, 5000, 5500, 6000, 6500
θ_1	40	5	25, 30, 35, 40, 45, 50, 55
$\beta_1^{\text{Transport}}$	5	0.5	3.5, 4, 4.5, 5, 5.5, 6, 6.5
m	25	5	15, 20, 25, 30, 35, 40, 45

As the result might be influenced by some or all parameters mentioned above, we only changed the value of one parameter each time and set the value of other parameters to their standard value, in order to evaluate their individual impact. The results are illustrated in Figure 4.

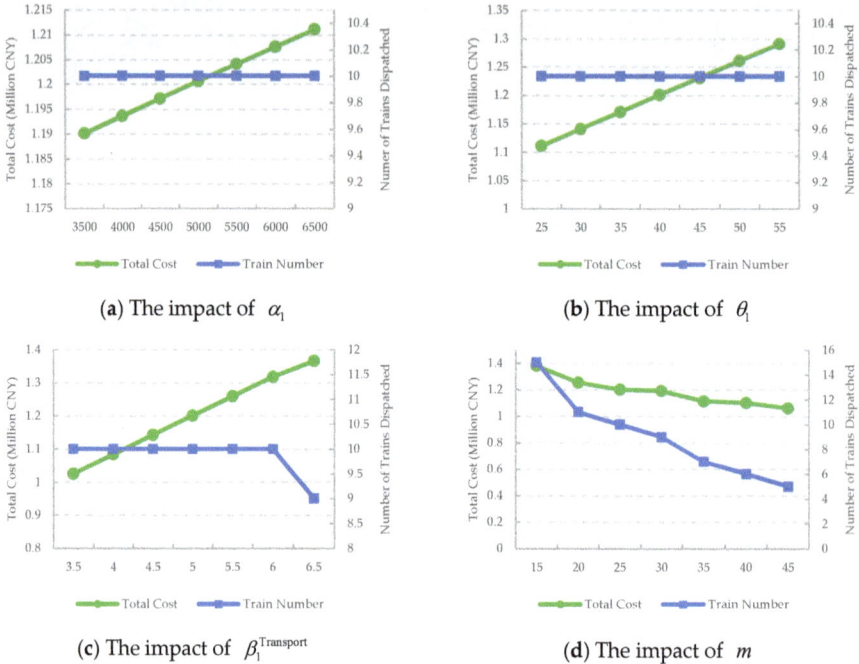

(a) The impact of α_1

(b) The impact of θ_1

(c) The impact of $\beta_1^{\text{Transport}}$

(d) The impact of m

Figure 4. The impact of some key parameters on the total cost and number of trains dispatched.

The impact of α_1, θ_1, $\beta_1^{\text{Transport}}$ and m on the total cost and number of trains dispatched are depicted by Figure 4a–d respectively.

As shown in Figure 4a, with the increase of α_1, the total cost steadily grows from 1,190,062 CNY to 1,211,062 CNY, while the number of trains stays the same, i.e., ten trains are dispatched each day no matter how the fixed cost of dispatching an express train at speed level I changes. It is demonstrated that α_1 is positively correlated with the total cost while it has no impact on the number of trains dispatched.

Similarly, as shown in Figure 4b, as θ_1 increases from 25 to 55, the total cost grows steadily from 1,110,727 CNY to 1,290,397 CNY while the number of trains dispatched remains stable no matter how the variable cost θ_1 changes. It seems that the total cost increases with θ_1 while the variable cost θ_1 has no impact on the number of trains dispatched.

It can be seen from Figure 4c that the total cost grows with $\beta_1^{\text{Transport}}$. On the contrary, the number of trains dispatched first remains unchanged when $\beta_1^{\text{Transport}}$ increases from 3.5 to 6, then decreases to nine trains when $\beta_1^{\text{Transport}}$ reaches 6.5. Thus, there is a negative correlation between the number of trains dispatched and $\beta_1^{\text{Transport}}$.

From Figure 4d, we can see that both the total cost and number of trains are negatively correlated with the train size. When m increases from 15 to 45, the total cost decreases from 1,379,512 CNY to 1,059,998 CNY, and the number of trains dispatched drops from 15 trains to five trains, respectively. This is because a train with larger size can accommodate more railcars at the same time, i.e., less train is needed to deliver the given shipments.

Symmetry **2018**, *10*, 391

6. Conclusions

In this paper, we formulated the express train service network design problem as a non-linear integer programming model which aims at finding a service network and shipment routing plan with minimum cost, while satisfying the transportation time constraints of shipments, carrying capacity constraints of train services, the flow conservation constraint and logical constraints among decision variables. Moreover, the model takes into account the speed levels and stop strategies of express trains, the block attachment/detachment operations in intermediate stations, as well as the time cost of express shipments. In order to obtain a global optimal solution, a linearization technique was adopted to transform our model into a linear one, which can be solved to optimality by using commercial software in small-scale cases. To evaluate the effectiveness and efficiency of our approach, the state-of-the-art mathematical programming solver Gurobi 7.5.2 was adopted to solve our model and an optimal solution was obtained rapidly. Furthermore, sensitivity analysis was carried out for several critical parameters to evaluate their impact on the total cost and number of trains dispatched. Currently, the express train service plan is made by a team of highly-experienced service designers, i.e., it requires painstaking manual effort. In this case, the model and solution approach we proposed can be used as an aid for railway operators in making express train service plans or improving their current train schedules, hence, achieving substantial savings in capital costs by running far fewer trains and significantly reducing block swaps (the transfer of a block from one train to another is called a block swap). In the long term, researchers can focus on the optimization of specific arrival/departure times of express trains when solving the RESSNDP. Moreover, the development of an exact solution approach for large-scale problems is also promising for future research.

Author Contributions: The authors contributed equally to this work.

Acknowledgments: This work was supported by the National Natural Science Foundation of China (Grant no. 51378056), and the science technology and legal department of the National Railway Administration of the People's Republic of China under Grant Number, KF2017-015.

Conflicts of Interest: The authors declare no conflict of interest.

References

1. Lin, B.L.; Liu, C.; Wang, J.X.; Liu, S.Q.; Wu, J.P.; Li, J. Modeling the railway network design problem: A novel approach to considering carbon emissions reduction. *Transp. Res. Part D Transp. Environ.* **2017**, *56*, 95–109. [CrossRef]
2. European Commission. *Roadmap to a Single European Transport Area—Towards a Competitive and Resource Efficient Transport System*; White Paper; European Commission: Brussels, Belgium, 2011; pp. 1–31.
3. Crainic, T.G.; Kim, K.H. *Transportation: Handbooks in Operations Research and Management Science*; Barnhart, C., Laporte, G., Eds.; Elsevier: Amsterdam, The Netherlands, 2006; Volume 14, pp. 189–284.
4. Barnhart, C.; Jin, H.; Vance, P. Railroad blocking: A network design application. *Oper. Res.* **2000**, *48*, 603–614. [CrossRef]
5. Campetella, M.; Lulli, G.; Pietropaoli, U.; Ricciardi, N. Freight service design for the Italian railways company. In Proceedings of the ATMOS 6th Workshop on Algorithmic Methods and Models for Optimization Railways, Zurich, Switzerland, 14 September 2006.
6. Ahuja, R.K.; Jha, K.C.; Liu, J. Solving real-life railroad blocking problems. *Interfaces* **2007**, *37*, 404–419. [CrossRef]
7. Lin, B.L.; Wang, Z.M.; Ji, L.J.; Tian, Y.M.; Zhou, G.Q. Optimizing the freight train connection service network of a large-scale rail system. *Transport. Res. B-Meth.* **2012**, *46*, 649–667. [CrossRef]
8. Zhu, E.; Crainic, T.G.; Gendreau, M. Scheduled service network design for freight rail transportation. *Oper. Res.* **2014**, *62*, 383–400. [CrossRef]
9. Barnhart, C.; Krishnan, N.; Kim, D.; Ware, K. Network design for express shipment delivery. *Comput. Optim. Appl.* **2002**, *21*, 239–262. [CrossRef]
10. Kim, D.; Barnhart, C.; Ware, K.; Reinhardt, G. Multimodal express package delivery: A service network design application. *Transp. Sci.* **1999**, *33*, 391–407. [CrossRef]

11. Grünert, T.; Sebastian, H.J. Planning models for long-haul operations of postal and express shipment companies. *Eur. J. Oper. Res.* **2000**, *122*, 289–309. [CrossRef]

12. Smilowitz, K.R.; Atamtürk, A.; Daganzo, C.F. Deferred item and vehicle routing within integrated networks. *Transp. Res. E-Logist. Transp. Rev.* **2002**, *39*, 305–323. [CrossRef]

13. Armacost, A.P.; Barnhart, C.; Ware, K.A. Composite variable formulations for express shipment service network design. *Transp. Sci.* **2002**, *36*, 1–20. [CrossRef]

14. Wang, D.Z.W.; Hong, K.L. Multi-fleet ferry service network design with passenger preferences for differential services. *Transp. Res. B-Meth.* **2008**, *42*, 798–822. [CrossRef]

15. Yu, S.N.; Yang, Z.Z.; Yu, B. Air express network design based on express path choices—Chinese case study. *J. Air Transp. Manag.* **2016**, *61*, 73–80. [CrossRef]

16. Quesada, J.M.; Tancrez, J.S.; Lange, J.C. A multi-hub express shipment service network design model with flexible hub assignment. *Louvain Sch. Manag. Work. Paper Ser.* **2016**, 1–26.

17. Zhao, J.; Zhang, J.J.; Yan, C.H. Research on hybrid hub-spoke express network decision with point-point direct shipment. *Chin. J. Manag. Sci.* **2016**, *24*, 58–65. [CrossRef]

18. Wang, B.H.; He, S.W. Resource planning optimization model and algorithm for multi-modal express shipment network. *J. China Railw. Soc.* **2017**, *39*, 10–15.

19. Ceselli, A.; Gatto, M.; Lübbecke, M.E.; Nunkesser, M.; Schilling, H. Optimizing the cargo express service of Swiss federal railways. *Transp. Sci.* **2008**, *42*, 450–465. [CrossRef]

MDPI

St. Alban-Anlage 66

4052 Basel

Switzerland

Tel. +41 61 683 77 34

Fax +41 61 302 89 18

www.mdpi.com

Symmetry Editorial Office

E-mail: symmetry@mdpi.com

www.mdpi.com/journal/symmetry